Lecture Notes in Mathematics

2118

Editors-in-Chief:
J.-M. Morel, Cachan
B. Teissier, Paris

Advisory Board:
Camillo De Lellis (Zürich)
Mario di Bernardo (Bristol)
Alessio Figalli (Austin)
Davar Khoshnevisan (Salt Lake City)
Ioannis Kontoyiannis (Athens)
Gabor Lugosi (Barcelona)
Mark Podolskij (Aarhus)
Sylvia Serfaty (Paris and NY)
Catharina Stroppel (Bonn)
Anna Wienhard (Heidelberg)

More information about this series at
http://www.springer.com/series/304

Claude Dellacherie • Servet Martinez •
Jaime San Martin

Inverse *M*-Matrices
and Ultrametric Matrices

 Springer

Claude Dellacherie
Laboratoire Raphael Salem, UMR 6085
Universite de Rouen
Rouen
France

Servet Martinez
CMM-DIM, FCFM
Universidad de Chile
Santiago
Chile

Jaime San Martin
CMM-DIM, FCFM
Universidad de Chile
Santiago
Chile

ISBN 978-3-319-10297-9   ISBN 978-3-319-10298-6 (eBook)
DOI 10.1007/978-3-319-10298-6
Springer Cham Heidelberg New York Dordrecht London

Lecture Notes in Mathematics ISSN print edition: 0075-8434
                              ISSN electronic edition: 1617-9692

Library of Congress Control Number: 2014953517

Mathematics Subject Classification (2010): 15B48; 60J45; 15B51; 05C50; 31C20

Printed on acid-free paper

Springer is part of Springer Science+Business Media (www.springer.com)

To our families

Preface

In this monograph we collect our work over the past 20 years linking ultrametric matrices with potential theory of finite state Markov chains and their consequences for the inverse M-matrix problem. This was triggered when Martínez, Michon and San Martín [44] proved that ultrametric matrices are inverse M-matrices. Nabben and Varga [51] provided a linear algebra proof of this fact, spreading ultrametricity towards linear algebra. Further developments were stated with the introduction of generalized ultrametric matrices by McDonald, Neumann, Schneider and Tsatsomeros [47] and Nabben and Varga [52].

Our presentation is grounded in a conceptual framework in which potential equilibrium and filtered matrices play a fundamental role. One of our main focal points is the study of the graph of connections associated with ultrametric and generalized ultrametric matrices. A fruitful line of research is to exploit the tree structure underlying these matrices, which provides a tool for understanding the associated Markov chain (see our papers [20] and [22]).

As it turns out, potential matrices perform well under Hadamard operations. We deal with an invariance of potentials under some Hadamard functions. Most notably powers and exponentials, see Neumann in [54], Chen in [11, 12] and our papers [23, 25].

This book is not intended to contain a complete discussion of the theory of inverse M-matrices and primarily reflects the interests of the authors in key aspects of this theory.

We acknowledge our coauthors Gérard Michon, Pablo Dartnell, Xiadong Zhang and Djaouad Taïbi. We also are indebted to Richard Varga, Reinhard Nabben and Miroslav Fiedler for the interest they have shown on our work and by helpful discussions. We thank anonymous referees for their recommendations and corrections that helped us to improve the presentation of this book. Finally, we express our gratitude to the support of Basal project PFB03 CONICYT and CNRS-UMI 2807.

Rouen, France   Claude Dellacherie
Santiago, Chile   Servet Martínez
Santiago, Chile   Jaime San Martín
October 29, 2014

Contents

1   Introduction 1

2   Inverse M-Matrices and Potentials 5
    2.1   Basic Notions 5
          2.1.1 M-Matrices in the Leontieff Model 6
          2.1.2   Potential Matrices.. 7
    2.2   Inverse M-Matrices and the Maximum Principle 8
    2.3   Probabilistic Interpretation of M-Matrices and Their Inverses 21
          2.3.1   Continuous Time Markov Chains 28
          2.3.2 M-Matrices and h-Transforms 32
    2.4   Some Basic Properties of Potentials and Inverse M-Matrices 37
    2.5   An Algorithm for CMP 42
    2.6   CMP and Non-Singularity 46
    2.7   Potentials and Electrical Networks 53

3   Ultrametric Matrices 57
    3.1   Ultrametric Matrices 57
    3.2   Generalized Ultrametric and Nested Block Form 63
    3.3   Graphical Construction of Ultrametric Matrices 66
          3.3.1   Tree Matrices 66
          3.3.2   A Tree Representation of GUM 72
          3.3.3   Gomory-Hu Theorem: The Maximal Flow Problem 75
          3.3.4   Potential Matrices on Graphs with Cycles
                  of Length Smaller Than 3 78
          3.3.5   Ultrametricity and Supermetric Geometry 81

4   Graph of Ultrametric Type Matrices 85
    4.1   Graph of an Ultrametric Matrix 85
          4.1.1   Graph of a Tree Matrix 90

    4.2   Graph of a Generalized Ultrametric Matrix.......................... 98
          4.2.1   A Dyadic Tree Supporting a GUM 99
          4.2.2   Roots and Connections 100
    4.3   Permutations That Preserve a NFB.................................. 113

5   **Filtered Matrices** 119
    5.1   Conditional Expectations 120
    5.2   Filtered Matrices 129
    5.3   Weakly Filtered Matrices 135
          5.3.1   Algorithm for Weakly Filtered Matrices...................... 141
          5.3.2   Sufficient Conditions for a Weakly Filtered
                  to be a Bi-potential.. 145
    5.4   Spectral Functions, M-Matrices and Ultrametricity 155
          5.4.1   Stability of M-Matrices Under Completely
                  Monotone Functions 161

6   **Hadamard Functions of Inverse M-Matrices** 165
    6.1   Definitions 166
    6.2   Hadamard Convex Functions and Powers 167
          6.2.1   Some Examples and a Conjecture 173
    6.3   A Sufficient Condition for Hadamard Invariance: Class $\mathscr{T}$ 175
          6.3.1   Class $\mathscr{T}$ and Weakly Filtered Matrices........................ 183
    6.4   Potentials of Random Walks and Hadamard Products 186
          6.4.1   Linear Ultrametric Matrices.................................. 188
          6.4.2   Hadamard Products of Linear Ultrametric Matrices 191
          6.4.3   Potentials Associated with Random Walks That
                  Lose Mass Only at the Two Ends 209

A   **Beyond Matrices** 215
    Extension of Z-Matrices, M-Matrices and Inverses 216
    Extension of Ultrametricity to Infinite Matrices 219
    Extension of Ultrametricity to L^2-Operators 220

B   **Basic Matrix Block Formulae** 223

C   **Symbolic Inversion of a Diagonally Dominant M-Matrix** 225

References.. 229

Index of Notations... 233

Index.. 235

Chapter 1
Introduction

This monograph deals with well established concepts in linear algebra and Markov chains: M-matrices, their inverses and discrete potential theory. A main focus of this monograph is the so called inverse M-matrix problem, which is the characterization of nonnegative matrices whose inverses are M-matrices. We present an answer given in terms of discrete potential theory. The primary drawback of this representation is the lack of an efficient algorithm for its implementation. The obstacles to securing a simple description have trigged research in subclasses of inverse M-matrices that are described easily. See Johnson and Smith [40] and references therein for more information about this problem.

The following diagram shows the primary connections among the concepts introduced in this monograph, and their relationship with the inverse M-matrix problem. Associated with each arrow, there is a reference to the main result where the equivalence or implication is shown. In this diagram we assume that the matrices involved are nonsingular.

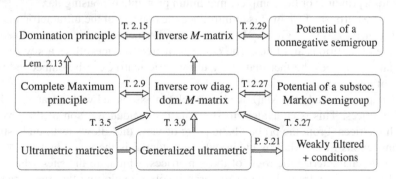

An important subclass of inverse M-matrices is given by the ultrametric matrices, which are given by elementary linear inequalities,

C. Dellacherie et al., *Inverse M-Matrices and Ultrametric Matrices*, Lecture Notes in Mathematics 2118, DOI 10.1007/978-3-319-10298-6_1

Initially introduced with regard to p-adic number theory, ultrametricity has gained particular attention over the last 20 years. Ultrametricity is an important notion in applications, like taxonomy [4], because of its relation with partitions. Moreover, strictly ultrametric matrices appear as covariance matrices of random energy models in statistical physics [10], as a generalization of the diagonal case. Relations between ultrametric matrices and filtrations were firstly developed by Dellacherie in [18]. A detailed study concerning ultrametric matrices, maximal filtrations and associated spectral decompositions for countable probability spaces can be found in [17].

The fact ultrametric matrices are inverse M-matrices was stated by Martínez, Michon and San Martín [44] and their proof used many probabilistic arguments. Shortly after, Nabben and Varga [51] provided a linear algebra proof of this fact and gave way to an interest in studying ultrametric matrices in both linear algebra and discrete potential theory. Further developments were stated with the introduction of generalized ultrametric matrices (GUM) by McDonald, Neumann, Schneider and Tsatsomeros [47] and Nabben and Varga [52]. Other developments can be found in Fiedler [30], Zhang [59].

In Chap. 2 we introduce the main concepts from probability and potential theory that characterize inverse M-matrix class as seen in the pioneering work by Choquet and Deny [14]. It seems that this fundamental work has not been sufficiently disseminated in the literature apart from potential theory. In this context, one characterization is based on the domination principle, see Theorem 2.15. A second stems from continuous positive-preserving semigroups, see Theorem 2.29. Both results are interesting from the linear algebra and probability perspectives providing a profound connection with the theory of Markov chains. Nevertheless, they are difficult to implement when seeking to ascertain whether a particular matrix is the inverse of an M-matrix, problem that will be tackled in Chap. 5. In the last section of Chap. 2, we characterize in simple terms, when a matrix that satisfies the domination principle or the complete maximum principle is nonsingular.

The core purpose of Chap. 3 is to introduce the concept of the ultrametric matrix and its generalization to GUM. We show how ultrametric matrices are related to sequence of partitions and trees. The concept of tree matrices plays a key role in this analysis as does the fact that every ultrametric matrix can be embedded in a matrix of this class.

In Chap. 4 we study the Markov chain skeleton associated with ultrametric and GUM matrices. This is equivalent to studying the incidence graph of the inverse of such matrices. From the probabilistic point of view this allows us to understand the connections in the Markov chain, whose potential is the given matrix. A key aspect is to characterize the roots of these matrices, which are the sites where the chain loses mass. A fruitful line of research involves exploiting the tree structure underlying these matrices, which in turn provides a means of understanding the Markov chain associated with them.

In Chap. 5 we introduce conditional expectations, which are a very special class of projections: they preserve positivity and constants. The linear combination of comparable conditional expectations define the class of filtered matrices.

We consider sums with constant coefficients as well as functional sums. We develop a backward algorithm to provide sufficient conditions for such sums to become inverse M-matrices. At the end of the chapter we relate filtered matrices and spectral functions of M-matrices.

In Chap. 6 we study stability properties under Hadamard functions for the class of inverse M-matrices. We prove that the class of GUM matrices is the largest class of bi-potential matrices stable under Hadamard increasing functions. We also study the conjecture performed by Neumann in [54] which states that the square function, in the sense of Hadamard, preserves the class of inverse M-matrices. This was solved for all integer powers by Chen in [11] and for general real powers greater than 1 in [12] (see also our article [23]). Our presentation relies on different ideas, where the concept of potential equilibrium plays a fundamental role. At the end of this chapter we include a study of potentials associated with random walks on the integers. We show they are Hadamard products of linear ultrametric matrices.

As a final remark, to give a probabilistic insight of the results of this monograph, a transient Markov chain can be described uniquely by its potential. In some applications one measures directly the number of visits between sites instead of the transition frequencies of the underlying Markov chain. With this information one could estimate the potential matrix and then the transition probabilities. The main drawback of this approach is that structural restrictions for potentials are difficult to state. For this reason it is interesting to know stability properties for potentials. Moreover, this approach is feasible for random walks as shown in the last chapter.

At the end of this monograph we include a section that places our presentation within the context of general potential theory. We mention our work [24] on infinite matrices and our collaboration with Taïbi [21] for filtered operators.

Chapter 2
Inverse M-Matrices and Potentials

In this chapter we investigate the relationship between M-matrices, their inverses and potentials of finite Markov chains. We supply the basic concepts from potential theory and put them in a linear algebra perspective. We highlight the concepts of potential matrices, equilibrium potentials and maximum and domination principles. The primary results are the characterizations of inverse M-matrices given in Theorems 2.15 and 2.29. The former is based on the domination principle while the latter is based on time continuous positive semigroups. We also illustrate the relationship between potential matrices and electrical networks.

2.1 Basic Notions

Throughout this monograph we shall assume that I is the index set of a matrix, unless the index set is otherwise specified. When J, K are nonempty subsets of I and A is a matrix, we denote by A_{JK} the matrix resulting from the restriction of A to the pairs in $J \times K$. When $J = K$ we denote A_J instead of A_{JJ}. We denote by $\mathbb{I}$ the identity matrix and by $\mathbb{1}$ the unit vector, whose dimensions will be clear from the context. When necessary we will use $\mathbb{1}_p$ to denote the unit vector of size p and even $\mathbb{1}_J$ the unit vector of size the cardinal of J. In general terms, we shall also use $\mathbb{1}_B$ for a set B, as the characteristic function of this set, which is the function equal to 1 in B and 0 otherwise. The elements of the canonical basis on $\mathbb{R}^n$ are denoted by $e(i) : i = 1, \cdots, n$. Given a matrix A, we denote by A' its transposed matrix. By $\#I$ we denote the number of elements of I.

We shall use the following standard convention. A real number r (a matrix, a vector or a function) is nonnegative if $r \geq 0$. We shall say that r is positive when $r > 0$. Similarly, we shall say that r is nonpositive if $r \leq 0$ and negative when $r < 0$.

Let us commence with the core concepts we shall use in this monograph.

© Springer International Publishing Switzerland 2014

C. Dellacherie et al., *Inverse M-Matrices and Ultrametric Matrices*, Lecture Notes in Mathematics 2118, DOI 10.1007/978-3-319-10298-6_2

Definition 2.1 A square matrix M is said to be a Z-**matrix** if the off diagonal elements of M are nonpositive.

A nonsingular matrix M is said to be a M-**matrix** if it is a Z-matrix, and its inverse is a nonnegative matrix.

As a general reference on this subject we use, among others, the books by Horn and Johnson [37] and [38]. We shall use Theorem 2.5.3 in [38] several times as it provides several necessary and sufficient conditions for a Z-matrix to be an M-matrix. In particular, a necessary condition is that the diagonal elements of an M-matrix are positive. For an account on the inverse M-matrix problem, that is to characterize the nonnegative matrices whose inverses are M-matrices, we refer to [40].

2.1.1 M-Matrices in the Leontieff Model

M-matrices appear in many applications such as economics, numerical procedures for partial differential equations, probability theory and potential theory. One good example is the Leontieff input-output model in economics (see for example [5]). In this context, $I = \{1, \cdots, n\}$ indexes the set of sectors of an economy. The interdependence of the sectors is given by the matrix $A = (a_{ij} : i, j \in I)$, where $a_{ij} \geq 0$ is the number of units of sector i necessary to produce one unit by sector j. The production level of sector i is denoted by x_i. Then $\sum_{j \in I} a_{ij} x_j$ is the production of sector i internally consumed by the entire economy.

In an open economy d_i denotes the external demand of sector i. Hence, the basic model for the production is the balance

$$x = Ax + d. \tag{2.1}$$

It is standard to assume that $M = \mathbb{I} - A$ is nonsingular. This is the case when A is irreducible and (2.1) has a nonnegative solution for some nonnegative $d \neq 0$. Indeed, let $\lambda > 0$ be the Perron-Frobenius eigenvalue of A and v be its left eigenvector, which is strictly positive. Then, from (2.1)

$$v'x = \lambda v'x + v'd,$$

and $v'd > 0$, one gets $\lambda < 1$. Since λ is the spectral radius of A we conclude $\mathbb{I} - A$ is nonsingular and $M^{-1} = \sum_{p \geq 0} A^p \geq 0$, showing that M is an M-matrix.

2.1.2 Potential Matrices

A row diagonally dominant matrix is a matrix in which each element in the diagonal dominates the sum of absolute values of the other elements in the corresponding row. This concept plays an important role in what follows and should not be confused with the domination in the sense that the diagonal elements dominate the other elements of the row (or column) as we seen in the next formal definition. Just as this is the case for rows a similar definition can be formulated for columns.

Definition 2.2 (i) A is said to be a **row diagonally dominant** if

$$\forall i \quad |A_{ii}| \geq \sum_{j \neq i} |A_{ij}|,$$

(ii) U is said to be **row pointwise diagonally dominant** if

$$\forall i, j \quad |U_{ii}| \geq |U_{ij}|.$$

In both definitions we use the prefix strictly when the inequalities are strict for all rows.

We introduce the main two concepts that play an important role in this monograph.

Definition 2.3 A nonnegative matrix U is said to be a **potential** if it is nonsingular and its inverse is a row diagonally dominant M-matrix. In addition, U is said to be a **bi-potential** if U^{-1} is also column diagonally dominant. This class is denoted by $bi \mathscr{P}$. We note that potentials and bi-potentials are inverse M-matrices.

Definition 2.4 Let V be a matrix. Any nonnegative vector that satisfies $V\lambda = \mathbb{1}$ is called a **right equilibrium potential** for V. Its total mass is $\bar{\lambda} = \mathbb{1}'\lambda$. Similarly any nonnegative vector μ that satisfies $\mu'V = \mathbb{1}'$ is called a **left equilibrium potential** for V. Its total mass $\mathbb{1}'\mu$ is denoted by $\bar{\mu}$.

When V is symmetric both concepts agree and we call them **equilibrium potentials**. If there exists a unique right equilibrium potential we denote it by λ^V (similarly for μ^V). A point $i \in I$ for which $\lambda_i^V > 0$ is called a **root** for V. The set $\mathscr{R}(V) = \{i : \lambda_i^V > 0\}$ is called the set of roots of V.

Sometimes we will need to relax the non-negativity of the equilibrium potentials and in this case we will call them **signed equilibrium potentials**.

It could happen that a matrix has a left equilibrium potential, but fails to have a right equilibrium potential, or vice versa. If the matrix is nonsingular and its inverse is a Z-matrix, the existence of a right equilibrium potential is equivalent to the fact that the inverse is row diagonally dominant (similarly the column diagonally dominance is equivalent to the existence of a left equilibrium potential).

Consider M, the inverse of a potential W. By definition M is a row diagonally dominant M-matrix. If $\kappa_0 = \max\{M_{ii} : i \in I\}$, then the matrix $P_\kappa = \mathbb{I} - \frac{1}{\kappa} M$ is substochastic for all $\kappa \geq \kappa_0$. That is, P is nonnegative and its row sums are bounded by one, with strict inequality for at least one row. Notice that $W^{-1} = \kappa(\mathbb{I} - P_\kappa)$. This decomposition of W^{-1} is not unique and if there is one for which we can take $\kappa = 1$, we say that W is a **Markov potential**. This is equivalent to having $\kappa_0 \leq 1$, that is $W_{ii}^{-1} \leq 1$ for all i. This distinction will be studied in depth when we link potential matrices with potentials of Markov process or potentials of continuous time Markov process (see Sect. 2.3). Essentially, any potential matrix is the potential of a continuous time Markov chain and any Markov potential, as presented above, is the potential of a discrete time Markov chain.

2.2 Inverse M-Matrices and the Maximum Principle

In this section we study a characterization of inverse M-matrices in terms of some domination properties related to the maximum principle. This section is inspired in the work of Choquet and Deny [14].

We will start with a strong concept that will characterize the inverses of row diagonally dominant M-matrices.

Definition 2.5 A nonnegative matrix U satisfies the **Complete Maximum Principle (CMP)** if for all $x \in \mathbb{R}^I$, $x_i \geq 0$ implies that $(Ux)_i \leq 1$, then also $Ux \leq \mathbb{1}$.

U satisfies the CMP if for all $x \in \mathbb{R}^I$, with at least one nonnegative coordinate, the maximum among the values $(Ux)_i : i \in I$ is attained in the set of coordinates where x is nonnegative.

Example 2.6 Consider the matrix $U = (1-\rho)\mathbb{I} + \rho \mathbb{1}\mathbb{1}'$, where $0 \leq \rho \leq 1$. Then, U satisfies the CMP. Indeed, take any $x \in \mathbb{R}^I$ and assume that $(Ux)_i \leq 1$ when $x_i \geq 0$. In particular, we have $(Ux)_i = \rho \mathbb{1}'x + (1-\rho)x_i \leq 1$ and a fortiori $\rho \mathbb{1}'x \leq 1$. Now, if $x_j < 0$, we get

$$(Ux)_j = \rho \mathbb{1}'x + (1-\rho)x_j \leq \rho \mathbb{1}'x \leq 1,$$

and U satisfies the CMP. Notice that for $\rho = 1$ this matrix is singular and still satisfies the CMP.

Moreover, when $\rho < 1$ we get

$$U^{-1} = \frac{1}{1-\rho}\left(\mathbb{I} - \frac{\rho}{1 + (n-1)\rho}\mathbb{1}\mathbb{1}'\right),$$

where $n = \#I$. Thus, U^{-1} is a Z-matrix and since $U \geq 0$, we get that U^{-1} is an M-matrix. Finally, $U^{-1}\mathbb{1} = \frac{1}{1+(n-1)\rho}\mathbb{1} \geq 0$ and U^{-1} is a row diagonally dominant M-matrix. Therefore, U is a potential.

Lemma 2.7 *Assume U is a nonnegative matrix that satisfies the CMP. Then*

(i) for all $x \in \mathbb{R}^I$ and all $c \geq 0$ we have

$$\left[\forall i \; [x_i \geq 0 \Rightarrow (Ux)_i \leq c] \right] \Rightarrow [Ux \leq c\mathbb{1}],$$

that is, if c is an upper bound for $(Ux)_i$ when x_i is nonnegative, then it is also an upper bound for the other coordinates;

(ii) for every nonnegative diagonal matrix L the matrix $L + U$ satisfies the CMP, in particular for all $a > 0$ the matrix $a\mathbb{1} + U$ satisfies the CMP;

(iii) U is column pointwise diagonally dominant;

(iv) if U is nonsingular then the diagonal is positive;

(v) if the diagonal of U is positive, then the equation $U\lambda = \mathbb{1}$ has a nonnegative solution.

Proof Part (i). The case $c > 0$ follows from homogeneity by replacing x by x/c. The special case $c = 0$ follows by a continuity argument. If $(Ux)_i \leq 0$ for all i where $x_i \geq 0$, then in the same set of coordinates and for all $\epsilon > 0$ it holds

$$(Ux)_i \leq \epsilon.$$

Then, from the CMP we deduce $Ux \leq \epsilon\mathbb{1}$. Letting ϵ decreases to 0 we deduce $Ux \leq 0$ as desired.

Part (ii). Consider $x \in \mathbb{R}^I$ such that $(Lx + Ux)_i \leq 1$ when $x_i \geq 0$. Then, $(Ux)_i \leq 1$ for these coordinates and since U satisfies the CMP, we conclude $Ux \leq \mathbb{1}$. This implies that

$$Lx + Ux \leq \mathbb{1}.$$

Hence, $L + U$ also satisfies the CMP.

In proving (iii) consider $x = (1, -\epsilon, \cdots, -\epsilon)'$ and $c = U_{11} \geq 0$, where $\epsilon > 0$ is arbitrary. Hence we have

$$(Ux)_1 \leq c,$$

where $i = 1$ is the unique coordinate where $x_i \geq 0$. The conclusion is that for all $j \in I$

$$(Ux)_j \leq c,$$

that is $U_{j1} - \epsilon \sum_{k \neq 1} U_{jk} \leq U_{11}$. Thus, for all j we have $U_{j1} \leq U_{11}$. Similar domination holds for the other columns and the property is proven.

When U is nonsingular, this domination implies the diagonal must be positive, proving (iv).

Part (v). Notice that for a small enough $a > 0$ the matrix $W(a) =: a\mathbb{I} + U$ is strictly column pointwise diagonally dominant, nonnegative, satisfies the CMP and is nonsingular (the matrix U has a finite number of eigenvalues). We shall prove that $\lambda(a)$, the unique solution of $W(a)\lambda(a) = \mathbb{1}$, is nonnegative.

Denote by $W = W(a)$. Assume that for coordinate i we have $\lambda(a)_i < 0$. Consider the vector $x = \lambda(a) + \alpha e$ where $e = e(i)$ is the i-th vector of the canonical basis. We assume that $\alpha > 0$ is small enough such that $x_i < 0$. On the other hand, we define $0 \le c = 1 + \alpha \max_{k \neq i} W_{ki}$. Since $Wx = W\lambda(a) + \alpha We = \mathbb{1} + \alpha We$, we conclude that for distinct i, j

$$(Wx)_j = 1 + \alpha W_{ji} \le c,$$

in particular $(Wx)_j \le c$ when $x_j \ge 0$. Then

$$(Wx)_i = 1 + \alpha W_{ii} \le 1 + \alpha \max_{k \neq i} W_{ki},$$

which is a contradiction since W is strictly column pointwise diagonally dominant. We conclude that $\lambda(a) \ge 0$. We now prove that $\lambda(a)$ remains bounded as $a \downarrow 0$. Take $\rho = \min\{U_{ii} : i \in I\}$, which by hypothesis is positive. Then

$$\mathbb{1}'(a\mathbb{I} + U)\lambda(a) = \mathbb{1}'\mathbb{1} = \#I \ge \sum_i (a + U_{ii})\lambda(a)_i \ge \rho \sum_i \lambda(a)_i.$$

The conclusion is that $\lambda(a)$ belongs to the compact set $K = \{x \in \mathbb{R}^I : x \ge 0, \mathbb{1}'x \le \#I/\rho\}$ and therefore there exists a sequence $a_n \downarrow 0$ and a vector $\lambda \in K$ such that $\lambda(a_n) \to \lambda$. From $(a_n\mathbb{I} + U)\lambda(a_n) = \mathbb{1}$ we deduce that $U\lambda = \mathbb{1}$ and the result follows.

We will need the following notion, taken from standard probability theory.

Definition 2.8 Let P be a nonnegative matrix. P is said to be a **stochastic matrix** if all the rows add up to 1. P is said to be a **substochastic matrix** if the row sums of P are bounded by one and at least one of them is strictly smaller than one. The rows where the sum are smaller than one are said to **lose mass** or be defective.

Let us state review the primary results of this section.

Theorem 2.9 *Assume U is a nonnegative nonsingular matrix. Then, U^{-1} is a row diagonally dominant M-matrix, that is U is a potential, if and only if U satisfies the CMP.*

Proof Assume first that U^{-1} is a row diagonally dominant M-matrix. This is equivalent to have $U^{-1} = \kappa(\mathbb{I} - P)$, with P substochastic and $\kappa > 0$ a constant. Consider $\lambda = U^{-1}\mathbb{1}$, which by hypothesis is nonnegative. Take any $x \in \mathbb{R}^I$ such that $(Ux)_i \le 1$ when x_i is nonnegative.

Define $y = U(\lambda - x) = \mathbb{1} - Ux$, which by construction is nonnegative on the coordinates where x is nonnegative. Assume that $y_i < 0$ for some coordinate i, optimal in the sense that $|y_i|$ is the largest possible value among all the negative terms. Notice that $\lambda - x = U^{-1}y = \kappa(y - Py)$, which gives us

$$\lambda_i - x_i = \kappa\left(y_i - \sum_j P_{ij}y_j\right) \leq \kappa\left(y_i - \sum_{j:y_j<0} P_{ij}y_j\right) \leq \kappa y_i\left(1 - \sum_{j:y_j<0} P_{ij}\right) \leq 0.$$

Furthermore, since $x_i < 0$ we get $\lambda_i - x_i > 0$, which is a contradiction. In conclusion $y = U(\lambda - x) \geq 0$ or equivalently $Ux \leq \mathbb{1}$ and U satisfies the CMP.

Conversely, assume that U satisfies the CMP. Take $\theta > 0, c \geq 0$ and $x \in \mathbb{R}^n$ such that $\theta x + Ux \leq c\mathbb{1}$. In particular, we have $(Ux)_i \leq c$ when x_i is nonnegative and from the CMP we deduce $Ux \leq c\mathbb{1}$.

Therefore, if $y = \theta x + Ux \leq c\mathbb{1}$, we conclude that $Ux \leq c\mathbb{1}$. On the other hand $Ux = U(\theta\mathbb{I} + U)^{-1}y = (\theta U^{-1} + \mathbb{I})^{-1}y$. Thus when $c = 0$ we have the property

$$y \leq 0 \Rightarrow (\theta U^{-1} + \mathbb{I})^{-1}y \leq 0,$$

or equivalently

$$y \geq 0 \Rightarrow (\theta U^{-1} + \mathbb{I})^{-1}y \geq 0.$$

For small $\theta > 0$, we have

$$(\theta U^{-1} + \mathbb{I})^{-1} = \mathbb{I} - \theta U^{-1} + \sum_{n\geq 2}(-\theta)^n U^{-n}.$$

Now, take $y = e(1)$ and $j \neq 1$ to get

$$-\theta(U^{-1})_{1j} + \sum_{n\geq 2}(-\theta)^n (U^{-n})_{1j} \geq 0.$$

From this inequality we conclude $U_{1j}^{-1} \leq 0$ for $j \neq 1$ (divide by θ and take a small enough θ). Similarly, $(U^{-1})_{ij} \leq 0$ for all different i, j, thus U^{-1} is a Z-matrix and therefore, since U is nonnegative we conclude U^{-1} is an M-matrix (see Theorem 2.5.3 in [38]).

Finally, from part (iv) in the previous Lemma, we have $\lambda = U^{-1}\mathbb{1} \geq 0$, which is equivalent to the fact that U^{-1} is row diagonally dominant and the result is proven.

In order to relax the row diagonally dominant condition we need to introduce new concepts, borrowed from [14]. Below, a nonnegative vector $x \in \mathbb{R}^I$ is said to be **carried** by $J \subset I$ if it vanishes outside J.

Definition 2.10 Let U be a nonnegative matrix. Then,

- U satisfies the **domination principle** if given two nonnegative vectors x, y such that $(Ux)_i \leq (Uy)_i$ when $x_i > 0$, then $Ux \leq Uy$.
- U satisfies the **balayage principle** if for any nonnegative vector y and any nonempty set $J \subseteq I$ there is a nonnegative vector x carried by J such that $Ux \leq Uy$ with equality on J, that is

$$\forall j \in J : \quad (Ux)_j = (Uy)_j.$$

- U satisfies the **maximum principle** if for any nonnegative $x \in \mathbb{R}^I$, $x_i > 0$ implies $(Ux)_i \leq 1$, then $Ux \leq 1$.
- U satisfies the **equilibrium principle** if for any nonempty set $J \subseteq I$ there exists a nonnegative vector λ carried by J, such that $U\lambda \leq 1$ with equality on J, that is

$$\forall j \in J : \quad (U\lambda)_j = 1$$

Remark 2.11 It is not difficult to see that if $J \subseteq I$ is a nonempty set and U satisfies any of the principles given above on I, then the sub matrix U_J satisfies the respective principle on J.

The following result is simple to prove and is quite useful.

Lemma 2.12 *Assume U satisfies the domination principle, and L_1, L_2 are two positive diagonal matrices. Then $V = L_1 U L_2$ also satisfies the domination principle. If U satisfies any of the principles introduced in Definition 2.10 or the CMP and Π is a permutation matrix, then $\Pi U \Pi'$ satisfies the same principle.*

Proof Take x, y nonnegative vectors such that $(Vx)_i \leq (Vy)_i$ when $x_i > 0$. Then, for these coordinates

$$(UL_2 x)_i \leq (UL_2 y)_i.$$

So, if we denote $z = L_2 x, w = L_2 y$, we get

$$(Uz)_i \leq (Uw)_i$$

for i in the set $\{k : x_k > 0\} = \{k : z_k > 0\}$. Since U satisfies the domination principle, we get $Uz \leq Uw$ and then

$$Vx \leq Vy,$$

proving the first part of the result. The second part is proven similarly.

The following result links the concepts introduced in Definition 2.10 with the *CMP* and shows interesting relations among them.

Lemma 2.13 *Assume U is a nonnegative matrix with a positive diagonal. Then*

(i) *U satisfies the maximum principle if and only if U satisfies the equilibrium principle;*

(ii) *U satisfies the domination principle if and only if U satisfies the balayage principle;*

(iii) *if U satisfies the CMP, then U satisfies the domination principle;*

(iv) *if U satisfies the CMP then U satisfies the maximum principle and the equilibrium principle;*

(v) *the following are equivalent*

(v.1) *U satisfies the CMP;*

(v.2) *U satisfies the domination and equilibrium principles;*

(v.3) *U satisfies the domination principle and it has a right equilibrium potential, that is, there exists a nonnegative vector λ solution of $U\lambda = \mathbb{1}$;*

(vi) *if U satisfies the domination principle, then there exists a diagonal matrix D, with positive diagonal elements, such that DU satisfies the CMP.*

Proof Part (i). Let us assume that U satisfies the maximum principle. We shall prove it satisfies the equilibrium principle by induction on $\#I$. Since the diagonal of U is positive the result is obvious when $\#I = 1$. So, assume the result holds whenever $\#I \leq n - 1$ and we shall prove that U satisfies the equilibrium principle if $\#I = n$. Consider $J \subseteq I$. If $\#J \leq n - 1$ then there exists a nonnegative vector $z \in \mathbb{R}^J$ such that

$$U_J z = \mathbb{1}_J.$$

Extend this vector by 0 on the coordinates of $I \setminus J$. We still denote by z this vector. Then for all $j \in J$

$$(Uz)_j = 1.$$

By the maximum principle we obtain $Uz \leq \mathbb{1}$ showing the inequality in this case.

The only case left is $J = I$, that is we need to show the existence of a nonnegative vector λ such that $U\lambda = \mathbb{1}$. Recall that $e(i)$ is the vector of the canonical basis associated with coordinate i. From the case already shown we have for all i, the existence a nonnegative vector $\lambda(i)$ supported on $I \setminus \{i\}$ such that

$$(U\lambda(i))_j = 1 \text{ for } j \neq i;$$
$$(U\lambda(i))_i \leq 1.$$

Then there exists $0 \leq \theta_i \leq 1$ such that

$$\mathbb{1} = U\lambda(i) + \theta_i e(i).$$

If it happens that $\theta_i = 0$, for some i, then the result is proved. On the contrary, since $\#I \geq 2$ the number $a = (\sum_i 1/\theta_i) - 1$ is positive and $\lambda = \frac{1}{a}\sum_i \lambda(i)/\theta_i$ is a nonnegative vector that satisfies $U\lambda = \mathbb{1}$ proving the desired result.

Conversely, we assume that U satisfies the equilibrium principle and need to prove that U satisfies the maximum principle. For that purpose consider a nonnegative vector x such that $(Ux)_i \leq 1$ for $i \in J = \{j : x_j > 0\}$. We must prove that $Ux \leq \mathbb{1}$, which is evident if $J = I$ or $J = \emptyset$ (in the latter because $x \leq 0$). The rest of the proof is done by induction on $\#J$, which is at least 1.

We denote by λ a nonnegative vector, carried by J, such that

$$(U\lambda)_j = 1 \quad \text{for } j \in J;$$
$$(U\lambda)_i \leq 1 \quad \text{for } i \in I \setminus J.$$

If $\#J = 1$, say $J = \{j\}$, we have that $x = a\,e(j)$ and $\lambda = b\,e(j)$ where $a > 0, b > 0$. We necessarily have $a \leq b$ and then

$$Ux \leq U\lambda \leq \mathbb{1},$$

showing the desired inequality in this case. Assume the result holds whenever $\#J \leq p$ and we must show the result for the case $\#J = p + 1$. Let us take $a = \min\{x_j/\lambda_j : j \in J\} \in [0, \infty]$. If a is greater than 1, we would have for all $j \in J$

$$(U\lambda)_j < (Ux)_j \leq 1,$$

which is a contradiction. Thus a is smaller or equal to 1 and therefore $y = x - a\lambda$ is a nonnegative vector supported by $K \subsetneq J$, a set that contains at most p points. If $y = 0$ the result is proved because

$$Ux = aU\lambda \leq \mathbb{1}.$$

So we may assume that $y \neq 0$. Let $b = \max\{(Uy)_j : j \in K\} > 0$ and $k_0 \in K \subseteq J$ be any coordinate such that $(Uy)_{k_0} = b$. We can apply the induction hypothesis to the vector $z = y/b$ which satisfies for $k \in K$

$$(Uz)_k \leq 1.$$

Since K is the support of z and this set has at most p elements we conclude that $Uz \leq \mathbb{1}$ and therefore

$$Uy \leq b\mathbb{1}.$$

This shows that

$$Ux = U(y + a\lambda) \leq (a + b)\mathbb{1}.$$

Finally, $(Ux)_{k_0} = a(U\lambda)_{k_0} + (Uy)_{k_0} = a + b$, which is smaller than one by hypothesis, proving the result.

The proof of part (ii) is similar to the proof of part (i).

Let us now prove (iii). So, we assume that U satisfies the CMP. In order to prove that U satisfies the domination principle, consider two nonnegative vectors x, y such that $(Ux)_i \leq (Uy)_i$ when $x_i > 0$. For all $\epsilon > 0$ and for those coordinates

$$(U(x - y - \epsilon\mathbb{1}))_i \leq 0.$$

Notice that $\{i : (x - y - \epsilon\mathbb{1})_i \geq 0\} \subseteq \{i : x_i > 0\}$ and therefore from the CMP we conclude

$$U(x - y - \epsilon\mathbb{1}) \leq 0.$$

Letting $\epsilon \downarrow 0$, we get the inequality $Ux \leq Uy$ and U satisfies the domination principle.

Part (iv). According to (i) it is sufficient to show that if U satisfies the CMP then U satisfies the maximum principle. For that purpose consider a nonnegative vector z such that

$$(Uz)_i \leq 1,$$

when $z_i > 0$. Recall that from Lemma 2.7 there exits a right equilibrium potential, that is, a nonnegative vector λ such that $U\lambda = \mathbb{1}$. Hence we obtain

$$(Uz)_i \leq (U\lambda)_i.$$

Since U satisfies the domination principle we have that $Uz \leq U\lambda = \mathbb{1}$, proving that U satisfies the maximum principle.

Now, we prove part (v). It is clear from (iii), (iv) that the CMP implies the domination and equilibrium principles. Then $(v.1)$ implies $(v.2)$, and the latter implies $(v.3)$.

So, to finish the proof of (v), we assume that U satisfies the domination principle and it possesses a right equilibrium potential. We shall prove that U satisfies the CMP. We denote by λ any nonnegative solution to the equation $U\lambda = \mathbb{1}$. Now, consider a vector $x \in \mathbb{R}^I$ such that $(Ux)_i \leq 1$ whenever $x_i \geq 0$, then we need to prove $Ux \leq \mathbb{1}$.

Set $J = \{j : x_j \geq 0\}$ and take $\epsilon > 0$ small enough such that for $x(\epsilon) = x + \epsilon \mathbb{1}$ we have $J = \{j : x(\epsilon)_j > 0\}$. Define $z = x(\epsilon)^+ \geq 0$ and $w = x(\epsilon)^- \geq 0$ the positive and negative parts of $x(\epsilon)$. Given that $x(\epsilon) = z - w$, $J = \{j : z_j > 0\}$ and for $j \in J$ we have

$$(Uz)_j = (U(w + \epsilon \mathbb{1}) + U(x))_j \leq (U(w + \epsilon \mathbb{1}))_j + 1 = (U(w + \epsilon \mathbb{1} + \lambda))_j,$$

we conclude from the domination principle that $Uz \leq U(w + \lambda + \epsilon \mathbb{1})$. Thus $Ux \leq U\lambda = \mathbb{1}$, as desired.

(vi). Assume that U satisfies the domination principle. Consider the diagonal matrix D, whose diagonal elements are defined as

$$D_{ii} = \left(\sum_{j \in I} U_{ij} \right)^{-1}.$$

According to Lemma 2.12 the matrix DU also satisfies the domination principle. Since $DU\mathbb{1} = \mathbb{1}$, we conclude that DU has a right equilibrium potential. Therefore, DU satisfies the CMP because (v.3).

Remark 2.14 Notice that in the proof that domination principle plus the existence of an equilibrium potential implies the CMP, we do not require the diagonal of U to be positive. This will be important in Sect. 2.5.

From part (iv) of the previous Lemma and Theorem 2.9, U is the inverse of a row diagonally dominant M-matrix if and only if U satisfies the domination and equilibrium principles. The equilibrium principle is related to the diagonally dominant property of U^{-1}. As such, the following result is natural.

Theorem 2.15 *Let U be a nonnegative and nonsingular matrix. Then, U^{-1} is an M-matrix if and only if U satisfies the domination principle.*

Proof Suppose that U satisfies the domination principle. We first prove that U has a positive diagonal. For that purpose assume that $U_{ii} = 0$ for some i. We have $(Ue(i))_i = 0 = (U0)_i$ and therefore from the domination principle $Ue \leq 0$ which implies that the i-th column of U is null, something not possible because U is nonsingular. Then, using Lemma 2.13 (ii) U also satisfies the balayage principle.

Now, for every fixed $j \in I$ consider $e = e(j)$. From the balayage principle there exists a nonnegative vector $\sigma = \sigma(j) \geq 0$ carried by $J = I \setminus \{j\}$ solution to the problem $U\sigma \leq Ue$ and

$$(U\sigma)_i = (Ue)_i \quad \text{for } i \neq j.$$

We denote by S the matrix given by $S_{kj} = \sigma(j)_k$ for $k, j \in I$. In particular S is a nonnegative matrix with a zero diagonal. We have the relation

$$(US)_{ij} = \sum_k U_{ik} S_{kj} = (U\sigma(j))_i = (Ue(j))_i - a(j)e(j)_i = U_{ij} - a(j)e(j)_i,$$

where $a(j) \in \mathbb{R}$ is nonnegative. If $a(j) = 0$ this means that $Ue(j) = U\sigma(j)$ and since U is nonsingular we conclude $e(j) = \sigma(j)$. This is not possible because $\sigma(j)_j = 0 \neq e(j)_j = 1$. The conclusion is that for all j it holds $a(j) > 0$ and then

$$US = U - A,$$

where $A \geq 0$ is the diagonal matrix with diagonal terms $a_{jj} = a(j) > 0$. Then

$$U(\mathbb{I} - S) = A.$$

From here we deduce that $\mathbb{I} - S$ is nonsingular and

$$A \sum_{n=0}^{N} S^n = U(\mathbb{I} - S) \sum_{n=0}^{N} S^n = U - US^{N+1} \leq U,$$

which implies that the series $\sum_{n=0}^{\infty} S^n$ converge. Therefore S^{N+1} tends to zero as $N \to \infty$ and a fortiori US^{N+1} converges to 0. In particular $(\mathbb{I} - S)^{-1} = \sum_{n=0}^{\infty} S^n$ and $U^{-1} = (\mathbb{I} - S)A^{-1}$, proving that U^{-1} is an M-matrix.

Conversely, assume that $M = U^{-1}$ is an M-matrix. It is well known that there is a diagonal matrix D, with positive diagonal elements, such that MD is a row diagonally dominant M-matrix (see Theorem 2.5.3 in [38], see also Sect. 2.3.2 below). It is not difficult to see that D given by $D_{ii} = \sum_j U_{ij}$ will work in this case, because $D^{-1}U\mathbb{1} = \mathbb{1}$ and therefore $MD\mathbb{1} = \mathbb{1}$. From Theorem 2.9 we conclude that $D^{-1}U$ satisfies the CMP, which then has positive diagonal elements according to Lemma 2.7. Thus, from Lemma 2.13 we obtain $D^{-1}U$ satisfies the domination principle. From here it is straightforward to show that U satisfies the domination principle as well.

The following result is a complement to Theorems 2.9 and 2.15.

Theorem 2.16 *Assume that U is a nonnegative matrix that satisfies the CMP and is strictly column pointwise dominant. Then U is nonsingular. Moreover $U^{-1} = \frac{1}{\Delta}(\mathbb{I} - P)$ for some substochastic matrix P, where $\Delta = \min\{U_{jj} - U_{ij} : i \neq j\}$ is positive by hypothesis,*

Proof Take $a > 0$ and $W = W(a) = a\mathbb{I} + U$. For small enough a the matrix W is nonsingular, nonnegative and satisfies the CMP. Therefore, its inverse is a row diagonally dominant M-matrix, that we denote by $M = M(a)$. In particular from $MW = \mathbb{I}$ we obtain for all j

$$1 = \sum_i M_{ji}W_{ij} = (a + \Delta)M_{jj} + (U_{jj} - \Delta)M_{jj} + \sum_{i \neq j} M_{ji}U_{ij}.$$

Since $U_{jj} - \Delta \geq U_{ij}$ and $M_{ji} \leq 0$, when $i \neq j$, we deduce

$$1 \geq (a + \Delta)M_{jj} + (U_{jj} - \Delta)\Big[M_{jj} + \sum_{i \neq j} M_{ji}\Big] \geq (a + \Delta)M_{jj},$$

because M is row diagonally dominant. We conclude for all j

$$M_{jj}(a) \leq \frac{1}{a + \Delta}.$$

This property and the fact that $M(a)$ is a row diagonally dominant M-matrix, show that $M(a)$ remains bounded as a decreases toward 0. Take a subsequence $a_n \downarrow 0$ such that $M(a_n)$ converges to a Z-matrix H. From

$$M(a_n)(a_n\mathbb{I} + U) = \mathbb{I},$$

we deduce that $HU = \mathbb{I}$ and therefore U is nonsingular, H is a row diagonally dominant M-matrix and moreover for all j

$$H_{jj} \leq \frac{1}{\Delta}.$$

From these inequalities we deduce that $P = \mathbb{I} - \Delta H$ is a substochastic matrix and the result is proved.

The next result is a generalization of the previous one.

Theorem 2.17 *Assume that U is a nonnegative matrix that satisfies the domination principle and is strictly column pointwise dominant. Then, U is nonsingular and its inverse is an M-matrix.*

In the proof of this result we need the following lemma.

Lemma 2.18 *Assume U satisfies the domination principle and has a positive diagonal, then for any nonnegative diagonal matrix L the matrix $L + U$ satisfies the domination principle.*

Proof Consider D the diagonal matrix given by $D_{ii} = \sum_j U_{ij} > 0$. Then it is obvious to show that $D^{-1}U$ also satisfies the domination principle and that $\lambda = \mathbb{1}$ is a right equilibrium potential, that is $D^{-1}U\mathbb{1} = \mathbb{1}$. Therefore from Lemma 2.13 we conclude that $D^{-1}U$ satisfies the CMP and then from Lemma 2.7 (ii) the matrix

$$D^{-1}L + D^{-1}U$$

also satisfies the CMP and then the domination principle by the Lemma 2.13. Thus $L + U$ satisfies the domination principle.

Proof (Theorem 2.17) We shall prove the result by induction on n the size of U. Clearly the result holds when $n = 1$, and furthermore it is quite simple to show when $n = 2$. So we assume that the desired property holds whenever the size of the matrix is smaller or equal than n. Consider a nonnegative matrix U of size $n + 1$ which is strictly column pointwise dominant and which satisfies the domination principle. Let us take $a > 0$ and define the matrix

$$W(a) = a\mathbb{I} + U,$$

which satisfies the domination principle (see Lemma 2.18) and which is nonsingular for small enough a. Then from Theorem 2.15 the inverse of $W(a)$ is an M-matrix. Now, consider $\lambda(a) = (W(a)^{-1})\mathbb{1}$. This vector may have negative components, but it nonetheless has at least one positive component since $W(a)\lambda(a) = \mathbb{1}$. Consider a sequence $a_p \downarrow 0$ for which there is a fixed index $i_0 \in I$ such that $\lambda(a_p)_{i_0} > 0$. For simplicity we assume that $i_0 = 1$. We decompose $W(a_p)$ and its inverse $M(a_p)$ in blocks as (see Appendix B for relations among these blocks)

$$W(a_p) = \begin{pmatrix} a_p + U_{11} & u' \\ w & a_p\mathbb{I}_n + V \end{pmatrix},$$

$$M(a_p) = \begin{pmatrix} M_{11}(a_p) & -\zeta(a_p)' \\ -\xi(a_p) & \Lambda(a_p) \end{pmatrix}$$

where u, w, are fixed nonnegative vectors, $\zeta(a_p)$ and $\xi(a_p)$ are also nonnegative and $\Lambda(a_p)$ is an M-matrix. In particular, we have

$$\lambda(a_p)_1 = M_{11}(a_p) - \zeta(a_p)'\mathbb{1}_n = -\sum_{j=1}^{n+1} M_{1j}(a_p) > 0.$$

From the fact $M(a_p)W(a_p) = \mathbb{I}_{n+1}$, we deduce

$$\begin{aligned}
1 &= (a_p + U_{11})M_{11}(a_p) + \sum_{i=2}^{n+1} M_{1i}(a_p)U_{i1} \\
&\geq (a_p + U_{11})M_{11}(a_p) + (U_{11} - \delta)\sum_{i=2}^{n+1} M_{1i}(a_p) \geq (a + \delta)M_{11}(a_p).
\end{aligned}$$

Here $\delta = \max\{U_{11} - U_{i1} : i \geq 2\} > 0$ according to the hypothesis that U is strictly column pointwise dominant. We deduce that $M_{11}(a_p) \leq 1/\delta$. Since for $i \geq 2$ we have $0 \leq -M_{1i}(a_p) \leq -\sum_{k\geq 2} M_{1k}(a_p) \leq M_{11}(a_p)$, we deduce the existence of a subsequence, that we still denote by (a_p), such that $M_{11}(a_p)$ and $\zeta(a_p)$ converge, as $p \to \infty$, to a nonnegative number M_{11} and to a nonnegative vector ζ. Again from $M(a_p)W(a_p) = \mathbb{I}_{n+1}$ we deduce that

$$1 = (a_p + U_{11})M_{11}(a_p) + \sum_{i=2}^{n+1} M_{1i}(a_p)U_{i1} \leq (a_p + U_{11})M_{11}(a_p),$$

which implies that

$$M_{11} \geq \frac{1}{U_{11}}.$$

Now we shall prove that $\xi(a_p)$ remains bounded. In fact from $W(a_p)M(a_p) = \mathbb{I}_{n+1}$ we deduce that for all $i \geq 2$

$$0 = U_{i1}M_{11}(a_p) + (a_p + U_{ii})M_{i1}(a_p) + \sum_{k \neq 1,i} U_{ik}M_{k1}(a_p)$$
$$\leq U_{i1}M_{11}(a_p) + (a_p + U_{ii})M_{i1}(a_p),$$

proving that

$$0 \leq -M_{i1}(a_p) \leq \frac{U_{i1}M_{11}(a_p)}{a_p + U_{ii}}.$$

Since $M_{11}(a_p)$ remains bounded and U_{ii} is positive, we deduce that $\xi(a_p)$ remains also bounded, and again via a subsequence, if necessary, it converges to a nonnegative vector ξ.

Furthermore, from Schur's formula, for the inverse by blocks, it holds that

$$(a_p\mathbb{I}_n + V)^{-1} = \Lambda(a_p) - \frac{1}{M_{11}(a_p)}\xi(a_p)\zeta(a_p)'.$$

Notice that V is nonnegative, strictly column pointwise dominant and satisfies the domination principle. Since the size of V is n, we conclude via induction hypothesis, that it is nonsingular. This shows that $\lim_p(a_p\mathbb{I}_n + V)^{-1} = V^{-1}$ exists and it is an M-matrix. From there, we conclude that $\Lambda(a_p)$ converges. This shows that $M(a_p)$ converges to a matrix M and the limit has to be a Z-matrix. Since $M(a_p)(a_p\mathbb{I} + U) = \mathbb{I}$, we deduce that

$$MU = \mathbb{I},$$

showing that U is nonsingular and $U^{-1} = M$. The fact that M is an M-matrix follows from the fact that U is nonnegative.

Corollary 2.19 *Assume that U is a nonnegative matrix which satisfies either the CMP or the domination principle. Then, for any positive diagonal matrix L the matrix $L + U$ is nonsingular and its inverse is an M-matrix, which is also row diagonally dominant when U satisfies the CMP.*

Proof Let us first consider the case U satisfies the CMP. Then from Lemma 2.7 the matrix $L + U$ satisfies the CMP and U is column pointwise dominant, which implies that $L + U$ is strictly column pointwise dominant. The result in this case follows from Theorem 2.16.

Now assume that U satisfies the domination principle. Consider the diagonal matrix D defined by $D_{ii} = \sum_j U_{ij}$ and $J = \{i : D_{ii} = 0\}$ which is exactly the index set of zero rows for U. We modify this diagonal by redefining $D_{ii} = 1$ for $i \in J$. Let us take $V = D^{-1}U$, which clearly satisfies the domination principle and $V\mathbb{1} \leq \mathbb{1}$ with equality on J^c. We shall prove that V satisfies the CMP. Let $x \in \mathbb{R}^I$ be such that

$$(Vx)_i \leq 1,$$

when $x_i \geq 0$. Clearly if $i \in J$ we have $(Vx)_i = 0$ and therefore we have

$$(Vx)_k \leq (V\mathbb{1})_k,$$

for those $k \in \{i : x_i > 0\} \subseteq \{i : x_i \geq 0\}$. Decompose $x = x^+ - x^-$ in its positive and negative part to get

$$(Vx^+)_k \leq (V(x^- + \mathbb{1}))_k,$$

for those k such that $x_k^+ > 0$. From the domination principle we get

$$Vx \leq V\mathbb{1} \leq \mathbb{1},$$

and V satisfies the CMP. Hence $D^{-1}L + D^{-1}U$ is nonsingular and its inverse is an M-matrix, from where $L + U$ is also nonsingular and its inverse is an M-matrix, proving the result.

2.3   Probabilistic Interpretation of M-Matrices and Their Inverses

In this section we introduce the basic facts needed on discrete and continuous Markov chains in order to interpret M-matrices in probabilistic terms. We shall see that the concept of M-matrix is intimately related to the notion of infinitesimal generator and the inverse of an M-matrix to the notion of Green potential (see Corollary 2.23). Our presentation will endeavor to be as self contained as possible and introduce only essential concepts. For a general treatment of Markov chains, please see the excellent books on the subject such as [1] and [33], among others.

Assume that P is a substochastic matrix over the set I. We consider the following stochastic extension of P. Take $\partial \notin I$ and define $\bar{I} = I \cup \{\partial\}$. This is like a compactification of I and ∂ will be an absorbing state. We define

$$\bar{P}_{ij} = \begin{cases} P_{ij} & i, j \neq \partial \\ 1 - \sum_{k \in I} P_{ik} & i \neq \partial, j = \partial \\ 1 & i = j = \partial \end{cases}$$

Consider (X_m) the Markov chain associated with $\bar{P}$. This means, for all $m \geq 0$ and all $i_0, \cdots, i_{m+1} \in \bar{I}$ we have

$$\mathbb{P}(X_0 = i_0, \cdots, X_m = i_m, X_{m+1} = i_{m+1}) = \mathbb{P}(X_0 = i_0) \prod_{k=0}^{m} \bar{P}_{i_k i_{k+1}}. \qquad (2.2)$$

Here $\mathbb{P}$ is the probability measure associated with the Markov chain. Also, $\mathbb{P}_i$ and $\mathbb{E}_i$ will denote the probability distribution and the expectation for this Markov chain, when the starting point is $i \in \bar{I}$. Notice that if $X_m = \partial$ then $X_{m+l} = \partial$ for all $l \geq 0$, that is ∂ is an absorbing state. In this situation $X = (X_m)$ is called a **killed Markov chain** on I.

Let us turn to the notion of irreducibility. In what follows P^m is the standard matrix powers of P whose elements are P_{ij}^m (as usual $P^0 = \mathbb{I}$). The state $i \in I$ is connected to state $j \in I$ if $P_{ij}^m > 0$ for some $m \geq 0$. We denote by $i \rightsquigarrow j$, when i is connected to j. This relation $\rightsquigarrow$ is reflexive and transitive. Notice that $i \rightsquigarrow j$ when i, j are different, is equivalent to the existence of a finite path $i_0 = i, \cdots, i_m = j$, in I, such that $P_{i_\ell i_{\ell+1}} > 0$ for $\ell = 0, \cdots, m - 1$. This implies that m can be chosen smaller or equal to the size of P.

In the theory of Markov chains, P is said to be **irreducible** whenever $\rightsquigarrow$ induces only one equivalence class, that is all states connect to each other. We point out that in many situations we can reduce the study of a particular property to the case where P is irreducible. Let us start with the following result.

Proposition 2.20 *Assume P is a nonnegative matrix, whose row sums are bounded by 1. Then*

(i) *If $\mathbb{I} - P$ is nonsingular then P is substochastic, that is, at least one row sum is smaller than one. Also, P^m tends to 0 at an exponential rate and $(\mathbb{I} - P)^{-1} = \sum_{m \geq 0} P^m$.*

(ii) *If P is irreducible and substochastic then $\mathbb{I} - P$ is nonsingular.*

(iii) *$\mathbb{I} - P$ is nonsingular if and only if every $i \in I$ is connected to a site were P loses mass or i itself loses mass.*

Proof As a general remark recall that the Perron-Frobenius eigenvalue of a substochastic matrix cannot exceed one.

(i). Since $\mathbb{I} - P$ is nonsingular then P is substochastic and the spectral radius of P must be smaller than one. This implies that the series $\sum_{m \geq 0} P^m$ converges and it is simple to prove that its limit is the inverse of $\mathbb{I} - P$. The exponential decay for P^m is given by the Perron-Frobenious eigenvalue.

(ii). If P is substochastic, all positive integer powers of P are also substochastic. Indeed, since $P\mathbb{1} \leq \mathbb{1}$ we obtain for $m \geq 2$

$$P^m\mathbb{1} \leq \cdots \leq P^2\mathbb{1} \leq P\mathbb{1} \leq \mathbb{1},$$

with at least one strict inequality at some coordinate.

We now use the irreducibility of P. Consider $n = \#I$, then

$$S = \frac{1}{n+1} \sum_{k=0}^{n} P^k$$

is positive and $S\mathbb{1} < \mathbb{1}$. This fact and the property $P\mathbb{1} \leq \mathbb{1}$ with strict inequality at least at one coordinate imply that $SP\mathbb{1} < \mathbb{1}$. Notice that $SP = PS$ is a positive matrix, since no row of P can be the zero row (except for the trivial case $n = 1$). The Perron-Frobenius principal eigenvalue θ of SP is simple and we shall prove that it is strictly smaller than 1. Indeed, take v a positive eigenvector associated with θ and let i_0 be a coordinate where v_{i_0} is maximal. Then,

$$\theta v_{i_0} = (SPv)_{i_0} - \sum_{j} (SP)_{i_0 j} v_j \leq v_{i_0} \sum_{j} (SP)_{i_0 j} < v_{i_0},$$

proving that $\theta < 1$. The conclusion is that the principal eigenvalue β of P, which is the spectral radius of P, is also strictly smaller than one. Since $Pv \neq 0$ the equalities $PSPv = \theta Pv = (SP)Pv$ imply that Pv is an eigenvector of SP associated with θ. The fact that this eigenvalue is simple, allows us to deduce that $Pv = \gamma v$, that is, v is also an eigenvector of P and $\gamma \leq \beta$ (actually $\gamma = \beta$ by the Perron-Frobenius's theorem). Now we prove that P^m decreases exponentially fast

$$P_{ij}^m \leq \frac{1}{v_j} \sum_{k} P_{ik}^m v_k = \frac{\gamma^m v_i}{v_j} \leq C\gamma^m,$$

where $C = \max\{\frac{v_s}{v_t} : s, t \in I\}$, and the result follows in this case because the series $\sum_m P^m$ is convergent and therefore $(\mathbb{I} - P)^{-1} = \sum_m P^m$. This can also be proved by means of the Gelfand's formula that states

$$\lim_{k \to \infty} \| P^k \|^{1/k} = \beta < 1.$$

(iii). Consider the equivalence relation $i \sim j$ if $i \leadsto j$ and $j \leadsto i$. We denote by $I_1, \cdots, I_r$ the respective equivalence classes. The extended relation $I_s \leadsto I_t$ given by: There exists a couple $(i, j) \in I_s \times I_t$ such that $i \leadsto j$; is a partial order relation. We also write $i \leadsto I_t$ when i is connected to some $j \in I_t$.

When $r = 1$, that is, all states communicate, every node is connected to a node that loses mass, showing the result in this case.

In what follows we assume $r > 1$. It is quite obvious to show that

$$\neg [I_s \rightsquigarrow I_t] \Rightarrow P_{I_s I_t} = 0.$$

So, after a permutation we can assume that P is a block upper triangular matrix.

Assume that every node $i \in I$ loses mass or it is connected to a one that does so. Since each block $P_{I_s I_s}$ is irreducible we obtain that at least on site in I_s loses mass for $P_{I_s I_s}$. Then each block $\mathbb{I} - P_{I_s I_s}$ is nonsingular and therefore $\mathbb{I} - P$ is also non-singular.

Conversely, if there is a node i which is not losing mass and it is not connected to anyone that loses mass then we can find a I_s with the properties: $i \rightsquigarrow I_s$ and I_s is not connected to any other class I_t for different s, t. In fact, it is enough to take any maximal element, with respect to $\rightsquigarrow$, in the partially ordered set $\{I_t : i \rightsquigarrow I_t\}$. Hence i is connected to I_s and every element in I_s connects only with elements in I_s. Thus $P_{I_s I_s}$ is a stochastic matrix and a fortiori

$$(\mathbb{I} - P_{I_s I_s})\mathbb{1} = 0,$$

proving that $\mathbb{I} - P_{I_s I_s}$ is singular and hence $\mathbb{I} - P$ is also singular.

We have proved that if $\mathbb{I} - P$ is nonsingular its inverse is $U = (\mathbb{I} - P)^{-1} = \sum_{m \geq 0} P^m$. We interpret now this inverse. We notice that

$$U_{ij} = \sum_{m \geq 0} P_{ij}^m = \sum_{m \geq 0} \mathbb{E}_i (\mathbb{1}_{X_m = j}) = \mathbb{E}_i \left(\sum_{m \geq 0} \mathbb{1}_{X_m = j} \right).$$

This means that U_{ij} is the expected amount of time spent by the chain at site j when starting from site i. This is finite because the chain is losing mass. In the language of potential theory $(\mathbb{I} - P)^{-1}$ is the **Green potential** of the Markov chain (X_m) on I killed at ∂. This matrix is used to solve the equation

$$\bar{Q} u = -\psi \text{ with boundary condition } u_\partial = 0.$$

Here $\bar{Q} = -(\mathbb{I} - \bar{P})$ is the **infinitesimal generator** of the Markov chain X on $\bar{I}$ and ψ is a function defined on $\bar{I}$. Clearly we need $\psi(\partial) = 0$ and in this case the solution on I is given by

$$u_I = U(w_I),$$

in other words for $i \in I$ we have

$$
\begin{aligned}
u_i &= \sum_{j \in I} U_{ij} \psi(j) = \sum_{j \in I} \sum_{m=0}^{\infty} \mathbb{E}_i(\mathbb{1}_{X_m=j}) \psi(j) = \sum_{m=0}^{\infty} \mathbb{E}_i \left(\sum_{j \in I} \mathbb{1}_{X_m=j} \psi(j) \right) \\
&= \sum_{m=0}^{\infty} \mathbb{E}_i(\psi(X_m)) = \mathbb{E}_i \left(\sum_{m=0}^{\infty} \psi(X_m) \right).
\end{aligned}
$$

Remark 2.21 This is analogous to solve in $\mathbb{R}^3$ the equation $\frac{1}{2} \Delta u = -\psi$, whose solution (under suitable conditions on ψ) is given by the integral operator

$$
u(x) = U \psi(x) = C \int_{\mathbb{R}^3} \frac{1}{\|x - y\|} \psi(y) dy
$$

for the constant $C = 1/(2\pi)$. In this situation $\frac{1}{2} \Delta$ is the infinitesimal generator of a Brownian Motion (B_t) and U is the Newtonian potential in $\mathbb{R}^3$ (up to a multiplicative constant). The solution is given by $u(x) = \mathbb{E}_x \int_0^{\infty} \psi(B_t) dt$.

In summary, a diagonally dominant M-matrix of the form $\mathbb{I} - P$ is the negative of the infinitesimal generator of a killed Markov chain and $(\mathbb{I} - P)^{-1}$ is the potential of this chain. Conversely we have

Proposition 2.22 *Assume that (Y_m) is a Markov chain on a set J, with transition matrix R and $I \subset J$ is a subset of transient states, that is for all $i, j \in I$*

$$
U_{ij} =: \mathbb{E}_i \left(\sum_{\ell \geq 0} \mathbb{1}_{Y_m=j} \right) < \infty.
$$

Then there exists a substochastic matrix P such that $U = (\mathbb{I} - P)^{-1}$.

Before proving this result we need an application of the **strong Markov property**. This is the fact that the Markov property (2.2) holds for some random times. The times we need are the hitting and entrance times of subsets of J. So, given a Markov chain (Y_m) taking values in J and $K \subset J$ we denote by

$$
\mathscr{T}_K = \inf\{n \geq 0 : Y_n \in K\}, \text{ and } \mathscr{T}_K^+ = \inf\{n \geq 1 : Y_n \in K\}
$$

the first time the chain hits K and enters to K, respectively. We take the convention that the infimum over the empty set is ∞. It is important to notice that it may happen that the chain starts at K, that is $Y_0 \in K$, and never returns to K in which case $\mathscr{T}_K = 0 < \mathscr{T}_K^+ = \infty$. When $K = \{i\}$ we denote $\mathscr{T}_{\{i\}}$ simply by $\mathscr{T}_i$, and $\mathscr{T}_{\{i\}}^+$ by $\mathscr{T}_i^+$.

So, assume that $\mathscr{T}$ is one of these random times. On the set where $\mathscr{T} < \infty$ we take $Y_{\mathscr{T}}$ as the random variable defined by $Y_{\mathscr{T}}(\omega) = Y_m(\omega)$, for $m = \mathscr{T}(\omega)$. The values of $Y_{\mathscr{T}}$ when $\mathscr{T} = \infty$ will play no role in what follows. The form of the strong Markov property that we need is

$$\mathbb{E}_i\left(f(Y_{\mathscr{T}+\ell}), \mathbb{1}_{\mathscr{T}<\infty}\right) = \mathbb{E}_i\left(\mathbb{E}_{Y_{\mathscr{T}}}(f(Y_\ell)), \mathbb{1}_{\mathscr{T}<\infty}\right), \qquad (2.3)$$

for all $i \in J$, all $\ell \geq 0$ and all functions f.

Now we have all the tools to prove the proposition.

Proof The matrix P is constructed with the aid of the strong Markov property. For that purpose take $\mathscr{T} = \mathscr{T}_I$ and define for $i, j \in I$

$$P_{ij} = \mathbb{P}_i(\mathscr{T} < \infty, Y_{\mathscr{T}} = j),$$

which is the probability that the first return to I, starting at i happens at site j. Notice that

$$U_{ij} = \mathbb{E}_i\left(\sum_{\ell\geq 0} \mathbb{1}_{Y_\ell=j}\right) = \mathbb{E}_i\left(\mathbb{1}_{\mathscr{T}<\infty}\sum_{\ell\geq\mathscr{T}} \mathbb{1}_{Y_\ell=j}\right)$$

$$= \sum_{k\in I} \mathbb{E}_i\left(\mathbb{1}_{\{\mathscr{T}<\infty, Y_{\mathscr{T}}=k\}}\sum_{\ell\geq\mathscr{T}} \mathbb{1}_{Y_\ell=j}\right)$$

From the strong Markov property we have

$$\mathbb{P}_i(\mathscr{T} < \infty, Y_{\mathscr{T}} = k, Y_{\mathscr{T}+m} = j) = \mathbb{P}_i(\mathscr{T} < \infty, Y_{\mathscr{T}} = k)\mathbb{P}_k(Y_m = j).$$

Since $\mathscr{T}$ is the first hitting time of I we have $\mathbb{P}_i(\mathscr{T} < \infty, Y_{\mathscr{T}} = k) = P_{ik}$ and therefore

$$U_{ij} = \sum_{k\in I} P_{ik}U_{kj} \quad \text{for } i \neq j,$$

and

$$U_{ii} = 1 + \sum_{k\in I} P_{ik}U_{ki}.$$

This is exactly the same as

$$PU = U - \mathbb{I},$$

or $\mathbb{I} = (\mathbb{I} - P)U$, from where the result follows.

We remark that $U = \sum_{m \geq 0} P^m$ and therefore the killed Markov chain, associated with the substochastic matrix P, has U as its potential matrix.

Thus, there is a one-to-one correspondence between potential matrices of transient Markov chains and inverses of row diagonally dominant M-matrices of the form $\mathbb{I} - P$ where P is a substochastic matrix. This is why we call these special inverse M-matrices Markov potentials.

We summarize this discussion in the next corollary.

Corollary 2.23 *A matrix U is an inverse M-matrix if and only if there exists a positive diagonal matrix $D > 0$ and a substochastic matrix P such that $U = ((\mathbb{I} - P)D)^{-1}$. If U is a potential we can take $D = \kappa \mathbb{I}$ for some constant $\kappa > 0$ and furthermore, if U is a Markov potential we can take $\kappa = 1$.*

Proof Clearly if $U = ((\mathbb{I} - P)D)^{-1}$ then U is an inverse M-matrix. Conversely, if $M = U^{-1}$ is an M-matrix, then there exists a diagonal matrix $E > 0$ such that ME is a row diagonally dominant M-matrix. Indeed, it is enough to take $E = U\mathbb{1}$ because $ME\mathbb{1} = \mathbb{1}$. Hence, there exists a constant $\kappa > 0$ and a substochastic matrix P such that $ME = \kappa(\mathbb{I} - P)$. It is enough to take $D = \kappa E^{-1}$. The rest of the corollary follows from definition.

The previous discussion justify the following definition.

Definition 2.24 The M-matrix M is said to be **irreducible** if for all different i, j there exists a finite path $i_0 = i, \cdots, i_k = j$, which we can assume are all different, such that for all $l = 0, \cdots k - 1$ we have $M_{i_l i_{l+1}} < 0$.

If we decompose $M = (\mathbb{I} - P)D$ as in the previous corollary, then M is irreducible if and only if P is irreducible. The next result characterizes irreducibility of M in terms of its inverse U.

Lemma 2.25 *M is irreducible if and only if $U > 0$.*

Proof Consider the decomposition of $M = (\mathbb{I} - P)D$ provided by Corollary 2.23. As noted before, M is irreducible if and only if P is irreducible: for all different i, j there exists $m \geq 0$ such that $P_{ij}^m > 0$. This fact is equivalent to

$$D_{ii}U_{ij} = (\mathbb{I} - P)_{ij}^{-1} = \sum_{k \geq 0} P_{ij}^k > 0$$

and result is proved.

In the next sections we provide a representation of U without the use of D or κ. The first thing we will do is to give a representation for a general potential.

2.3.1 Continuous Time Markov Chains

A representation of a general row diagonally dominant M-matrix will be given by
continuous time Markov chains. There are several ways to construct these processes.
The most natural way to do so, for a non-probabilist, is to use semigroup theory.

Consider the M-matrix M indexed by I. The semigroup associated with
$Q = -M$ is given by

$$P_t = e^{Qt} = \sum_{n=0}^{\infty} \frac{t^n}{n!} Q^n = \lim_{m \to \infty} \left(\mathbb{I} + \frac{t}{m} Q \right)^m \quad \text{for } t \geq 0.$$

Q is the generator of this semigroup, that is $Q = \frac{dP_t}{dt}\big|_{t=0}$, and P_t solves the so-
called Chapman-Kolmogorov equations

$$\frac{dP_t}{dt} = QP_t = P_t Q, \quad P_0 = \mathbb{I}.$$

For large m the matrix $\mathbb{I} + \frac{t}{m} Q$ is nonnegative and therefore for each $t \geq 0$ the
matrix P_t is also nonnegative. Notice that $P_t \mathbb{1}$ satisfies the equation

$$\frac{dP_t \mathbb{1}}{dt} = P_t Q \mathbb{1}.$$

By assumption $Q\mathbb{1} = -M\mathbb{1} \leq 0$ and P_t preserves positivity, then we conclude that
each coordinate of $P_t \mathbb{1}$ is a decreasing function of t. In particular

$$P_t \mathbb{1} \leq P_0 \mathbb{1} = \mathbb{1}$$

with equality at all indexes if and only if $M\mathbb{1} = 0$, which is not allowed since M is
assumed to be nonsingular. In this context (P_t) is called a substochastic semigroup.
We shall discuss later on how to construct a Markov process on a enlarged space $\bar{I}$
whose trace on I induces the semigroup (P_t).

We introduce the following definition for future references.

Definition 2.26 A semigroup $(P_t)_{t \geq 0}$ is said to be a **substochastic Markov
semigroup** if P_t are nonnegative matrices and $P_t \mathbb{1} \leq \mathbb{1}$, with strict inequality at
some coordinate. If $P_t \mathbb{1} = \mathbb{1}$, for all t, we say it is a **Markov semigroup**.

Now, we prove the following representation of M and its inverse.

Theorem 2.27 *Assume that M is a nonsingular matrix. Then, M is row diagonally
dominant M-matrix if and only if $P_t = e^{-Mt}$ is a substochastic Markov semigroup.*

Moreover, under any of these equivalent conditions, we have the relations

(i) $\int_0^\infty P_t dt = M^{-1}$;

(ii) $P_t \mathbb{1}$ *is a strictly decreasing function of t, in particular for all $t > 0$ we have*
$P_t \mathbb{1} < \mathbb{1}$;

(iii) *there exists a constant $0 < \rho < 1$ such that for all $v \in \mathbb{R}^I$ and $t \geq 0$*

$$|P_t v| \leq \|v\|_\infty \, \rho^{t-1}.$$

Remark 2.28 In Chap. 5, Theorem 5.34, we shall study the equivalence between generators of semigroups and Z-matrices.

Proof Assume first that M is a row diagonally dominant M-matrix. As before we denote by $Q = -M$. Then, for all $t \geq 0$

$$P_t = \lim_{n \to \infty} (\mathbb{I} + \frac{t}{n} Q)^n, \tag{2.4}$$

is a positive semigroup, because for large n the matrix $(\mathbb{I} - \frac{t}{n} M)^n$ is nonnegative.

A simple computation shows that $\frac{d}{dt} P_t \mathbb{1} = P_t(Q\mathbb{1}) \leq 0$, proving that $P_t \mathbb{1}$ is a decreasing function of t (on each coordinate). Since $P_0 \mathbb{1} = \mathbb{1}$, we conclude that (P_t) is a substochastic Markov semigroup (we prove below that $P_t \mathbb{1} < \mathbb{1}$).

Conversely, assume that (P_t) is a substochastic Markov semigroup. The fact that P_t is nonnegative and (2.4) implies that M is a Z-matrix. We shall prove now that in fact is row diagonally dominant. Since $P_t \mathbb{1} \leq \mathbb{1} = P_0 \mathbb{1}$, for all t, we deduce

$$M\mathbb{1} = -\frac{d}{dt} P_t \mathbb{1}|_{t=0} \geq 0,$$

proving the claim.

In what follows we assume that M is a row diagonally dominant M-matrix and denote $Q = -M$. Consider a fixed $\theta > 0$ and the semigroup $S_t = e^{-\theta t} P_t = e^{(Q-\theta\mathbb{I})t}$. The matrix $Q - \theta\mathbb{I}$ has negative row sums, that is $(Q - \theta\mathbb{I})\mathbb{1} < 0$. Since

$$\frac{dS_t}{dt} = (Q - \theta\mathbb{I})S_t = S_t(Q - \theta\mathbb{I}),$$

we obtain that $\frac{d}{dt} S_t \mathbb{1} < 0$ (notice that each row of the nonsingular and nonnegative matrix S_t has at least one positive term). Therefore $S_t \mathbb{1}$ is strictly decreasing. In particular S_1 is a nonnegative nonsingular matrix and $\beta = \max_i (S_1 \mathbb{1})_i \leq e^{-\theta} < 1$.

Then for all $v \in \mathbb{R}^I$ and $t > 1$, we obtain

$$|S_t v| \leq S_t |v| \leq \|v\|_\infty S_t \mathbb{1} \leq \|v\|_\infty S_{[t]} \mathbb{1} = \|v\|_\infty S_1^{[t]} \mathbb{1} \leq \|v\|_\infty \beta^{[t]} \mathbb{1},$$

where $[t]$ denotes the integer part of t. We get that S_t is integrable in $\mathbb{R}_+$ and moreover

$$(S_u - \mathbb{I}) = \int_0^u \frac{dS_t}{dt} dt = \int_0^u (Q - \theta\mathbb{I}) S_t dt,$$

which converges towards $-\mathbb{I}$ when $u \to \infty$. We deduce that

$$\int_0^\infty e^{-\theta t} P_t dt = -(Q - \theta\mathbb{I})^{-1}.$$

Using the Monotone Convergence Theorem (here we use that $P_t \geq 0$) and the fact that $(Q - \theta\mathbb{I})^{-1}$ converges toward Q^{-1} when $\theta \to 0$, we obtain

$$\int_0^\infty P_t dt = -(Q)^{-1} = M^{-1},$$

proving part (i).

Let us now prove part (ii). We know that $P_t \mathbb{1}$ is decreasing on t, nonnegative and integrable on $\mathbb{R}_+$. Thus $P_t \mathbb{1}$ tends to 0 as $t \to \infty$. The function of $z \in \mathbb{C}$ given by

$$(P_z \mathbb{1})_i = \sum_{n \geq 0} \frac{(Q^n \mathbb{1})_i}{n!} z^n,$$

is analytic on $\mathbb{C}$, because Q^n has at most exponential growth. Then, for each coordinate i, the function $\left(\frac{dP_t}{dt} \mathbb{1}\right)_i$ is nonnegative and not identically 0. So it has isolated zeros. The conclusion is that $P_t \mathbb{1}$ is strictly decreasing. In particular $P_1 \mathbb{1} < \mathbb{1}$.

Part (iii) follows from (ii) by considering $\rho = \max_i (P_1 \mathbb{1})_i < 1$. Indeed

$$|P_t v| \leq P_t |v| \leq \|v\|_\infty P_t \mathbb{1} \leq \|v\|_\infty P_{[t]} \mathbb{1} \leq \|v\|_\infty \rho^{[t]} \mathbb{1} \leq \|v\|_\infty \rho^{t-1} \mathbb{1},$$

and the result is shown.

Relation (i) in the previous Theorem gives a representation of a potential U of a transient continuous time Markov chain with generator $Q = -U^{-1}$ as

$$U_{ij} = \int_0^\infty (e^{-U^{-1}t})_{ij} \, dt.$$

To interpret this formula it is important to enlarge the space and to add an absorbing state ∂ as we did for discrete time Markov chains. Consider $\bar{M}$ a matrix indexed by $\bar{I} = I \cup \{\partial\}$ where $\bar{M}_{II} = M$ and

$$\bar{M}_{i\partial} = M_{ii} - \sum_{j \in I, j \neq i} M_{ij} \text{ for } i \neq \partial,$$

$$\bar{M}_{\partial i} = 0 \text{ for all } i \in \bar{I}.$$

Then $\bar{M}$ is a Z-matrix and it is singular because $\bar{M}\mathbb{1} = 0$. Let us take $\bar{Q} = -\bar{M}$. The semigroup $\bar{P}_t = e^{\bar{Q}t}$ is positive-preserving and $\bar{P}_t\mathbb{1} = \mathbb{1}$. Associated with this semigroup there is a continuous time Markov chain (Y_t) with values on $I \cup \{\partial\}$ such that:

- the trajectories of (Y_t) are right continuous with left limits;
- (Y_t) satisfies the strong Markov property;
- $(\bar{P}_t v)_i = \mathbb{E}_i(v(Y_t))$ for all $i \in \bar{I}$ and $v \in \mathbb{R}^{\bar{I}}$;
- ∂ is an absorbing state, that is $\mathbb{P}_\partial(Y_t = \partial \text{ all } t \geq 0) = 1$;
- if $\mathbb{1}_I$ denotes the vector equal to 1 on I and 0 on ∂, then for all $i \in I$ we have $(\bar{P}_t \mathbb{1}_I)_i = \mathbb{P}_i(\mathscr{T}_\partial > t) < 1$. This means that under our conditions for any positive t there exists a set of trajectories, of positive probability $\mathbb{P}_i$ that are absorbed at ∂ before t;
- $U_{ij} = \int_0^\infty \mathbb{P}_i(Y_t = j)dt$ is the mean total time expended by the chain at site j starting from i.

A probabilistic construction of (Y_t) is usually done as follows. An exponential variable (of parameter one) is a random element Z whose distribution is exponential: $\mathbb{P}(Z > t) = e^{-t}$. For the construction we need a countable set $(Z_m)_{m \geq 1}$ of independent identically distributed exponential random variables. These variables regulate the times at which the process (Y_t) jump.

We first construct the **skeleton** $(X_n)_{n \geq 0}$ of (Y_t), which is a Markov chain that records the visited sites. Its transition matrix indexed by I is given by

$$P_{ij} = \frac{-M_{ij}}{M_{ii}} = \frac{Q_{ij}}{-Q_{ii}} \text{ if } i \neq j, \text{ and } P_{ii} = 0.$$

The fact that M is a row diagonally dominant M-matrix implies that P is substochastic. Extend as before P to $\bar{I} = I \cup \{\partial\}$ and consider a Markov chain $(X_n)_{n \geq 0}$ with transition probability matrix $\bar{P}$ on $\bar{I}$. This process (X_n) is chosen independent of the process (Z_m).

Now we can construct (Y_t). At the initial state $X_0 = i \in I$ the process sojourns a random time $T_1 = \frac{1}{-Q_{ii}} Z_1$ and then jumps to $j = X_1 \neq i$ where the process (Y_t) waits a random time $T_2 = \frac{1}{-Q_{jj}} Z_2$ and then jumps again to X_2 and so on. Then $Y_t = X_0$ for $0 \leq t < T_1$, $Y_t = X_1$ for $T_1 \leq t < T_1 + T_2$ and so on. This is done until the chain (X_n) jumps over ∂ where (X_n) is absorbed, and (Y_t) is absorbed too. This is a version of the continuous time Markov chain with generator $\bar{Q}$ and semigroup $(\bar{P}_t)$.

The strong Markov property satisfied by (Y_t) allows us to show that the following restriction on I is effectively a semigroup, for $i \in I$ and $v \in \mathbb{R}^I$

$$(P_t v)_i = \mathbb{E}_i (v(Y_t), \mathcal{T}_\partial > t),$$

which is the mean average of $v(Y_t)$ on the set of trajectories that are not absorbed before t.

It is evident that $P_t v = (\bar{P}_t \bar{v})|_I$, where $\bar{v}$ is the extension of v to $\bar{I}$ with $\bar{v}(\partial) = 0$. Since $Qv = (\bar{Q}\bar{v})|_I$ then P_t is the semigroup associated with the generator Q on I. The semigroup (P_t) is called substochastic because it loses mass $P_t \mathbb{1} < \mathbb{1}$.

Then the representation of $U = M^{-1}$ as $U_{ij} = \int_0^\infty (P_t)_{ij} \, dt$ is interpreted as

$$U_{ij} = \int_0^\infty \mathbb{E}_i (Y_t = j, \mathcal{T}_\partial > t) \, dt = \mathbb{E}_i \left(\int_0^\infty \mathbb{1}_{Y_t = j, \mathcal{T}_\partial > t} \, dt \right)$$

$$= \mathbb{E}_i \left(\int_0^\infty \mathbb{1}_{Y_t = j} \, dt \right),$$

which is the mean expected total time spent by the chain (Y_t) at site j, starting from i, before absorption. This is the meaning of the potential U of the transient continuous Markov chain (Y_t).

To get a discrete time interpretation of M and $U = M^{-1}$ take $K = \max_i \{M_{ii}\}$. For any $\kappa \geq K$ the matrix $P = \mathbb{I} - \frac{1}{\kappa} M$ is nonnegative, substochastic and

$$M = \kappa (\mathbb{I} - P).$$

Notice that P depends on κ.

We can associate a Markov chain to P taking values again on $\bar{I}$. This Markov chain is not necessarily the skeleton (X_n) and for this reason we denote it by $(\hat{X}_n)$. The potential of $(\hat{X}_n)$ is then

$$W = (\mathbb{I} - P)^{-1} = \sum_{n \geq 0} P^n = \kappa M^{-1} = \kappa U.$$

That is the potential of the continuous time Markov chain (Y_t) is proportional to the potential of the discrete time Markov chain $(\hat{X}_n)$. An interesting case happens when it is satisfied that $M_{ii} \leq 1$ for all i. In this situation we can take $\kappa = 1$ and then U is obviously the potential of a Markov chain.

2.3.2 *M-Matrices and h-Transforms*

Thus far we have provided a representation of any row diagonally dominant M-matrix as the generator of a continuous time Markov chain and the special form $M = \mathbb{I} - P$ in the setting of discrete time Markov chains. This is a partial representation of inverses M-matrices as potential matrices for these processes. We will now remove the row diagonally dominance condition and we assume that M is an M-matrix. The probabilistic interpretation of M and its inverse can be effected in several ways.

One of them is as follows. Let D be a positive diagonal matrix such that MD is diagonally dominant. For example, take $U = M^{-1}$ which is a nonnegative matrix and consider the diagonal matrix D defined by

$$D_{ii} = \sum_{j \in I} U_{ij}.$$

Then clearly $D^{-1}U\mathbb{1} = \mathbb{1}$ and then $MD = (D^{-1}U)^{-1}$ is an M-matrix and satisfies $MD\mathbb{1} = \mathbb{1}$, thus it is a row diagonally dominant M-matrix (see Theorem 2.5.3 in [38]). This new matrix has a representation in terms of a Markov chain, but the initial matrix M does not have a simple interpretation.

A more standard transformation used in probability theory is to consider

$$N = D^{-1}MD,$$

where as before we assume that MD is row diagonally dominant. Obviously, N is again a row diagonally dominant M-matrix. Then $Q = -N$ is the generator of a continuous time substochastic Markov semigroup (P_t), such that $V = N^{-1}$ is its potential. The original matrix $U = DVD^{-1}$ has a representation

$$U_{ij} = \frac{D_{ii}}{D_{jj}} V_{ij} = \frac{D_{ii}}{D_{jj}} \int_0^\infty (P_t)_{ij}\, dt.$$

The family $S_t = DP_t D^{-1}$ form also a semigroup of positive-preserving transformations which is not necessarily Markovian or substochastic because the inequality $S_t \mathbb{1} \le \mathbb{1}$ does not necessarily hold. Nevertheless (S_t) is integrable and

$$U = \int_0^\infty S_t dt.$$

This gives a representation of U as the potential of a semigroup. Here we have used very few properties of D, namely that it is a positive diagonal matrix and $MD\mathbb{1} > 0$. We claim this representation characterizes the inverse M-matrix class.

Theorem 2.29 *U is an inverse M-matrix if and only if there exists a positive-preserving continuous and integrable semigroup (S_t) such that $S_0 = I$ and*

$$U = \int_0^\infty S_t\, dt.$$

Remark 2.30 For an extension of this result for Z-matrices, see Theorem 5.34 in Chap. 5.

Proof The only part to be proved is that $U = \int_0^\infty S_t \, dt$, then U is an inverse M-matrix. By the Hille-Yosida theorem there exists a matrix $\tilde{Q}$ such that $S_t = e^{\tilde{Q}t}$. We give a proof of this fact here (see for example [36] for a proof in a general setting).

From continuity of (S_t) we have $\lim_{h \to 0} \frac{1}{h} \int_t^{t+h} S_u \, du = S_t$, for all $t \geq 0$. In particular, there exists a $a > 0$ such that $\|\frac{1}{a} \int_0^a S_u \, du - \mathbb{I}\| < 1$ which implies that $\frac{1}{a} \int_0^a S_u \, du$ is nonsingular.

Let us continue with $\int_0^a S_{h+u} \, du - \int_0^a S_u \, du = (S_h - \mathbb{I}) \int_0^a S_u \, du$, and then we get

$$\frac{1}{h} \left(\int_a^{a+h} S_u \, du - \int_0^h S_u \, du \right) = \frac{1}{h}(S_h - \mathbb{I}) \int_0^a S_u \, du.$$

Hence we have proved the identity

$$\frac{1}{h}(S_h - \mathbb{I}) = \frac{1}{h} \left(\int_a^{a+h} S_u \, du - \int_0^h S_u \, du \right) \left(\int_0^a S_u \, du \right)^{-1}.$$

The right side converges to $(S_a - \mathbb{I}) \left(\int_0^a S_u \, du \right)^{-1}$ when $h \downarrow 0$ and then

$$\lim_{h \downarrow 0} \frac{1}{h}(S_h - \mathbb{I}) = (S_a - \mathbb{I}) \left(\int_0^a S_u \, du \right)^{-1}.$$

Let us take $\tilde{Q} = \lim_{h \downarrow 0} \frac{1}{h}(S_h - \mathbb{I})$ which satisfies $\tilde{Q} = \frac{dS_t}{dt}\big|_{t=0} = (S_a - \mathbb{I}) \left(\int_0^a S_u \, du \right)^{-1}$ and therefore $(S_a - \mathbb{I}) = \tilde{Q} \int_0^a S_u \, du$.

Using the semigroup property, we get $\frac{dS_t}{dt}\big|_t = \tilde{Q} S_t = S_t \tilde{Q}$, which implies for all $t > 0$

$$S_t - \mathbb{I} = \tilde{Q} \int_0^t S_u \, du.$$

Now, we prove that $S_t = e^{\tilde{Q}t}$. Indeed, consider $R_t = e^{\tilde{Q}t} - S_t$ which satisfies $R_0 = 0$ and

$$\frac{dR_t}{dt} = \tilde{Q} R_t.$$

Hence for any $v \in \mathbb{R}^I$ the function $\psi(t) = \|R_t v\|^2$ is differentiable and

$$\frac{d\psi}{dt} = 2\langle \frac{dR_t}{dt} v, R_t v \rangle = 2\langle \tilde{Q} R_t v, R_t v \rangle \leq 2\|\tilde{Q}\| \, \|R_t v\|^2 = 2\|\tilde{Q}\| \psi(t).$$

Since $\psi(0) = 0$, we get from Gronwall's inequality that $\psi \equiv 0$ and the claim is proved.

In summary $\tilde{Q}$ is the generator of S_t and then

$$\tilde{Q}U = \int_0^\infty \tilde{Q} S_t \, dt = \int_0^\infty \frac{dS_t}{dt} \, dt = -S_0 = -\mathbb{I},$$

which implies that U is nonsingular and $U^{-1} = -\tilde{Q}$.

Now we need to prove that $M = -\tilde{Q}$ is an M-matrix. Since its inverse is U, a nonnegative matrix, this is equivalent to prove that M is a Z-matrix, which in this case is to prove that for all different i, j it holds $\tilde{Q}_{ij} \geq 0$. From $S_t = e^{\tilde{Q}t}$ we have for all $t > 0$

$$S_t = \mathbb{I} + \sum_{m \geq 1} \tilde{Q}^m \frac{t^m}{m!}.$$

Finally, since S_t is positive-preserving, we get $(S_t)_{ij} \geq 0$ for all different i, j. Therefore

$$0 \leq \tilde{Q}_{ij} + \sum_{m \geq 2} (\tilde{Q}^m)_{ij} \frac{t^{m-1}}{m!},$$

implying that $\tilde{Q}_{ij} \geq 0$ and the result is proved.

Remark 2.31 Take a constant $\kappa > 0$ and define $Q = \tilde{Q} - \kappa\mathbb{I}$, which is also a generator of a positive-preserving semigroup $P_t = e^{-\kappa t} S_t$. Since for large κ we have $Q\mathbb{1} < 0$ we obtain that $P_t \mathbb{1} \leq \mathbb{1}$, which indicates it is the semigroup of a killed Markov chain. Also we obtain that

$$U = \int_0^\infty e^{\kappa t} P_t \, dt,$$

which represents U as an element of the resolvent for a killed Markov chain.

To finish this section, we will give a natural representation of M in terms of what is called an h-**transform** in potential theory and probability theory. Here we shall assume that M is irreducible, in other words that $U = M^{-1} > 0$.

Notice that $M = a\mathbb{I} - T$ where T is a nonnegative irreducible matrix and a is a real number strictly greater than the spectral radius of T. The Perron-Frobenius theorem applied to T gives the existence of a maximum eigenvalue $\theta \in (0, a)$ and a positive eigenvector h that we can normalize to $h'\mathbb{1} = 1$. We have $Th = \theta h$ and moreover $Mh = (a - \theta)h$. Define the nonnegative matrix P as

$$P_{ij} = a^{-1} \frac{h_j}{h_i} T_{ij},$$

that is $P = a^{-1}H^{-1}TH$, where H is the diagonal matrix define by h, that is $H_{ii} = h_i$. Then, we have $P\mathbb{1} = \frac{\theta}{a}\mathbb{1}$ and therefore a substochastic irreducible matrix. The potential induced by P is

$$V = (\mathbb{I} - P)^{-1} = \sum_{m \geq 0} P^m.$$

Since $P^m = \frac{1}{a^m}H^{-1}T^m H$, we obtain

$$V = H^{-1} \sum_{m \geq 0} \left(\frac{T}{a}\right)^m H = H^{-1}(\mathbb{I} - T/a)^{-1}H = aH^{-1}UH,$$

or $U = \frac{1}{a}HVH^{-1}$ which means

$$U_{ij} = \frac{1}{a}\frac{h_i}{h_j}V_{ij}.$$

We notice that $Uh = \frac{1}{a-\theta}h$ and the Perron-Frobenius eigenvalue of U is $\frac{1}{a-\theta}$.

Finally, the potential $W = \frac{1}{a}V$ corresponds to a continuous time Markov chain (Y_t) whose generator is $Q = -a(\mathbb{I} - P)$ and its semigroup $P_t = e^{Qt}$. Then, $U = HWH^{-1}$ and we get

$$U_{ij} = \frac{h_i}{h_j}\mathbb{E}_i \int_0^\infty \mathbb{1}_{\{Y_t = j\}}\,dt = \int_0^\infty \frac{h_i(P_t)_{ij}}{h_j}dt.$$

Consider the semigroup (not necessarily Markovian) $S_t(\bullet) = hP_t\left(\frac{1}{h}\bullet\right)$, where

$$S_t(f)_i = h_i P_t(f/h)_i = h_i \sum_j (P_t)_{ij}\frac{f(j)}{h(j)},$$

and its generator $\tilde{Q}(\bullet) = hQ(\frac{1}{h}\bullet)$ (here we identify vectors, diagonal matrices and functions over I). Notice that $\tilde{Q}(h) = Qh = -(a - \theta)h$, that is h is a eigenvector associated with a negative eigenvalue of $\tilde{Q}$ (the largest in absolute value). One obtains the representation

$$U = \int_0^\infty S_t\,dt,$$

and $U^{-1} = -\tilde{Q} = M$.

We notice that $h > 0$ is an eigenvector of the semigroup (S_t) and moreover $S_t h = e^{-(a-\theta)t}h$ converges exponentially fast to 0.

2.4 Some Basic Properties of Potentials and Inverse M-Matrices

Here we present some basic properties of potential matrices, which are simple to prove from their representation as potentials of Markov processes. Recall that a potential U can be represented as $U^{-1} = \kappa(\mathbb{I} - P)$, for a constant $\kappa > 0$ and a substochastic matrix P.

Lemma 2.32 *Assume that U is a potential (bi-potential, an inverse M-matrix). Let $J = \{\ell_1, \cdots, \ell_p\}$ be a subset of I and $A = U_J$ the corresponding principal submatrix. Then, A is a potential (respectively a bi-potential, an inverse M-matrix) and the right equilibrium potentials λ^U, λ^A satisfy*

$$\lambda^A \geq \lambda_J^U.$$

In particular, the set of roots of A and U satisfy $\{s : \ell_s \in \mathcal{R}(U)\} \subseteq \mathcal{R}(A)$ (see Definition 2.4). A similar relation holds for the left equilibrium potentials when U is a bi-potential. Finally, if U is a Markov potential then A is also a Markov potential.

Proof The fact that any principal submatrix of an M-matrix is still an M-matrix is well known (see [38] p. 119]).

Let us assume that U is a potential of size n. Using induction and a suitable permutation, it is enough to prove the result when A is the restriction of U to $\{1, \ldots, n-1\} \times \{1, \ldots, n-1\}$. Assume that

$$U = \begin{pmatrix} A & b \\ c' & d \end{pmatrix} \text{ and } U^{-1} = \begin{pmatrix} \Lambda & -\zeta \\ -\varrho' & \theta \end{pmatrix}. \tag{2.5}$$

where $b, c, \zeta, \varrho \in \mathbb{R}^{n-1}$ are nonnegative and $d, \theta \in \mathbb{R}$ are positive. Since

$$A^{-1} = \Lambda - \frac{1}{\theta}\zeta\varrho', \tag{2.6}$$

we obtain that the off diagonal elements of A^{-1} are non-positive, that is A is a Z-matrix. In order to prove that A is a potential is enough to show that $\lambda^A = A^{-1}\mathbb{1} \geq 0$.

Since U is a potential we have that $\Lambda\mathbb{1} - \zeta \geq 0$ and $\theta \geq \varrho'\mathbb{1}$. Therefore,

$$\lambda^A = A^{-1}\mathbb{1} = \Lambda\mathbb{1} - \frac{1}{\theta}\zeta\varrho'\mathbb{1} = \Lambda\mathbb{1} - \frac{\varrho'\mathbb{1}}{\theta}\zeta \geq \Lambda\mathbb{1} - \zeta = \lambda_J^U \geq 0,$$

showing also the desired relation between the right equilibrium potentials.

Now, we have to prove that if U is a Markov potential then A is also a Markov potential. We need to show the diagonal elements of A^{-1} are bounded by 1. This follows from the formula (2.6) because $A_{ii}^{-1} = A_{ii} - \frac{\zeta_i \rho_i}{\theta} \leq A_{ii} \leq 1$.

Remark 2.33 The probabilistic insight of this result is the following. Assume that U is the Markov potential of the chain (X_n) killed at ∂. Now, for $J \subset I$ we induce a process (Y_m) by recording only the visits to J. The strong Markov property shows that (Y_m) is also a killed Markov chain and it is not difficult to see that for $i, j \in J$ the mean number of visits to j starting from i is the same for both chains. That is $A = U_J$ is the potential of the Markov chain (Y_m). On the other hand the transition kernel Q of (Y_m) is obtained from the Schur decomposition

$$A = (\mathbb{I}_J - P_{JJ} - P_{JJ^c}(\mathbb{I}_{J^c} - P_{J^c J^c})^{-1} P_{J^c J})^{-1} = (\mathbb{I}_J - Q)^{-1}$$

Hence, a formula for Q is

$$Q = P_{JJ} + P_{JJ^c}(\mathbb{I}_{J^c} - P_{J^c J^c})^{-1} P_{J^c J}.$$

For $i \in J$ it is clear that

$$\lambda_i^U = 1 - \sum_{k \in I} P_{ik} = P_{i\partial} = \mathbb{P}_i(X_1 = \partial) \leq \mathbb{P}_i(Y_1 = \partial) = Q_{i\partial} = \lambda_i^A.$$

The inequality occurs because there could be trajectories for (X_n) that are not killed in the first step, but nevertheless only visit sites in J^c until finally absorbed at ∂. We conclude that both equilibrium potentials agree at site $i \in J$ if and only if every path of positive probability and length $n \geq 2$: $X_0 = i, X_1 \in I, \cdots, X_{n-1} \in I, X_n = \partial$ must enters in J, that is there exists $1 \leq m \leq n - 1$ such that $X_m \in J$. A sufficient condition for this to happen is $P_{ik} > 0 \Rightarrow k \in J$.

Some parts of the next lemma can be proved using the maximum principle Theorem 2.9 and Lemma 2.7, but we prefer to give a self contained proof.

Lemma 2.34 *Assume U is an inverse M-matrix. Then, $(\mathbb{I} + tU)$ is an inverse M-matrix, for all $t \geq 0$. Moreover, if U is a potential so is $(\mathbb{I} + tU)$ and its right equilibrium potential is strictly positive. In particular if U is a bi-potential, then so is $\mathbb{I} + tU$ and its equilibrium potentials are strictly positive.*

Similarly, if for some $t > 0$ the matrix $\mathbb{I} + tU$ is a bi-potential, then for all $0 < s \leq t$ the matrix $\mathbb{I} + sU$ is also a bi-potential and its equilibrium potentials are strictly positive.

Proof For some $\kappa > 0$ large enough, $U^{-1} = \kappa(\mathbb{I} - P)$ where $P \geq 0$ (and $P\mathbb{1} \leq \mathbb{1}$ in the row diagonally dominant case). In what follows we can assume that $\kappa = 1$, because it is enough to consider κU instead of U.

From the equality $(\mathbb{I} - P)(\mathbb{I} + P + P^2 + \cdots P^r) = \mathbb{I} - P^{r+1}$ we get that

$$\mathbb{I} + P + P^2 + \cdots P^r = U(\mathbb{I} - P^{r+1}) \le U,$$

and we deduce that the series $\sum\limits_{l=1}^{\infty} P^l$ is convergent and its limit is U.

Consider now the matrix

$$P_t = t\left(\left(\mathbb{I} - \frac{1}{1+t}P\right)^{-1} - \mathbb{I}\right) = t\sum_{l=1}^{\infty}\left(\frac{1}{1+t}\right)^l P^l. \tag{2.7}$$

We have that $P_t \ge 0$ (and $P_t \mathbb{1} \le \mathbb{1}$ whenever $P\mathbb{1} \le \mathbb{1}$). Therefore the matrix $\mathbb{I} - P_t$ is an M-matrix (which is row diagonally dominant when M is so). On the other hand we have

$$\mathbb{I}+tU = \mathbb{I}+t(\mathbb{I}-P)^{-1} = (t\mathbb{I}+\mathbb{I}-P)(\mathbb{I}-P)^{-1} = (1+t)\left(\mathbb{I} - \frac{1}{1+t}P\right)(\mathbb{I}-P)^{-1},$$

from where we deduce that $\mathbb{I} + tU$ is nonsingular and its inverse is

$$\begin{aligned}
(\mathbb{I} + tU)^{-1} &= \tfrac{1}{1+t}(\mathbb{I} - P)\left(\mathbb{I} - \tfrac{1}{1+t}P\right)^{-1} \\
&= \tfrac{1}{1+t}\left(\left(\mathbb{I} - \tfrac{1}{1+t}P\right)^{-1} - P\left(\mathbb{I} - \tfrac{1}{1+t}P\right)^{-1}\right) \\
&= \tfrac{1}{1+t}\left(\sum_{l=0}^{\infty}(1+t)^{-l}P^l - \sum_{l=0}^{\infty}(1+t)^{-l}P^{l+1}\right) \\
&= \tfrac{1}{1+t}(\mathbb{I} - P_t).
\end{aligned}$$

The only thing left to prove is that $P_t\mathbb{1} < \mathbb{1}$ in the row diagonally dominant case, that is when $P\mathbb{1} \le \mathbb{1}$. For that it is enough to prove that $P^l\mathbb{1} < \mathbb{1}$ for large l according to formula (2.7). From the equality $U = \sum\limits_{l=1}^{\infty} P^l$, we deduce that $\sum\limits_{l=1}^{\infty} P^l\mathbb{1} < \infty$, and therefore $P^l\mathbb{1}$ tend to zero as $l \to \infty$. This proves the claim.

When κ is not 1 we have the following equality

$$(\mathbb{I} + tU)^{-1} = \frac{\kappa}{t + \kappa}(\mathbb{I} - \frac{t}{\kappa}\sum_{l=1}^{\infty}(\frac{\kappa}{t + \kappa})^l P^l),$$

where $P = \mathbb{I} - \frac{1}{\kappa}U^{-1}$.

Finally, assume that $\mathbb{I} + tU$ is a bi-potential. Hence $\mathbb{I} + \beta(\mathbb{I} + tU)$ is a bi-potential for all $\beta \geq 0$. This implies that for any nonnegative β

$$\mathbb{I} + \frac{\beta}{1 + \beta} t\, U,$$

is a bi-potential. Now, it is enough to take $\beta \geq 0$ such that $s = \frac{\beta}{1+\beta} t$.

The following result will play an important role in some of our arguments. In particular, it implies that a potential is a column pointwise dominant matrix (see [28] for a relation between pointwise diagonally dominance and Hadamard products).

Proposition 2.35 *Assume that U is a potential, with an inverse $U^{-1} = \kappa(\mathbb{I} - P)$. Assume that (X_n) is a killed Markov chain associated with the substochastic matrix P. Then for all i, j*

$$U_{ij} = \mathbb{P}_i(\mathscr{T}_j < \infty)U_{jj}. \tag{2.8}$$

So, U_{ij} is a fraction of U_{jj}, and this fraction is the probability that the chain (X_n) ever visits j, when starting from i.

Notice that P is not uniquely defined, because the representation $U^{-1} = \kappa(\mathbb{I} - P)$ is not unique. Nevertheless the probability $\mathbb{P}_i(\mathscr{T}_j < \infty)$ does not depend on the selection of P. In the language of Markov chains, $\mathbb{P}_i(\mathscr{T}_j < \infty)$ is usually denoted by f_{ij}. For this reason, in what follows we denote it by f_{ij}^U, to emphasizes its dependence on U (we recall that $f_{ii}^U = 1$). Thus, for any potential matrix it holds that

$$U_{ij} = f_{ij}^U U_{jj}.$$

Proof Consider $V = (\mathbb{I} - P)^{-1}$. Then, as before we have

$$V_{ij} = \mathbb{E}_i\left(\sum_{m \geq 0} \mathbb{1}_{X_m = j}\right) = \mathbb{E}_i\left(\mathbb{1}_{\mathscr{T}_j < \infty} \sum_{m \geq \mathscr{T}_j} \mathbb{1}_{X_m = j}\right)$$

$$= \mathbb{P}_i(\mathscr{T}_j < \infty)\mathbb{E}_j\left(\sum_{m \geq 0} \mathbb{1}_{X_m = j}\right).$$

The last equality follows from the strong Markov property and we obtain

$$V_{ij} = \mathbb{P}_i(\mathscr{T}_j < \infty)V_{jj},$$

proving the desired equality for V. Since U and V are proportional, the result follows.

Lemma 2.36 *(i) If U is a potential then it is a column pointwise diagonal dominant matrix, that is $U_{jj} \geq U_{ij}$ for all different j, i, with strict inequality at those coordinates j where the right equilibrium potential $\lambda = U^{-1}\mathbb{1}$ is positive. Moreover, consider $\kappa_0 = \max\{(U^{-1})_{ll} : l \in I\}$, then for all i we have $\lambda_i \leq \kappa_0$, with strict inequality unless the i-th row of U^{-1} is $\kappa_0(e(i))'$. Also this implies that the i-th row of U is equal to $\frac{1}{\kappa_0}e(i)$. In general we have the inequality for all i, j different*

$$U_{ij} \leq \left(1 - \frac{\lambda_i}{\kappa_0}\right) U_{jj}.$$

(ii) If U is the inverse of an M-matrix, then its diagonal is positive.

Proof (i). As before consider $U^{-1} = \kappa_0(\mathbb{I} - P)$ and the killed Markov chain (X_n) associated with P, whose potential matrix is $V = (\mathbb{I} - P)^{-1}$.

From the previous Proposition, we have $U_{ij} = \mathbb{P}_i(\mathscr{T}_j < \infty)U_{jj} \leq U_{jj}$, proving that U is a column pointwise dominant.

Moreover, we have $U_{ij} = U_{jj}$ if and only if in the Markov chain associated with P every path starting from i must visit j before it is absorbed. This is not the case if $\theta_i = P_{i\partial} = 1 - \sum_{k \in I} P_{ik} > 0$, because in this situation there is a positive probability that the chain starting from i is absorbed in the first step. Then we have

$$\mathbb{P}_i(\mathscr{T}_j = \infty) \geq P_{i\partial} = \theta_i.$$

Then, we obtain the inequality $V_{ij} \leq (1 - \theta_i)V_{jj}$ and therefore

$$U_{ij} \leq (1 - \theta_i)U_{jj}.$$

Since $(\mathbb{I} - P)\mathbb{1} = \theta$, we get $V\theta = \mathbb{1}$ and then $\theta = \lambda/\kappa_0$, proving the desired inequality. Also, we find $\theta_i \leq 1$ with equality if and only if $P_{ik} = 0$ for all $k \in I$, which proves that the i-th row of U^{-1} is $\kappa_0 e(i)'$.

Let us prove that the i-th row of U is $\frac{1}{\kappa_0}e(i)'$. Indeed, since $P_{ik} = 0$ for all $k \in I$ we obtain the same is true for any power of P, that is, $P_{ik}^n = 0$ for all $n \geq 1$. Hence, $V_{ik} = 0$ for k different from i and $V_{ii} = 1$, proving the claim.

(ii). Now, assume that U is the inverse of an M-matrix. Then, $D^{-1}U$ is a potential where the diagonal matrix D is given by

$$D_{ii} = \sum_{j \in I} U_{ij} > 0.$$

Since $D^{-1}U\mathbb{1} = \mathbb{1}$, we have $(D^{-1}U)^{-1}$ is a row diagonally dominant M-matrix, Therefore $D^{-1}U$ is a column pointwise diagonally dominant matrix. Since no column of $D^{-1}U$ can be the zero column we deduce the result.

2.5   An Algorithm for CMP

In this section we present an algorithm for deciding when a nonnegative matrix satisfies the CMP. Of course if the matrix is nonsingular, we could compute its inverse and verify if it is a row diagonally dominant M-matrix. The algorithm we propose is recursive, which is probably slower than computing the inverse. Nevertheless, this algorithm may be helpful in applications when describing what restrictions a potential matrix must satisfy. Another advantage of this algorithm is that it works for singular matrices.

For a nonnegative matrix U, we recall that domination principle and the existence of a right equilibrium potential implies CMP (see Remark 2.14 after Lemma 2.13). The basis of the algorithm is given in the following proposition.

Theorem 2.37 *Assume that U is an $n \times n$ nonnegative matrix, with positive diagonal elements. Consider the following block decomposition of U*

$$U = \begin{pmatrix} V & v \\ u' & z \end{pmatrix},$$ (2.9)

where V is the principal submatrix of U given by the $(n-1)^{st}$ rows and columns.

(I) If U satisfies the CMP then

(i) U is a column pointwise dominant matrix and it has a right equilibrium potential, that is, a nonnegative vector λ solution of $U\lambda = 1$;

(ii) V and $V - \frac{1}{z}vu'$ satisfy the CMP;

(iii) u is a barycenter of the rows of V, that is, there exists a nonnegative vector $\eta \in \mathbb{R}^{n-1}$ with total mass $\bar{\eta} \leq 1$, such that $u' = \eta'V$;

(iv) v is a nonnegative potential of V, that is, there exists a nonnegative vector $\tau \in \mathbb{R}^{n-1}$ such that $v = V\tau$;

(v) $\eta'v \leq z$;

Moreover, U is nonsingular if and only if V is nonsingular and

(v') $\eta'v < z$.

(II) Conversely, if $(i) - (iv)$, (v') are fulfilled then U satisfies the CMP and $V, V - \frac{1}{z}vu'$ have right equilibrium potentials.

First, we shall prove a stability property under limits for CMP, and the other principles.

Lemma 2.38 *Assume that $(U_p)_{p \in \mathbb{N}}$ is a sequence of nonnegative matrices converging to U as p grows to infinity. Then,*

(i) if each U_p satisfies the CMP (respectively maximum principle), then U satisfies the CMP (respectively maximum principle);

(ii) *if each U_p satisfies the domination principle (respectively equilibrium principle, balayage principle) and U has a positive diagonal, then U satisfies the domination principle (respectively equilibrium principle, balayage principle).*

Proof (i). Consider a vector $x \in \mathbb{R}^I$ such that $(Ux)_i \leq 1$ when $x_i \geq 0$. Fix some $\epsilon > 0$. Then, for all large p and all those coordinates we have

$$(U_p x)_i \leq 1 + \epsilon.$$

Using that each U_p satisfies the CMP, we deduce that $U_p x \leq (1 + \epsilon)\mathbb{1}$. Thus, passing to the limit we deduce $Ux \leq (1 + \epsilon)\mathbb{1}$ and since ϵ is arbitrary we conclude $Ux \leq \mathbb{1}$, showing that U satisfies the CMP. The stability for the maximum principle is shown in the same way.

(ii). The hypothesis about the diagonal of U implies the existence of a positive constant C, such that for large p the diagonal of U_p is bounded from below by C. We take D_p (respectively D) the diagonal matrix, whose diagonal elements are the reciprocal of the row sums of U_p (respectively U). So, for large p the matrix $D_p U_p$ satisfies the CMP. From (i) we conclude that DU satisfies the CMP and then U satisfies the domination principle. The stability for the other principles is shown similarly.

Proof (Theorem 2.37) Assume that U satisfies the CMP and its diagonal is positive. (i) follows from Lemma 2.7. Also, since V is a principal submatrix of U, it satisfies the CMP, proving the first part of (ii) (see Lemmas 2.32 and 2.38).

In order to prove (ii)–(v), we consider $U_a = U + a\mathbb{I}$, for $a > 0$. U_a is nonsingular and satisfies the CMP, so it is a potential (see Theorem 2.9). Hence the inverse of U_a has the form

$$U_a^{-1} = \begin{pmatrix} \Omega_a & -\epsilon_a \\ -\zeta_a' & \alpha_a \end{pmatrix},$$

where Ω_a is a row diagonally dominant M-matrix, ϵ_a, ζ_a are nonnegative vectors, the total mass of ζ_a is bounded by α_a and $\epsilon_a \leq \Omega_a \mathbb{1}$. Now, it is simple to check that

$$V + a\mathbb{I} - \tfrac{vu'}{z+a} = \Omega_a^{-1}, \ (V + a\mathbb{I})\left(\tfrac{1}{\alpha_a}\epsilon_a\right) = v, \ \tfrac{1}{\alpha_a}\zeta_a'(V + a\mathbb{I}) = u';$$
$$0 < \det(U_a) = \det(V + a\mathbb{I})(z + a - u'(V + a\mathbb{I})^{-1}v)$$
$$= \det(V + a\mathbb{I})(z + a - \tfrac{1}{\alpha_a}\zeta_a'v) = \det(V + a\mathbb{I})(z + a - u'\tfrac{1}{\alpha_a}\epsilon_a).$$

We conclude that $V - \tfrac{1}{z}vu'$ satisfies the CMP (see previous lemma). Since $\tfrac{1}{\alpha_a}\zeta_a$ is a nonnegative vector and its total mass is at most one, we deduce, by a compactness argument, the existence of a sequence $a_n \downarrow 0$, such that

$$\eta = \lim_{n \to \infty} \frac{1}{\alpha_{a_n}}\zeta_{a_n}.$$

This vector is nonnegative and its total mass is at most 1. Moreover, $\eta'V = u'$, showing (*iii*).

Let us take $C = \min\{U_{ii} : i \in I\}$, a lower bound for the diagonal of U. Then

$$C\frac{1}{\alpha_a}\epsilon_a \le (V + a\mathbb{I})\left(\frac{1}{\alpha_a}\epsilon_a\right) = v,$$

which implies that $\frac{1}{\alpha_a}\epsilon_a \in K = \{x \in \mathbb{R}^{n-1} : 0 \le x \le \frac{1}{C}v\}$. Given that K is a compact set, we deduce again the existence of $\tau \in K$ solution to the equation $V\tau = v$, proving that (*iv*) holds. Finally, since $0 < z + a - \frac{1}{\alpha_a}\zeta_a'v$ we obtain that $0 \le z - \eta'v$, showing that (*v*) is satisfied. We point out that $\eta'v = u'\tau$.

To conclude the first part, notice that if U is nonsingular then V is nonsingular (it is a principal submatrix of a potential) and $0 < \det(U) = \det(V)(z - u'V^{-1}v)$, showing that (*v'*) holds. On the other hand, if V is nonsingular and (*v'*) holds, we deduce $\det(U) > 0$.

(*II*). We assume that (*i*)–(*iv*), (*v'*) hold. U is assumed to have a right equilibrium potential. In order to prove that it satisfies the CMP, it is enough to show U satisfies the domination principle (see Lemma 2.13).

For that reason take two nonnegative vectors $x, y \in \mathbb{R}^I$ and assume that $(Ux)_i \le (Uy)_i$ when $x_i > 0$. We decompose x, y in blocks as $x = (p, q)'$, $y = (r, s)'$, with $q, s \in \mathbb{R}$. Notice that

$$Ux = \begin{pmatrix} Vp + vq \\ u'p + qz \end{pmatrix} \text{ and } Uy = \begin{pmatrix} Vr + vs \\ u'r + sz \end{pmatrix}.$$

If $q = 0$, then $(Vp)_i \le (Vr+vs)_i = (V(r+\tau s))_i$ when $p_i > 0$ and $i \le n-1$. Since V satisfies the domination principle we deduce that $Vp \le Vr+vs$. Now, multiplying this inequality by $\eta' \ge 0$ we deduce

$$u'p + qz = \eta'Vp \le \eta'(Vr + vs) = u'r + \eta'vs \le u'r + zs,$$

showing in this case that $Ux \le Uy$.

Finally, we need to analyze the case $q > 0$. By hypothesis $(Ux)_n \le (Uy)_n$, which is exactly the inequality $u'p + qz \le u'r + sz$ or $f = u'r + sz - (u'p + qz) \ge 0$. Let us show that

$$\left(\left[V - \frac{1}{z}vu'\right]p\right)_i \le \left(\left[V - \frac{1}{z}vu'\right](r + f\tau)\right)_i,$$

holds when $p_i > 0$. Indeed, since $z - u'\tau = z - \eta'V\tau = z - \eta'v > 0$, we deduce that

$$\left(V - \frac{1}{z}vu'\right)\left(\frac{z}{z - u'\tau}\tau\right) = v,$$

and then

$$Vp + vq = \left[V - \frac{1}{z}vu' \right] p + v \left(\frac{1}{z}u'p + q \right) = \left[V - \frac{1}{z}vu' \right] \cdot \left(p + \frac{u'p + qz}{z - u'\tau}\tau \right)$$

Hence, for the coordinates i where $p_i > 0$ we have

$$\left(\left[V - \frac{1}{z}vu' \right] p \right)_i \le \left(\left[V - \frac{1}{z}vu' \right] w \right)_i ,$$

where w is the nonnegative vector

$$w = r + \frac{u'r + sz - (u'p + qz)}{z - u'\tau}\tau = r + \frac{f}{z - u'\tau}\tau.$$

Using that $V - \frac{1}{z}vu'$ satisfies the CMP we conclude the inequality

$$\left[V - \frac{1}{z}vu' \right] p \le \left[V - \frac{1}{z}vu' \right] w,$$

or equivalently

$$Vp + vq \le Vr + vs.$$

This shows that $Ux \le Uy$ and then U satisfies the domination principle.

Let us prove that V and $V - \frac{1}{z}vu'$ have right equilibrium potentials. Denote by $\lambda = (\chi, \delta)'$, where $\delta \in \mathbb{R}$, a block decomposition of a right equilibrium potential of U. Since

$$\mathbb{1} = V\chi + \delta v = V(\chi + \delta\tau)$$

and $\gamma = \chi + \delta\tau \ge 0$, we get V has a right equilibrium potential. Now, let us take $s \ge 0$ and compute

$$\left(V - \frac{1}{z}vu' \right)(\gamma + s\tau) = \mathbb{1} + v \left(s \left(1 - \frac{u'\tau}{z} \right) - \frac{u'\gamma}{z} \right).$$

So, for $s = \frac{u'\gamma}{z - u'\tau}$ the vector $\gamma + s\tau \ge 0$ is a right equilibrium potential of $V - \frac{1}{z}vu'$.

We can use this result to obtain a similar characterization for a matrix that satisfies the domination principle. This is important because it gives a recursive algorithm to recognize when a nonnegative matrix is the inverse of an M-matrix. Recall that if U is a nonnegative matrix, with a positive diagonal and U satisfies the domination principle then there exists a diagonal matrix D, with positive diagonal elements, such that DU satisfies the CMP (see Lemma 2.15 (vi)). Moreover, we can take the diagonal of D as the reciprocals of the row sums of U. We state the following corollary only for inverse M-matrices.

Corollary 2.39 *Assume that U is an $n \times n$ nonnegative and nonsingular matrix, with a block structure $U = \begin{pmatrix} V & v \\ u' & z \end{pmatrix}$, where V is the principal submatrix of U given by the $(n-1)^{st}$ rows and columns. Define the diagonal matrix D of size n, whose diagonal elements are $D_{ii} = \left(\sum_{j \in I} U_{ij} \right)^{-1}$, and consider $E = D_{\{1, \cdots, n-1\}}$.*

(I) U is an inverse M-matrix then

> *(i) for all i, j it holds $D_{ii} U_{ij} \le D_{jj} U_{jj}$;*
> *(ii.1) V is an inverse M-matrix;*
> *(ii.2) $V - \frac{1}{z} v u'$ is an inverse M-matrix;*
> *(iii.1) there exists a nonnegative vector $\rho \in \mathbb{R}^{n-1}$ such that $u' = \rho' V$;*
> *(iii.2) $\eta = D_{nn} E^{-1} \rho$ has total mass $\bar{\eta} \le 1$;*
> *(iv) there exists a nonnegative vector $\tau \in \mathbb{R}^{n-1}$ such that $V\tau = v$;*
> *(v) $z - u'\tau > 0$.*

(II) Conversely, $(ii.2)$, $(iii.1)$, (iv) and (v) imply that U is an inverse M-matrix.

The proof of this corollary is based on the formula of U^{-1} by blocks (see for example Appendix B)

$$
U^{-1} = \begin{pmatrix} \left(V - \frac{1}{z} v u' \right)^{-1} & -\frac{1}{z - u'\tau} \tau \\[2ex] -\frac{1}{z - u'\tau} \rho' & \frac{1}{z - u'\tau} \end{pmatrix}.
$$

We also notice that if U is an inverse M-matrix then EV, $E(V - \frac{1}{z} v u')$ satisfy the CMP. This follows from the domination principle and

$$
EV(\mathbb{1} + \tau) = EV\mathbb{1} + Ev = \mathbb{1}, \text{ and } E\left(V - \frac{1}{z} v u' \right) \left(\mathbb{1} + \frac{\bar{u} + z}{z - u'\tau} \tau \right) = \mathbb{1}.
$$

2.6 CMP and Non-Singularity

The purpose of this section is to give a characterization of matrices satisfying CMP (or the domination principle) that are nonsingular. This characterization is inspired by Theorem 4.4 in [47], where the authors studied nonsingular generalized ultrametric matrices (see also Theorem 1 in [20] and Theorem 3.9 in Chap. 3). Generalized ultrametric matrices satisfy CMP and for them there is a simple criterion for non-singularity: no row is the zero row and all rows are different. We shall extend this characterization to the class of CMP, and surprisingly enough, the condition is the same. In a first stage we assume that U is positive. We prefer to state the result for singular matrices because it is simple to do it.

Theorem 2.40 *Assume that U is a positive matrix and it satisfies CMP. Then, the following are equivalent*

(i) *U is singular;*
(ii) *there exist different i, j, such that rows i and j of U are equal and columns i and j of U are proportional.*

Proof Clearly condition (ii) is sufficient for U to be singular. In what follows we show (ii) is also necessary. We shall prove this by induction on n, the size of U. For $n = 2$, we have that U has the form

$$U = \begin{pmatrix} a & b \\ c & d \end{pmatrix},$$

where $0 < c \le a, 0 < b \le d$, because U is a column diagonally dominant matrix. Hence, U is singular if and only if $ad - bc = 0$, which implies that $c = a, b = d$, showing in this case the rows of U are equal and the columns are proportional.

Now, we perform the inductive step. We assume that (i) and (ii) are equivalent for matrices with size at most $n - 1$ for $n \ge 3$ and we show the equivalence for any matrix $U > 0$ that satisfies CMP of size n. For the inductive step, we consider two different situations. The first is that all principal submatrices of U of size $n - 1$, are nonsingular. The second, is the presence of at least one principal submatrix of size $n - 1$, which is singular.

Let us show that the first case is not possible when $n \ge 3$. To do this, we decompose U in blocks as

$$U = \begin{pmatrix} V & v \\ u' & z \end{pmatrix},$$

where V has size $n - 1$ and it is nonsingular. Since we assume that U is singular then necessarily $z = u'V^{-1}v$, because $\det(U) = \det(V)(z - u'V^{-1}v)$. Recall that, from Theorem 2.37 we have $\eta' = u'V^{-1}$ is nonnegative and its total mass is $\bar{\eta} = \mathbb{1}'\eta \le 1$. We also have $\tau = V^{-1}v$ is nonnegative. The fact that U is column pointwise dominant implies that $v \le z$.

The only way $z = u'V^{-1}v$ holds is that $\bar{\eta} = 1$ and $v_i = z$, for those $i \le n - 1$ in the support of η, that is, for all $i \le n - 1$

$$\eta_i > 0 \Rightarrow v_i = z. \tag{2.10}$$

Recall that, for any $a > 0$ the matrix $U_a = U + a\mathbb{I}$ is nonsingular and satisfies the CMP. So, U_a^{-1} is a row diagonally dominant M-matrix, which implies

$$U_a^{-1} = \begin{pmatrix} \Omega_a & -\epsilon_a \\ -\zeta_a' & \theta_a \end{pmatrix}.$$

Here Ω_a is a row diagonally dominant M-matrix, ϵ_a, ζ_a are nonnegative and $\theta_a > 0$. From Cramer's rule we have that

$$\mathrm{adjoint}(U + a\mathbb{I}) = \det(U + a\mathbb{I})U_a^{-1},$$

showing that $\mathrm{adjoint}(U)$ is a Z-matrix of the form

$$\mathrm{adjoint}(U) = \begin{pmatrix} \Omega & -\alpha \\ -\beta' & \rho \end{pmatrix}.$$

The hypothesis that every principal submatrix of U of size $n - 1$ is nonsingular shows that $\Omega_{ii} > 0$ for all $i \leq n - 1$ and $\rho > 0$. On the other hand we have $U\,\mathrm{adjoint}(U) = \mathrm{adjoint}(U)U = 0$ and then

$$V\alpha = \rho v, \ \beta'V = \rho u', \ V\Omega = v\beta'.$$

In particular, we get

$$\beta = \rho\eta, \ \Omega = \tau\beta' = \rho\tau\eta',$$

and therefore $0 < \Omega_{ii} = \rho v_i \eta_i$, for all $i \leq n - 1$. The important conclusion is that $\eta > 0$ and then $v = z\mathbb{1}$ (see (2.10)). Thus, the last column of U is constant.

Summarizing, if every principal submatrix of U of size $n - 1$ is nonsingular, then the columns of U are constant: $U_{\bullet j} = U_{jj}\mathbb{1}$ for all j. This is not possible if $n \geq 3$, because every principal submatrix V of U (of size $n - 1$) contains at least two proportional columns, so V is singular, contradicting the assumption.

The only possible case is that some principal submatrix of size $n - 1$ is singular. Without loss of generality we assume that $V = U_{\{1,\cdots,n-1\}}$ is singular. Again, from Theorem 2.37 we have $v = V\tau, u' = \eta'V$. The induction hypothesis implies the existence of different $i, j \leq n - 1$ such that rows i, j of V are equal and columns i, j of V are proportional. Since $u_i = \eta'V_{\bullet i}$ and $u_j = \eta'V_{\bullet j}$, we conclude the columns i, j of U are proportional. Similarly, since $v_i = V_{i\bullet}\tau$ and $v_j = V_{j\bullet}\tau$, we deduce $v_i = v_j$ and then rows i, j of U are equal, showing the result.

As a matter of completeness we state the result for nonsingular matrices.

Corollary 2.41 *Assume that $U > 0$ satisfies the CMP. Then, the following are equivalent*

 (i) *U is nonsingular;*
 (ii) *no two rows of U are equal;*
 (iii) *no two columns of U are proportional.*

The characterization of non-singularity for positive matrices satisfying the domination principle is obtained from the previous result. Indeed, if U satisfies the domination principle and we consider D the diagonal matrix whose diagonal elements are the reciprocals of the row sums of U, then DU satisfies the CMP. This transformation allows us to conclude.

Corollary 2.42 *Assume that $U > 0$ satisfies the domination principle. Then, the following are equivalent*

 (i) *U is nonsingular;*
 (ii) *no two rows of U are proportional;*
 (iii) *no two columns of U are proportional.*

Remark 2.43 Notice the symmetry between rows and columns in the previous corollary. This is a consequence of the fact that the domination principle is closed under transposition of matrices. This is not the case for CMP. Indeed, recall that a nonsingular matrix satisfying the CMP has an inverse, which is row diagonally dominant and it can happen that the inverse is not column diagonally dominant.

When U has some 0 entries the analysis of non-singularity becomes more complicated. A key observation is given by the following lemma.

Lemma 2.44 *Assume that U is nonnegative and it satisfies the CMP. If $U_{ij} = 0$ for some i, j then there exists a permutation matrix Π such that $\Pi U \Pi'$ has the block structure*

$$\Pi U \Pi' = \begin{pmatrix} A & 0 \\ B & C \end{pmatrix},$$

where A, C are square nonnegative matrices that satisfies the CMP.

Proof For any permutation matrix Π, the matrix $\Pi U \Pi'$ satisfies the CMP (see Lemma 2.12). The fact that A, C also satisfy the CMP follows from Theorem 2.37. So, we only have to prove the existence of a set $L \subsetneq I$ such that $U_{LL^c} = 0$. Given this set L, a possible permutation matrix supplying the required block structure is the one associated with σ^{-1}, where σ is a permutation of I such that $\sigma(L) = \{1, \cdots, \#L\}$ and $\sigma(L^c) = \{\#L + 1, \cdots, n\}$. That is

$$\Pi_{pq} = 1 \text{ if and only if } \sigma(p) = q.$$

There are two different cases. The first is $i = j$ and of course the second is $i \neq j$. In the first case, the column dominance of U implies that $U_{ki} = 0$ for all k. Hence, in this situation we have $L = I \setminus \{i\}$.

For the second case we can assume $U_{ij} = 0$ for some $i \neq j$ and all the diagonal elements of U are positive. We consider the set $L = \{k \in I : U_{kj} = 0\}$. Obviously $i \in L$ and $j \notin L$. We point out that $r \in L^c$ if and only if $U_{rj} > 0$. We have to show that $U_{kr} = 0$ whenever $k \in L, r \in L^c$.

For that purpose we perturb U to make it nonsingular. So, for $a > 0$ we know that $W_a = U + a\mathbb{I}$ is the inverse of a row diagonally dominant M-matrix. Therefore, $W_a^{-1} = \kappa(a)(\mathbb{I} - P(a))$, for some real number $\kappa(a) > 0$, and a substochastic matrix $P(a)$. Consider a killed Markov chain associated with this substochastic kernel $P(a)$. We denote by $\mathbb{P}^{(a)}$ the probability measure induced by $P(a)$. We have for all different k, ℓ (see formula (2.8))

$$U_{k\ell} = (W_a)_{k\ell} = \mathbb{P}_k^{(a)}(\mathscr{T}_\ell < \infty)(W_a)_{\ell\ell} = \mathbb{P}_k^{(a)}(\mathscr{T}_\ell < \infty)(U_{\ell\ell} + a),$$

where we recall that $\mathscr{T}_\ell$ is the random time of first visit to ℓ.

So, for any $k \in L$ and any $a > 0$ the probability of ever visit j is 0, that is $\mathbb{P}_k^{(a)}(\mathscr{T}_j = \infty) = 1$. On the other hand, since $U_{rj} > 0$, for any $r \in L^c$, we deduce that $\mathbb{P}_r^{(a)}(\mathscr{T}_j < \infty) > 0$. These two facts should imply that the probability of ever visits r, when starting from k is 0. Indeed,

$$0 = \mathbb{P}_k^{(a)}(\mathscr{T}_j < \infty) \geq \mathbb{P}_k^{(a)}(\mathscr{T}_r < \mathscr{T}_j < \infty) = \mathbb{P}_k^{(a)}(\mathscr{T}_r < \mathscr{T}_j)\mathbb{P}_r^{(a)}(\mathscr{T}_j < \infty).$$

The last equality is obtained from the strong Markov property. The conclusion is that

$$\mathbb{P}_k^{(a)}(\mathscr{T}_r < \mathscr{T}_j) = 0.$$

Since $\mathbb{P}_k^{(a)}(\mathscr{T}_j = \infty) = 1$, we conclude that

$$\mathbb{P}_k^{(a)}(\mathscr{T}_r < \infty) = 0,$$

and then $(W_a)_{kr} = 0$, that is, $U_{kr} = 0$. The result is shown.

If we repeat the above algorithm in each of the blocks A, C we will end up, with the help of a possible extra permutation, with a matrix of the following type

$$\Pi U \Pi' = \begin{pmatrix} A_1 & 0 & 0 & \cdots & 0 \\ B_{21} & A_2 & 0 & \cdots & 0 \\ \vdots & \vdots & \ddots & \vdots & \vdots \\ B_{p1} & B_{p2} & \cdots & A_p & 0 \\ B_{p+1,1} & B_{p+1,2} & \cdots & B_{p+1,p} & T \end{pmatrix}, \qquad (2.11)$$

where $A_1, \cdots, A_p$ are positive square matrices satisfying CMP and $T = 0$. In some cases it could happen that T is not present or $p = 0$, but both conditions cannot occur simultaneously. The matrices $B_{p+1,1}, \cdots, B_{p+1,p}$ are not necessarily 0 as it is shown by the following example

$$U = \begin{pmatrix} a & 0 \\ b & 0 \end{pmatrix},$$

where $0 \leq b \leq a$. In this example, U satisfies the CMP, it is singular, but nevertheless can have different rows.

The block structure in (2.11) allows us to show the next result.

Proposition 2.45 *Assume that U is nonnegative and satisfies CMP. Then the following are equivalent*

 (i) *U is singular;*
 (ii) *U has a column of zeroes or there are i, j different such that columns i, j are proportional and rows i, j are equal.*

Proof Obviously (ii) implies (i). So, for the converse we may assume that U is nonnegative, satisfies the CMP, it is singular, it has some 0 entries, but there is not a column of zeroes. We must show there are two proportional columns and the corresponding rows are equal.

Without loss of generality we assume that U has the form given in (2.11). The assumption is that $p \geq 2$ and that T is not present in this form. So, one of the blocks $A_1, \cdots, A_p$ is singular. Since each block contains no zeroes, we deduce that at least one of them has two proportional columns and the corresponding rows are equal.

If the block A_p is singular then there exist i, j different such that the columns associated with i, j in A_p are proportional and the corresponding rows in A_p are equal (see Theorem 2.40).

Obviously the columns i, j in U are proportional. We need to show that the rows i, j of U are equal. Let us assume that $A_p = U_L$ for some subset $L \subset I$, such that $i, j \in L$. We can assume that i, j correspond to the first and second rows in A_p. As before we consider $U + a\mathbb{I}$ where $a > 0$ and deduce that (after a compactness argument)

$$U_{L \times L^c} = A_p H,$$

for some nonnegative matrix H of size $\#L \times \#L^c$. Hence,

$$U_{\{i\} \times L^c} = (A_p)_{1\bullet} H, \, U_{\{j\} \times L^c} = (A_p)_{2\bullet} H.$$

Since by assumption $(A_p)_{1\bullet} = (A_p)_{2\bullet}$, we conclude the rows i, j in U are equal.

To finish the proof let us introduce some notation. Consider the set of index $L_\ell \subset I$, for $\ell = 1, \cdots, p$, that defines each block $A_\ell = U_{L_\ell}$. Also, define $L_\ell^+ = \overset{p}{\underset{s=\ell}{\cup}} L_s$ and $L_\ell^- = \overset{\ell-1}{\underset{s=1}{\cup}} L_s$.

The matrix $W_\ell = U_{L_\ell^+}$ contains all the blocks $\ell, \cdots, p$ and it has a block form as

$$W_\ell = \begin{pmatrix} A_\ell & 0 \\ G & W_{\ell+1} \end{pmatrix}.$$

This matrix satisfies CMP and again we get that

$$U_{L_{\ell+1}^+ \times L_\ell} = G = A_\ell R,$$

for some nonnegative matrix R of the appropriate size. Thus, if the columns s, t of A_ℓ are proportional, then the corresponding columns in U are proportional. On the other hand, the rows s, t in A_ℓ are equal and the corresponding rows in U are also equal because

$$U_{L_\ell \times L_\ell^-} = F A_\ell,$$

for some nonnegative matrix F. This implies the result.

Corollary 2.46 *Assume that U is nonnegative and it satisfies the CMP. Then, the following are equivalent:*

(i) *U is nonsingular;*
(ii) *there is not a column of zeroes and no two rows of U are equal;*
(iii) *no two columns of U are proportional.*

Corollary 2.47 *Assume that U is nonnegative and it satisfies the domination principle. Then, the following are equivalent:*

(i) *U is nonsingular;*
(ii) *there is not a column of zeroes and no two rows of U are proportional;*
(iii) *no two columns of U are proportional.*

Notice that if U satisfies the CMP, then the column j of U is zero if and only if $U_{jj} = 0$. So the condition of not having a column of zeroes is equivalent to have a positive diagonal. The same property holds for U that satisfies the domination principle (use that DU satisfies the CMP for some nonsingular diagonal matrix D).

2.7 Potentials and Electrical Networks

In this section we summarize some of the well-known results that provide a electrical network interpretation of potential matrices. For a more detailed discussion see the online monograph by Lyons with Peres [42] and references therein. For this section we shall assume that P is a stochastic matrix on $\bar{I}$, that satisfies the **reversibility condition**

$$\forall i, j \in \bar{I}, \quad \pi_i P_{ij} = \pi_j P_{ji}, \tag{2.12}$$

where π is a positive vector. Notice that if P is symmetric we can take $\pi = \mathbb{1}$ in the reversibility condition. For the sake of simplicity, we also assume that P is irreducible.

Consider $\mathcal{G}$ the incidence graph of P, that is (i, j) is an edge of this graph if and only if $P_{ij} > 0$. To every edge (i, j) in $\mathcal{G}$ we associate a weight $C_{ij} = \pi_i P_{ij}$ and call this nonnegative number the **conductance** of this edge. Notice that reversibility of P is equivalent to the symmetry of C. As it is usual in electrical networks $R_{ij} = \frac{1}{C_{ij}}$ is called the **resistance** of this edge. For a fixed node i, the nodes j such that $(i, j) \in \mathcal{G}$ are called the neighbors of i. Reversibility shows that the relation of being neighbors is symmetric and we denote this relation by $i \sim j$.

An important concept in what follows is the notion of harmonic function.

Definition 2.48 A function $f : \bar{I} \to \mathbb{R}$ is said to be **harmonic** at i if

$$f(i) = \sum_{j:i \sim j} P_{ij} f(j),$$

that is, $f(i)$ is the average of f in the neighborhood of i. We also say f is harmonic on $I \subset \bar{I}$ if it is harmonic at every $i \in I$.

We also need the notion of a voltage in a network.

Definition 2.49 Let K, L be disjoint subsets of $\bar{I}$. A function $v : \bar{I} \to \mathbb{R}$ is said to be a **voltage** if it is harmonic on $\bar{I} \setminus (K \cup L)$, it is 1 at K and 0 at L. We shall say that v is a voltage between K and L.

In most cases we shall assume that K is a singleton. The existence of a voltage is guaranteed by the next proposition. Recall that T_{i_0} is the random time where a Markov chain visits for the first time the node i_0.

Proposition 2.50 Let $K = \{i_0\}$ and $L \neq \emptyset$ be disjoint subsets of $\bar{I}$. Consider $I = \bar{I} \setminus L$ and assume that $Q = P_{II}$ satisfies that $\mathbb{I} - Q$ is nonsingular. Take $U = (\mathbb{I} - Q)^{-1}$ the potential associated with Q. Then, the unique voltage $v = v^{i_0,L}$ between K and L is given by $v = 0$ in L and for all $i \in I$

$$v_i = \frac{\pi_{i_0}}{U_{i_0 i_0}} \frac{U_{i_0 i}}{\pi_i} = \frac{U_{i i_0}}{U_{i_0 i_0}} = \mathbb{P}_i(\mathcal{T}_{i_0} < \infty). \tag{2.13}$$

Proof By definition v takes the value 1 at K and 0 at L. So, we need to prove that v is harmonic in $\bar{I} \setminus (K \cup L)$. This follows quite easily from the facts that $U(\mathbb{I} - Q) = \mathbb{I}$ and that P is reversible. Indeed, for any i different from i_0 we have

$$U_{i_0 i} = \sum_{j \in I} U_{i_0 j} Q_{ji} = \sum_{j \in I} U_{i_0 j} P_{ji} = \sum_{j \in I} U_{i_0 j} \frac{\pi_i}{\pi_j} P_{ij} = \frac{\pi_i U_{i_0 i_0}}{\pi_{i_0}} \sum_{j \in I} P_{ij} v_j$$

$$= \frac{\pi_i U_{i_0 i_0}}{\pi_{i_0}} \sum_{j \in \bar{I} : i \sim j} P_{ij} v_j,$$

where in the last equality we have used the fact that v is zero on L. Hence v is harmonic in $\bar{I} \setminus (K \cup L)$. Uniqueness of a voltage is also obvious.

Notice that the reversibility condition (2.12) imposed on P is also valid for U, that is for all $i, j \in I$

$$\pi_i U_{ij} = \pi_j U_{ji}.$$

Hence, the unique voltage v between $\{i_0\}$ and L, is for all $i \in I$

$$v_i = \frac{U_{i i_0}}{U_{i_0 i_0}} = \mathbb{P}_i(\mathcal{T}_{i_0} < \infty),$$

where the last probability is computed under the Markov chain on I whose transition probabilities are given by Q.

Remark 2.51 L is an absorbing set for the killed Markov chain induced by Q. In particular, for all $i \in L$ we have $\mathbb{P}_i(\mathcal{T}_{i_0} < \infty) = 0$.

We now introduce the **current** $\mathcal{I}$ associated with a voltage v, which is given by Ohm's law, for all $i, j \in \bar{I}$

$$\mathcal{I}_{ij} = (v(i) - v(j)) C_{ij}.$$

Consider the voltage $v = v^{i_0, L}$ given by Proposition 2.50. The **total current** at node i_0 is defined as

$$\overline{\mathcal{I}_{i_0}} = \sum_{j \in \bar{I}} \mathcal{I}_{i_0, j}.$$

Let us compute this quantity. By definition

$$\overline{\mathscr{I}_{i_0}} = \sum_{j \in \bar{I}} (v(i_0) - v(j))C_{i_0,j} = \sum_{j \in \bar{I}} (1 - \mathbb{P}_j(\mathscr{I}_{i_0} < \infty))\pi_{i_0} P_{i_0 j}$$

$$= \pi_{i_0} \left(1 - \sum_{j \in I} P_{i_0 j} \mathbb{P}_j(\mathscr{I}_{i_0} < \infty) \right) = \pi_{i_0} \left(1 - \frac{1}{U_{i_0 i_0}} \sum_{j \in I} P_{i_0 j} U_{j i_0} \right)$$

$$= \frac{\pi_{i_0}}{U_{i_0 i_0}} \left(U_{i_0 i_0} - \sum_{j \in I} Q_{i_0 j} U_{j i_0} \right) = \frac{\pi_{i_0}}{U_{i_0 i_0}},$$

$$(2.14)$$

where we have used $U(\mathbb{I} - Q) = \mathbb{I}$.

Proposition 2.52 *Under the same hypothesis of Proposition 2.50, the current associated with $v^{i_0, L}$ has the probabilistic representation: For all $i, j \in I$ such that $i \sim j$, we have*

$$\mathscr{I}_{ij} = \overline{\mathscr{I}_{i_0}} \mathbb{E}_{i_0}(S_{ij} - S_{ji}),$$

where S_{ij} is the number of transitions from i to j for the chain associated with Q. Moreover, we also have

$$\mathbb{E}_{i_0}(S_{ij}) = U_{i_0 i} P_{ij}.$$

Proof Let (X_m) be the killed Markov chain associated with Q. Then

$$S_{ij} = \sum_{m \geq 0} \mathbb{1}_{X_m = i, X_{m+1} = j}.$$

Then, we obtain from Fubini's theorem and the Markov property

$$\mathbb{E}_{i_0}(S_{ij}) = \sum_{m \geq 0} \mathbb{E}_{i_0}(\mathbb{1}_{X_m = i, X_{m+1} = j}) = \sum_{m \geq 0} \mathbb{E}_{i_0}(\mathbb{1}_{X_m = i}) P_{ij} = U_{i_0 i} P_{ij}.$$

Thus, if we consider the matrix Θ defined on I by $\Theta_{ij} = \mathbb{E}_{i_0}(S_{ij} - S_{ji})$, we obtain from the reversibility of U and formula (2.13) for the voltage

$$\Theta_{ij} = U_{i_0 i} P_{ij} - U_{i_0 j} P_{ji} = \frac{U_{i_0 i_0}}{\pi_{i_0}} (v_i \pi_i P_{ij} - v_j \pi_j P_{ji}) = \frac{U_{i_0 i_0}}{\pi_{i_0}} (v_i - v_j)C_{ij} = \frac{U_{i_0 i_0}}{\pi_{i_0}} \mathscr{I}_{ij}.$$

The result follows from Eq. (2.14).

We notice that $\mathscr{I}$, v, R satisfy the standard basic laws of electrical networks such as: Ohm's law, Kirchhoff's node and cycle laws. For details on these laws, the reader may consult [42]. The concepts discussed in Sect. 2.2, such as the domination principle and balayage principle, have its appropriate interpretation in electrical networks (see for example Chapter 7 in [41]). Further results can be found in [29] and [6].

Chapter 3
Ultrametric Matrices

This chapter is devoted to the study of ultrametric matrices introduced by Martínez, Michon and San Martín in [44], where it was proved that the inverse of an ultrametric matrix is a row diagonally dominant Stieltjes matrix (a particular case of an M-matrix). We shall include this result in Theorem 3.5 and give a proof in the lines done by Nabben and Varga in [51]. One of the important aspects of ultrametric matrices is that they represent a class of inverse M-matrices described in very simple combinatorial terms.

We will subsequently discuss the generalization of ultrametric matrices to the nested block form introduced by McDonald, M. Neumann, H. Schneider and Tsatsomeros in [47] and Nabben and Varga in [52]. Again, the inverse of a nested block form matrix is an M-matrix. We state this result in Theorem 3.9. A proof of this result can be effected by extending the one performed in the ultrametric case. Nevertheless, we prefer to postpone such extension to Chap. 5 where we look at a further generalization of ultrametric matrices called weakly filtered. Other developments around these concepts can be found in [53] and [45]

Ultrametric matrices are intimately related to tree structures and we exploit this fact to give a geometric characterization of ultrametric matrices. This is seen in Theorem 3.16, where we show that every ultrametric matrix can be embedded into a tree matrix, and in Theorem 3.26 where ultrametricity is described by a natural notion of isosceles triangles.

3.1 Ultrametric Matrices

Ultrametricity was firstly introduced in relation with p-adic number theory. In applications, like taxonomy [4], ultrametricity is an important notion because its relation with partitions. On the other hand, strictly ultrametric matrices appear as covariance matrices of random energy models in statistical physics [10], as a

© Springer International Publishing Switzerland 2014
C. Dellacherie et al., *Inverse M-Matrices and Ultrametric Matrices*, Lecture Notes in Mathematics 2118, DOI 10.1007/978-3-319-10298-6_3

generalization of the diagonal case. Since most of the relevant quantities depend on the inverse of the covariance matrix, knowledge on the inverse of a strictly ultrametric matrix might be useful in this theory.

In this chapter, we also study relations between ultrametric matrices and chains of partitions, which were firstly developed by Dellacherie in [18]. A detailed study concerning ultrametric matrices, maximal chains and associated spectral decompositions for countable probability spaces was made in [17]. Since partitions is a major concept in this monograph, in the next definition we fix some notions related to it.

Definition 3.1 A **partition** $\mathfrak{R} = \{R_1, \cdots, R_p\}$ of a set I is a collection of nonempty disjoint subsets of I, which cover I. The elements of $\mathfrak{R}$ are called **atoms**.

A partition $\mathfrak{S}$ is **finer** than $\mathfrak{R}$ if each atom of $\mathfrak{R}$ is a union of atoms of $\mathfrak{S}$. We denote this partial order as $\mathfrak{R} \preccurlyeq \mathfrak{S}$. Associated with this partial order there are two extreme partitions: The coarsest one $\mathfrak{N} = \{I\}$ and the finest one $\mathfrak{F} = \{\{i\} : i \in I\}$.

A **chain of partitions** $\mathbf{F}$ is a sequence $\mathfrak{R}_1 \preccurlyeq \mathfrak{R}_2 \preccurlyeq \cdots \preccurlyeq \mathfrak{R}_s$, of comparable partitions. Notice that repetition of partitions is allowed.

We say that $\mathbf{F}$ is **dyadic** if every atom $J \in \mathfrak{R}_q \in \mathbf{F}$, that is not a singleton, is the union of two different atoms $Q, S \in \mathfrak{R}_{q+1}$, for $q = 1, \cdots, s-1$.

The definition, which generated some ideas and results that are part of this monograph is the following.

Definition 3.2 A nonnegative matrix U is said to be **ultrametric** if

(i) U is symmetric;
(ii) for all j, $U_{jj} \geq \max\{U_{kj} : k \in I\}$, that is U is column pointwise diagonal dominant;
(iii) for all $j, k, l \in I$ it is verified the **ultrametric inequality**

$$U_{jk} \geq \min\{U_{jl}, U_{lk}\}.$$

If the inequalities are strict in (ii) we say U is **strictly ultrametric**.

We observe that (ii) is a consequence of (i) and (iii), but we included it to highlight the strictly ultrametric case.

The next example is one of the simplest ultrametric matrices and its importance is due to its relation to the simple random walk.

Example 3.3 For $n \geq 2$ consider the matrix

$$U = \begin{pmatrix} 1 & 1 & 1 & \cdots & 1 \\ 1 & 2 & 2 & \cdots & 2 \\ 1 & 2 & 3 & \cdots & 3 \\ \vdots & \vdots & \vdots & \ddots & \vdots \\ 1 & 2 & 3 & \cdots & n \end{pmatrix}.$$

Notice that for all $1 \leq i, j \leq n$ we have $U_{ij} = i \wedge j$ and it is straightforward to show that U is ultrametric. This is a particular case of a D-matrix introduced by Markham [43] (see also Sect. 6.4.1).

U is proportional to the potential of a simple random walk on $\{1, \cdots, n\}$, which is reflected at n and absorbed at 0. Indeed, the inverse of $2U$ is $M = \mathbb{I} - P$, where

$$
P = \begin{pmatrix}
0 & 1/2 & 0 & \cdots & 0 & 0 & 0 \\
1/2 & 0 & 1/2 & \cdots & 0 & 0 & 0 \\
& \ddots & \ddots & \ddots & & & \\
& & \ddots & \ddots & \ddots & & \\
& & & \ddots & \ddots & \ddots & \\
0 & 0 & 0 & \cdots & 1/2 & 0 & 1/2 \\
0 & 0 & 0 & \cdots & 0 & 1/2 & 1/2
\end{pmatrix}
$$

is the transition matrix of that random walk.

The matrix U has a block structure

$$
U = \begin{pmatrix} 1 & \mathbb{1}' \\ \mathbb{1} & V \end{pmatrix},
$$

where V is an ultrametric matrix and the block $U_{\{1\} \times \{2, \cdots, n\}}$ is the constant 1, which is the minimum value of U. This fact is the basis of the proof that U is the inverse of a row diagonally dominant M-matrix.

In Sect. 3.3 we shall give a relation between ultrametric matrices and ultrametric distances, that inspired the name of these matrices. For $i \neq j$, U_{ij} is the reciprocal of d_{ij}, where d is an ultrametric distance. This explains the reversal inequality in the definition of ultrametric matrices. We shall explain better this reciprocal relation between a potential (inverse of M-matrix) and a distance, in much the same spirit as the Euclidian distance is the reciprocal of the Newtonian potential in $\mathbb{R}^3$.

One of the fundamental properties associated with ultrametricity is that among the three numbers U_{ij}, U_{ik}, U_{kj} at least two of them are equal. This is the well known property of an ultrametric distance: Every triangle is isosceles. Indeed, assume that $U_{ij} = \max\{U_{ij}, U_{ik}, U_{kj}\}$. Then, the ultrametric inequality gives

$$
U_{ik} \geq U_{kj} \text{ and } U_{kj} \geq U_{ik},
$$

showing the claim. This remark is the beginning of a much deeper geometric relation we shall study in Sect. 3.3.5.

Proposition 3.4 *Assume that U is an ultrametric matrix of size n, then there exists a permutation Π such that*

$$\Pi U \Pi' = \begin{pmatrix} V_1 & \alpha \mathbb{1}_p \mathbb{1}'_{n-p} \\ \alpha \mathbb{1}_{n-p} \mathbb{1}'_p & V_2 \end{pmatrix},$$

where V_1, V_2 are ultrametric matrices of sizes $p, n-p \in [1, n-1]$ and $\alpha = \min\{U\}$.

Proof The first observation is that any principal submatrix W of U is also ultrametric. So, the only thing left to prove is the existence of a block with the constant value α. If $U = \alpha$ is the constant matrix, then the result is obvious. Then, in what follows, we assume there exists a couple (i_0, j_0) (not necessarily different) such that $U_{i_0 j_0} > \alpha$. Let us define the relation

$$i \overset{\mathfrak{R}}{\sim} j \Leftrightarrow [i = j] \text{ or } U_{ij} > \alpha.$$

Ultrametricity and symmetry of U imply this is an equivalence relation. Indeed, if i, j, k are all different and $i \overset{\mathfrak{R}}{\sim} k, k \overset{\mathfrak{R}}{\sim} j$, then

$$U_{ij} \geq \max\{U_{ik}, U_{kj}\} > \alpha.$$

Denote by $\mathfrak{R} = \{A_1, \cdots, A_r\}$ the set of atoms for this equivalence relation. By definition these atoms are disjoint and they cover I. Our extra assumption $U_{i_0 j_0} > \alpha$ ensures that $r \geq 2$. Now, consider $K = \overset{r-1}{\underset{\ell=1}{\cup}} A_\ell$ and $L = A_r$. For any $i \in K, j \in L$ we have $U_{ij} = U_{ji} = \alpha$. So, U_{KL} is the desired constant block with value α.

Let $p = \#K$, $K = \{i_1, \cdots, i_p\}$ and σ a permutation on I such that $\sigma(i_\ell) = \ell$ for $\ell = 1, \cdots, p$. The permutation matrix Π defined as

$$\Pi_{ij} = 1 \Leftrightarrow \sigma(j) = i,$$

satisfies the required property.

The following is the main result of this section and it originates most of our interest in the inverse M-matrix problem. We recall that a right equilibrium potential of U is a nonnegative vector λ such that $U\lambda = \mathbb{1}$. Its total mass is denoted by $\bar{\lambda} = \mathbb{1}'\lambda$.

Theorem 3.5 *Let U be an ultrametric matrix. Assume that no row of U is the zero vector (which is equivalent to having only positive elements on the diagonal of U). Let $\alpha = \min\{U\}$. Then*

  (i) U has a right equilibrium potential. All of them has the same total mass, $\alpha\bar{\lambda} = \alpha \mathbb{1}'\lambda \leq 1$, with equality only if U has a constant row equal to $\alpha\mathbb{1}$;
  (ii) U is nonsingular if and only if no two rows are equal;

(iii) *for all* $t > 0$ *the ultrametric matrix* $U(t) = U + t\mathbb{I}$ *is nonsingular and there exist a real number* $c(t) > 0$ *and a doubly substochastic symmetric matrix* $P(t)$ *such that*

$$(U + t\mathbb{I})^{-1} = c(t)(\mathbb{I} - P(t)).$$

In particular, if U *is nonsingular then* U^{-1} *is a row and column diagonally dominant* M-*matrix. Thus, there exist a constant* $c > 0$ *and a doubly substochastic symmetric matrix* P *such that* $U^{-1} = c(\mathbb{I} - P)$. U *is a positive definite matrix because it is a symmetric inverse* M-*matrix. If* U *is singular then* adjoint(U) *is a* Z-*matrix whose diagonal terms are nonnegative. Also, we deduce that every ultrametric matrix is positive semi-definite.*

Proof We shall prove the result by induction on n, the size of the matrix U. For $n = 2$ the result is obvious. Assume the result holds for any ultrametric matrix of size at most $n - 1$, where $n \geq 3$. All the properties to be shown are invariant under permutations of rows and columns and therefore we can assume without loss of generality that

$$U = \begin{pmatrix} A & \alpha \mathbb{1}_p \mathbb{1}'_q \\ \alpha \mathbb{1}_q \mathbb{1}'_p & B \end{pmatrix},$$

where A, B are ultrametric matrices of sizes p and $q = n - p \in [1, n - 1]$ respectively and $\alpha = \min\{U\}$. Notice that neither A nor B can have a row of zeros, because on the contrary $\alpha = 0$ and then U would have a row of zeros.

Let λ_A, λ_B be equilibrium potentials for A, B. For $a, b \in \mathbb{R}$, to be fixed, consider

$$\lambda = \begin{pmatrix} a\lambda_A \\ b\lambda_B \end{pmatrix}.$$

The conditions on a, b in order that λ is an equilibrium potential for U are

$$a \quad\;\; + b\,\alpha\bar{\lambda}_B = 1$$
$$a\,\alpha\bar{\lambda}_A + b \quad\;\;\; = 1,$$

where $\bar{\lambda}_A = \mathbb{1}'_p \lambda_A, \bar{\lambda}_B = \mathbb{1}'_q \lambda_B$. The determinant of this 2×2 linear system is $\Delta = 1 - \alpha^2 \bar{\lambda}_A \bar{\lambda}_B$. From $\alpha \leq \min\{A\} \wedge \min\{B\}$ and the induction hypothesis, we get $\alpha\bar{\lambda}_A \leq 1$, $\alpha\bar{\lambda}_B \leq 1$, which implies that $\Delta \geq 0$.

Notice that the unique possibility for $\Delta = 0$ is $\alpha = \min\{A\} = \min\{B\}$ and both A, B have a constant row of α values. In this case U has at least one constant row and therefore also a constant column of α values, say the i-th column, which by the hypothesis made on U is not zero. In this situation $\lambda = \alpha^{-1} e(i)$ is an equilibrium potential where $e(i)$ is the i-th vector of the canonical basis. Then in what follows we may assume $\Delta > 0$, which implies

$$a = \frac{1 - \alpha \bar{\lambda}_B}{\Delta} \geq 0, \ b = \frac{1 - \alpha \bar{\lambda}_A}{\Delta} \geq 0,$$

and the existence of an equilibrium potential is proven for U. If λ is any equilibrium potential then for all i

$$\sum_j U_{ij} \lambda_j = 1 \geq \alpha \bar{\lambda}.$$

Assume that $\alpha \bar{\lambda} = 1$ and take i any index for which $\lambda_i > 0$. Then, from the previous inequality, we have $U_{ii} = \alpha = \min\{U\}$, which by the ultrametricity of U implies that the i-th row is constant α. This finishes part (i).

In proving (ii) the non trivial statement is that: If U has no two equal rows then it is nonsingular. We notice that A, B cannot have two equal rows and therefore are nonsingular, by the induction hypothesis. Using Schur's formula (see for example [37] and (3.1) below) U is nonsingular if B is nonsingular and $A - \alpha^2 \mathbb{1}_p \mathbb{1}'_q B^{-1} \mathbb{1}_q \mathbb{1}'_p$ is nonsingular. This matrix is equal to

$$C = A - \alpha^2 \bar{\lambda}_B \mathbb{1}_p \mathbb{1}'_p,$$

which is the matrix obtained from A by subtracting the constant $\alpha^2 \bar{\lambda}_B \leq \alpha \leq \min\{A\}$. Therefore C is nonnegative and ultrametric. The unique way this matrix C has a zero row is that $\alpha = \min\{A\}$, A has a constant row α and $\alpha \bar{\lambda}_B = 1$. This last equality happens only if B has a constant row α which is impossible since we assume that U has no two equal rows. Therefore, no row of C is the zero vector. Since C cannot have two equal rows (otherwise A would have two equal rows) the inductive step implies that C is nonsingular and then U is also nonsingular. In particular the inverse of C is a row diagonally dominant M-matrix.

In a similar way it is deduced that $D = B - \alpha^2 \bar{\lambda}_A \mathbb{1}_q \mathbb{1}'_q$ is nonsingular. Let $\lambda_C = C^{-1} \mathbb{1}_p$ and $\lambda_D = D^{-1} \mathbb{1}_q$ be the equilibrium potentials of C and D respectively (we recall that both are nonnegative vectors). From Schur's decomposition we get

$$U^{-1} = \begin{pmatrix} C^{-1} & -\alpha B^{-1} \mathbb{1}_q \mathbb{1}'_p C^{-1} \\ -\alpha C^{-1} \mathbb{1}_p \mathbb{1}'_q B^{-1} & D^{-1} \end{pmatrix} = \begin{pmatrix} C^{-1} & -\alpha \lambda_B \lambda'_C \\ -\alpha \lambda_C \lambda'_B & D^{-1} \end{pmatrix}.$$
$$(3.1)$$

(iii) It is clear that $U(t) = U + t\mathbb{I}$ is a strictly ultrametric matrix and so it is row and column strictly pointwise diagonally dominant. Then, it cannot have two equal rows and therefore is nonsingular. Consider $C(t) = C + t\mathbb{I}$, $D(t) = D + t\mathbb{I}$. By induction their inverses are M-matrices. According to formula (3.1) $U(t)^{-1}$ is a Z-matrix and since $U(t)$ is nonnegative, we conclude $U(t)^{-1}$ is an M-matrix.

Moreover, from (i) we deduce that $U(t)^{-1}\mathbb{1} = \lambda_{U(t)} \geq 0$, which implies that $U(t)^{-1}$ is row diagonally dominant (being symmetric it is also column diagonally dominant). Take $c(t) = \max\{U(t)_{ii}^{-1} : i \in I\}$, then it is not difficult to prove that

$$P(t) = \mathbb{I} - \frac{1}{c(t)}U(t)^{-1}$$

is a doubly substochastic nonnegative matrix.

The last part of the theorem, is obtained by letting t converges to 0, in the above reasoning.

3.2   Generalized Ultrametric and Nested Block Form

The main concepts and results of this section are taken from: McDonald, Neumann, Schneider and Tsatsomeros [47]; and from Nabben and Varga [52]. Let us start with the basic definitions we shall use in this section.

Definition 3.6 A matrix V of size n is said to be a **constant block form** matrix (CBF) if $n = 1$ or there exists a non trivial partition $\{R, R^c\}$ of I such that

$$V = \begin{pmatrix} A & \alpha\mathbb{1}_p\mathbb{1}_q' \\ \beta\mathbb{1}_q\mathbb{1}_p' & B \end{pmatrix} \tag{3.2}$$

where $A = V_{RR}$, $B = V_{R^c R^c}$ are constant block form matrices. Here $\#R = p, \#R^c = n - p = q$, and $\alpha, \beta \in \mathbb{R}$.

In addition, we say that V is a nonnegative CBF if $V \geq 0$.

The following is a generalization of ultrametricity to non-symmetric matrices given in [47, 52].

Definition 3.7 A positive CBF matrix V of size n is said to be a **nested block form** matrix (NBF) if $n = 1$ or there is a decomposition like in (3.2)

$$V = \begin{pmatrix} A & \alpha\mathbb{1}_p\mathbb{1}_q' \\ \beta\mathbb{1}_q\mathbb{1}_p' & B \end{pmatrix}, \tag{3.3}$$

where

(1) $0 \leq \alpha = \min\{V\}$;
(2) $\beta \leq \min\{A\}$ and $\beta \leq \min\{B\}$;
(3) A, B are in nested block form of the respective sizes.

Moreover a nonnegative matrix U is called a **generalized ultrametric matrix**
(GUM) if there exists a permutation Π such that $V = \Pi U \Pi'$ is in nested block
form.

We notice that a nested block form matrix is in general nonsymmetric. On the
other hand, the symmetric GUM are exactly the ultrametric matrices. We state here
a combinatorial description of a GUM given in Lemma 4.1 in [47].

Theorem 3.8 *Assume that U is a nonnegative matrix of size n. Then, U is GUM if
and only if U satisfies*

  (i) *for all i, j we have $U_{ii} \geq \max\{U_{ij}, U_{ji}\}$;*
 (ii) *$n \leq 2$ or $n > 2$ and every subset $\mathscr{A} = \{i, j, k\} \subseteq I$, of size 3, has a preferred
       element $i \in \mathscr{A}$, which means*

     (ii.1) $U_{ij} = U_{ik}$;
     (ii.2) $U_{ji} = U_{ki}$;
     (ii.3) $\min\{U_{jk}, U_{kj}\} \geq \min\{U_{ji}, U_{ij}\}$;
     (ii.4) $\max\{U_{jk}, U_{kj}\} \geq \max\{U_{ji}, U_{ij}\}$.

Proof We give just an idea of the proof, which is done by induction. Assume that
U is in NBF and $n > 2$. Then, there exists a decomposition of the form

$$U = \begin{pmatrix} A & \alpha \mathbb{1}_p \mathbb{1}_q' \\ \beta \mathbb{1}_q \mathbb{1}_p' & B \end{pmatrix}$$

where $A = U_{KK}, B = U_{LL}$ and $K \cup L = I$. Consider now $\mathscr{A} = \{i, j, k\} \subseteq I$ of
size 3. If $\mathscr{A} \subseteq K$ or $\mathscr{A} \subseteq L$ the existence of a preferred element is obtained by
the induction hypothesis. So, essentially the only case left to analyze is when $i \in K$
and $j, k \in L$. In this situation i is preferred and

$$U_{ij} = U_{ik} = \alpha, U_{ji} = U_{ki} = \beta,$$

showing $(ii.1)$ and $(ii.2)$. On the other hand, $(ii.3)$ follows from $\alpha = \min\{U\}$, and
$(ii.4)$ follows from part (2) in Definition 3.7.

The converse is obtained by considering first $\alpha = \min\{U\}$ and to show the
existence of a partition $I = K \cup L$ such that $U_{KL} = \alpha$ or $U_{LK} = \alpha$. Then, in the
former case one has to show that U_{LK} is constant, call it β and then the result follows
by induction after showing (2) in the same definition. This follows at once if $\alpha = \beta$.
When $\alpha < \beta$ the result follows from $(ii.4)$.

We now state the main result for GUM matrices.

Theorem 3.9 *Let U be an GUM. Assume that no row of U is the zero vector (which
is equivalent to have no zero on the diagonal). Let $\alpha = \min\{U\}$. Then*

  (i) *U has a right (and a left) equilibrium potential. All these vectors λ satisfy
       $\alpha \bar{\lambda} = \alpha \mathbb{1}' \lambda \leq 1$, with equality only if U has a constant column equal to $\alpha \mathbb{1}$;*

(ii) *U is nonsingular if and only if no two rows are equal (equivalently no two columns are equal);*

(iii) *for all $t > 0$ the matrix $U(t) = U + t\mathbb{I}$ is nonsingular and there exists a real number $c(t) > 0$ and a doubly substochastic matrix $P(t)$ such that*

$$(U + t\mathbb{I})^{-1} = c(t)(\mathbb{I} - P(t)).$$

In particular, if U is nonsingular then U^{-1} is a row and column diagonally dominant M-matrix. Thus, there exist a constant $c > 0$ and a doubly substochastic matrix P such that $U^{-1} = c(\mathbb{I} - P)$.

This theorem can be proved using induction, much in the same lines as the proof of Theorem 3.5. We shall prove (*iii*) in Chap. 5 (see Corollary 5.30) where we study a generalization of GUM.

The decomposition $U^{-1} = \kappa(\mathbb{I} - P)$, where P is a doubly substochastic matrix, is not unique. For every $\kappa \geq \kappa(U) := \max\{(U^{-1})_{ii} : i \in I\}$, the matrix $P = \mathbb{I} - \kappa^{-1}U^{-1}$ satisfies the required property. The next result gives a priori estimate of $\kappa(U)$ in terms of U

Proposition 3.10 *Let U be a non-singular GUM. Define*

$$\epsilon(U) := \min\{U_{ii} - \max\{U_{ij}, U_{ji} : j \neq i\} : i \in I\}.$$

Then $\kappa(U) \leq \epsilon(U)^{-1}$.

Proof The result is trivial if $\epsilon(U) = 0$, so in what follows we assume $\epsilon(U) > 0$. Take $0 < \epsilon < \epsilon(U)$ and denote by $V = -\epsilon\mathbb{I} + U$. The diagonal elements of V dominate their respective columns and rows, and since U^{-1} is a row and column diagonally dominant M-matrix we get

$$(\epsilon + V_{ii})(U^{-1})_{ii} = 1 - \sum_{j \neq i} V_{ij}(U^{-1})_{ij} \leq 1 - V_{ii}\sum_{j \neq i}(U^{-1})_{ij} \leq 1 + V_{ii}(U^{-1})_{ii}.$$

Then $(U^{-1})_{ii} \leq \epsilon^{-1}$ and the result follows.

For $J \subseteq I$ the principal submatrix U_{JJ} is also a GUM and moreover $\kappa(U_{JJ}) \leq \kappa(U)$. Then, the estimation given in the previous result also works for any principal submatrix of U. Notice that if $\epsilon(U) \geq 1$ means that the diagonal of U dominates pointwise, at least by one unit, each row and column. In this situation we get $\kappa(U) \leq 1$ and then $P^U = \mathbb{I} - U^{-1}$ is a doubly substochastic matrix, as well as it is any principal sub matrix $P^{U_{JJ}} = \mathbb{I} - (U_{JJ})^{-1}$.

3.3   Graphical Construction of Ultrametric Matrices

In this section we give some graphical constructions of ultrametric matrices, which have interest on their own and also in the study of the Markov chain associated with an ultrametric matrix.

We first state ultrametric matrices in terms of ultrametric distances. We remind that a metric space (I, d) is ultrametric if the metric d satisfies the ultrametric inequality

$$d_{ij} \leq \max\{d_{ik}, d_{kj}\} \text{ for any } i, j, k \in I.$$

Hence, it is straightforward to show that the matrix U, indexed by I, is ultrametric if and only if there exists an ultrametric distance d on I such that for all i, j, k one has $d_{ij} \leq d_{ik}$ implies $U_{ij} \geq U_{ik}$. Moreover, U is strictly ultrametric if and only if for all i, j, k one has $d_{ij} \leq d_{ik}$ is equivalent to $U_{ij} \geq U_{ik}$.

Since an ultrametric matrix is diagonally dominant it follows that $U_{ii} = 0$ implies that the i-th row $U_{i\bullet}$ and the i-th column $U_{\bullet i}$ vanish. Then $U_{ii}^n = 0$ for all $n > 0$ and so i is called an inessential state, following a terminology of [15]. Without loss of generality we can assume that I contains no inessential states.

For i, j consider the relation $i \leftrightarrow j$ if and only if $U_{ij} > 0$. Ultrametricity implies that this is an equivalence relation. Let us denote by $\mathfrak{R} = \{R\}$ the set of equivalence classes. Then,

$$I = \bigcup_{R \in \mathfrak{R}} R$$

and $U_{ij} = 0 = U_{ji}$ if and only if $(i, j) \in C \times C'$, where C and C' are different classes in $\mathfrak{R}$. In studying properties of ultrametric matrices, this decomposition of the space in equivalence classes allows us to restrict ourselves to the case of positive ultrametric matrices.

3.3.1   Tree Matrices

Let us introduce some basic definitions about graphs needed for the coming sections. Given a finite set I and a set $E \subset I \times I$ the couple $\mathscr{G} = (I, E)$ is called a graph. The elements of I are called points or vertices and the elements of E are called edges or arcs. Most of the graphs we consider are non-oriented which means that E is symmetric.

A **path** γ in $\mathscr{G}$ is a finite sequence $i_0, \cdots, i_m \in I$ such that $(i_\ell, i_{\ell+1}) \in E$ for all $\ell = 0, \cdots, m-1$. We say that γ joins i_0 and i_m. The **length** of γ is m. By convention the sequence with one element $i \in I$ is a path that joins i with itself and has length 0. A **geodesic** between $i, j \in I$ is a path that joins i and j with minimal length (among all paths that join i and j). We put $d_{\mathscr{G}}(i, j)$ the length of any such optimal path, if there exists one, and $d_{\mathscr{G}}(i, j) = \infty$ otherwise. $d_{\mathscr{G}}$ is called the geodesic distance on the graph. $\mathscr{G}$ is said connected if $d_{\mathscr{G}}$ is bounded, or equivalently given two points in I there is a path that joins them. A **loop** at i is the path $i_0 = i_1 = i$, which simple means that $(i, i) \in E$. A path γ given by $i = i_0, \cdots, i_m = i$, with $m \geq 1$ is called a **cycle** if it is not a loop and $i_0, \cdots, i_{m-1}$ are all different.

The following special class of graphs will play an important role in the study of ultrametric matrices.

Definition 3.11 A graph $T = (I, E)$ is a **tree** if

(i) it is non-oriented and connected;
(ii) it has no cycles.

Given a tree T and two different vertices i, j in it, there is a unique geodesic joining i and j, which we denote by $\mathrm{Geod}(i, j)$. We take by convention $\mathrm{Geod}(i, i) = \{i\}$ and its length is 0. The tree T may have a loop at a node i, that is the pair (i, i) could belong to E.

In what follows, to simplify notation we confound T with I and E, so sometimes we denote by $i \in T$ instead of $i \in I$ and $(i, j) \in T$ instead of $(i, j) \in E$. Fix a point $r \in T$, which will be called the **root of the tree** and T will be called a **rooted tree**. In what follows, most of our trees are rooted and when it is clear from the context we omit to mention it explicitly. This root induces a partial order and a level function in the following way. We write

$$j \preceq i \quad \text{if} \quad j \in \mathrm{Geod}(i, r),$$

and say j is an **ancestor** of i. This is a partial order relation on T. The root r is the common ancestor to all points in T. For any two points i and j we denote by $i \wedge j$ the nearest common ancestor of i, j, that is

$$(i \wedge j) \preceq i, \ (i \wedge j) \preceq j \ \text{and if} \ (k \preceq i, k \preceq j) \Rightarrow k \preceq (i \wedge j).$$

We define the **level function of the tree** by $L(i) = $ length of $\mathrm{Geod}(i, r)$. Observe that L is an increasing function with respect to $\preceq$. We call the **height** of T the value $H(T) = \max\{L(i) : i \in T\}$.

A point i is said to be a **leaf** of the tree if $(i, j) \in E, j \neq i$ implies $j \prec i$. We denote by $\mathscr{L}(T)$ the set of leaves of T. Notice that a leaf is a local maxima for the level function L. A **branch** of T is any geodesic from a leaf to r.

Every node $i \neq r$ possesses a unique **immediate predecessor** i^- which is characterized as $(i^-, i) \in E$ and $i^- \prec i$. On the other hand if i is not a leaf we denote by $\mathfrak{S}(i) = \{j : (i, j) \in E, i \prec j\}$ the set of **immediate successors** of i.

Finally, a **linear tree** is a rooted tree that has just one leaf. It is clear that any branch of T is identified with a maximal linear subtree of T.

Definition 3.12 A nonnegative matrix U is said to be a **tree matrix** if there exist a rooted tree T with root r, a level function L and a nonnegative nondecreasing sequence of real numbers $w_0 \leq w_1 \leq \cdots \leq w_{H(T)}$, called the **weight function**, such that

$$U_{ij} = w_{L(i \wedge j)}.$$

We call the triplet (T, r, w) a **weighted tree** and say U is induced by it.

Example 3.13 Consider the matrix

$$W = \begin{pmatrix} w_0 & w_0 & w_0 & w_0 & w_0 & w_0 & w_0 & w_0 & w_0 \\ w_0 & w_1 & w_1 & w_1 & w_0 & w_0 & w_0 & w_0 & w_0 \\ w_0 & w_1 & w_2 & w_1 & w_0 & w_0 & w_0 & w_0 & w_0 \\ w_0 & w_1 & w_1 & w_2 & w_0 & w_0 & w_0 & w_0 & w_0 \\ w_0 & w_0 & w_0 & w_0 & w_1 & w_1 & w_1 & w_1 & w_1 \\ w_0 & w_0 & w_0 & w_0 & w_1 & w_2 & w_1 & w_1 & w_1 \\ w_0 & w_0 & w_0 & w_0 & w_1 & w_1 & w_2 & w_1 & w_1 \\ w_0 & w_0 & w_0 & w_0 & w_1 & w_1 & w_2 & w_3 & w_2 \\ w_0 & w_0 & w_0 & w_0 & w_1 & w_1 & w_2 & w_2 & w_3 \end{pmatrix}.$$

This matrix is a tree matrix, whose associated tree is given below.

A tree matrix U is ultrametric. Indeed, U is clearly symmetric and it satisfies the ultrametric inequality because given any points i, j, k, we have

$$i \wedge k \succcurlyeq (i \wedge j) \wedge (j \wedge k)$$

which together with the fact that L, w are increasing implies

$$U_{ik} \geq \min\{U_{ij}, U_{jk}\}.$$

Also it is straightforward to conclude that U has no inessential states if and only if $w_0 > 0$, and U is strictly ultrametric if and only if w is strictly increasing.

We summarize some of the basic properties of tree matrices in the next result.

Proposition 3.14 *Consider a tree matrix U induced by the weighted tree (T, r, w). Then*

(i) U is ultrametric;
(ii) U has no inessential states if and only if $w_0 > 0$;
(iii) U is nonsingular if and only if $w_0 > 0$ and w is strictly increasing.

The next notion gives a partial order among weighted trees which is helpful to define a minimal tree structure associated with an ultrametric matrix.

Definition 3.15 The weighted tree (T', r', w') is said to be an **extension** of the weighted tree (T, r, w) if there exists a one to one mapping $\varphi : T \to T'$ such that

$$w'_{L'(\varphi(i))} = w_{L(i)},$$

for all $i \in T$.

It is not difficult to see that not all ultrametric matrices are tree matrices, but all of them are obtained by restriction of tree matrices.

Theorem 3.16 *A matrix U defined in $I \times I$ is an ultrametric matrix if and only if there exists a weighted tree $(\hat{T}, \hat{r}, \hat{w})$ on an extension $\hat{I} \supseteq I$ such that $\hat{U}_{II} = U$, where $\hat{U}$ is the matrix associated with $(\hat{T}, \hat{r}, \hat{w})$. Moreover, there exits a minimal weighted tree matrix $(\tilde{T}, \tilde{r}, \tilde{w})$, defined on some $\tilde{I}$, with this property. This minimal extension $\tilde{U}$ is nonsingular if U is nonsingular and positive.*

Proof The condition is clearly sufficient since ultrametricity is preserved under restriction. Now we prove the condition is necessary. In the first part of the proof we assume that U has no repeated rows.

We denote by $w_0 < \ldots < w_p$ the ordered set of values $\{U_{k\ell} : k, \ell \in I\}$. For each $0 \le q \le p$ and $i \in I$ we denote by $[i]_q = \{j \in I : U_{ij} \ge w_q\}$. Observe that $[i]_0 = I$ and put $q_i = \max\{q : [i]_q \ne \emptyset\}$, where $\emptyset$ denotes the empty set.

The extension is obtained on the set $\tilde{I} = \{([i]_q, q) : i \in I \text{ and } q = 0, \ldots, q_i\}$ where we identify i and $\bar{i} = ([i]_{q_i}, q_i)$. We also denote $\bar{I} = \{\bar{i} : i \in I\}$.

Consider the graph $\tilde{T} \subseteq \tilde{I} \times \tilde{I}$ with connections

$$(([i]_{q-1}, q-1), ([i]_q, q)) \text{ and } (([i]_q, q), ([i]_{q-1}, q-1)),$$

when these couples are defined. The graph $\tilde{T}$ is a tree, and we call it the tree associated with U. Take the root $\tilde{r} = ([I], 0)$, which induces a level function $\tilde{L}$ given by $\tilde{L}(([i]_q, q)) = q$.

The order $\preceq$ on $\tilde{T}$ is described as follows. The first case is

$$([i]_q, q) \wedge ([i]_s, s) = ([i]_{q \wedge s}, q \wedge s),$$

and we write $t(i, i, q, s) = q \wedge s$. When

$$[i]_q \cap [j]_s = \emptyset,$$

we denote by $t(i, j, q, s)$ the largest value u for which $[i]_u = [j]_u$. In this way we have

$$([i]_q, q) \wedge ([j]_s, s) = ([i]_t, t).$$

The matrix defined by $\tilde{U}_{([i]_q,q),([j]_s,s)} = w_t$ if $t = t(i,j,q,s)$ is a tree matrix associated with the tree $\tilde{T}$. It is straightforward to check the equality:

$$\tilde{U}_{\bar{i}\,\bar{j}} = U_{ij} \text{ for every } i,j \in I. \tag{3.4}$$

By identifying $\bar{i}$ and i and then $\bar{I}$ with I, (3.4) asserts that $\tilde{U}_{II} = U$.

Let us show that $\tilde{U}$ is minimal. For that purpose assume $\hat{U}$ induced by the weighted tree $(\hat{T}, \hat{r}, \hat{w})$ is an extension of U that is $I \subseteq \hat{I}$ and $\hat{U}_{II} = U$. We must show the existence of a one-to-one function $\varphi : \tilde{I} \to \hat{I}$ such that $\hat{w}_{\hat{L}(\varphi(\kappa))} = w_{\tilde{L}(\kappa)}$ for all $\kappa \in \tilde{I}$. First of all remark that the values taken by $\hat{w}$ cover the values of w.

Let us construct φ. Fix $q \in \{0,\dots,p\}$ and consider $i \in I$ such that $q_i \geq q$. Find the largest value s such that $\hat{w}_s = w_q$ and define $\hat{I}(s) = \{\ell \in \hat{I} : \hat{L}(\ell) = s\}$, where $\hat{L}$ is the level function of the tree $\hat{T}$.

The fact that $\hat{T}$ is a tree implies the existence of a unique $\ell(i,q) \in \hat{I}(s) \cap \widehat{\mathrm{Geod}}(i,\hat{r})$. Now, define

$$\varphi([i]_q, q) = \ell(i,q).$$

It is obvious that φ is monotone increasing with respect to the partial orders $\preccurlyeq$ in T and $\hat{\preccurlyeq}$ in $\hat{T}$. Let us show φ is one-to-one. First we prove that φ is strictly increasing on each branch of T. This is equivalent to prove that

$$\varphi(([i]_q, q)) \,\hat{\succcurlyeq}\, \varphi(([i]'_q, q')),$$

whenever $q' > q$. Since $w_q < w_{q'}$ we conclude that $\ell(i,q) \,\hat{\succcurlyeq}\, \ell(i,q')$ proving the desired property. To finish with the proof that φ is one to one, we consider the situation where $i,j \in I$ are such that $q_i \wedge q_j \geq q$ and $[i]_q \cap [j]_q = \emptyset$. Then, we need to prove $\ell(i,q) \neq \ell(j,q)$.

If we assume the contrary we have the existence of $\ell \in \hat{I}(s)$ satisfying $\ell \in \widehat{\mathrm{Geod}}(i,\hat{r}) \cap \widehat{\mathrm{Geod}}(j,\hat{r})$. Hence, $\ell \,\hat{\preccurlyeq}\, \tau = i \hat{\wedge} j$ and so

$$U_{ij} = \hat{U}_{ij} = \hat{w}_{\hat{L}_\tau} \geq \hat{w}_s = w_q.$$

But this last condition implies $[i]_q = [j]_q$, which is a contradiction. This shows that $\varphi(([i]_q, q)) \neq \varphi(([j]_q, q))$ and φ is one to one.

On the other hand by construction φ satisfies

$$\hat{w}_{\hat{L}(\varphi(\kappa))} = w_{L(\kappa)},$$

for all $\kappa \in \tilde{I}$. This finishes the first part of the proof in the case U has no repeated rows.

We notice that if U is nonsingular and positive then $w_0 > 0$ and w is strictly increasing, therefore $\tilde{U}$ is nonsingular according to Proposition 3.14.

We now assume that U has repeated rows. This forces any tree representing U to have repeated weights for different levels and the construction of a minimal tree is more involved. As before, let $w_0 < w_1 < \cdots < w_p$ be the different values $U_{ij} : i, j \in I$, properly ordered. Assume that U has $k \geq 1$ groups of repeated rows. We perturb U on the diagonal of each group of repeated rows by a small parameter $\epsilon > 0$, that we choose smaller than $\min\{w_i - w_{i-1} : i = 1, \cdots, p\}$, in the following way: If the group of repeated rows consists of the rows $i_1 < i_2 < \cdots < i_m$ we modify the diagonal as

$$U^\epsilon_{i_1 i_1} = U_{i_1 i_1}$$
$$U^\epsilon_{i_1 i_1} = U_{i_2 i_2} + \epsilon$$
$$\vdots$$
$$U^\epsilon_{i_m i_m} = U_{i_m i_m} + \epsilon,$$

that is the first diagonal term remains the same and the rest are enlarged by ϵ. This is done in each group. The matrix U^ϵ is ultrametric and has no repeated rows (notice that in the modified rows the diagonal strictly dominates the respectively row). Then there exists a minimal weighted tree $(T^\epsilon, r^\epsilon, w^\epsilon)$ associated with U^ϵ. It is quite straightforward to prove that T^ϵ does not depend on ϵ. We call this tree $\tilde{T}$. Since $\tilde{U}^\epsilon_{II} = U^\epsilon$, we can pass to the limit to get that $(\tilde{T}, \tilde{r}, \tilde{w})$ is the desired tree, where $\tilde{w} = \lim_{\epsilon \downarrow 0} w^\epsilon$.

Example 3.17 Consider the tree matrix W given in Example 3.13 and take $V = W_{II}$ where $I = \{3, 4, 5, 6, 8, 9\}$.
   Then

$$V = \begin{pmatrix} w_2 & w_1 & w_0 & w_0 & w_0 & w_0 \\ w_1 & w_2 & w_0 & w_0 & w_0 & w_0 \\ w_0 & w_0 & w_1 & w_1 & w_1 & w_1 \\ w_0 & w_0 & w_1 & w_2 & w_1 & w_1 \\ w_0 & w_0 & w_1 & w_1 & w_3 & w_2 \\ w_0 & w_0 & w_1 & w_1 & w_2 & w_3 \end{pmatrix}$$

and $W = \tilde{V}$ is the tree extension of V, whose tree $\tilde{T}$ is given in Fig. 3.1.

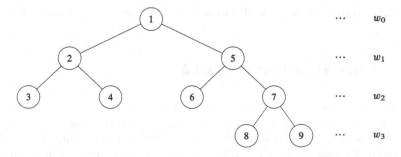

Fig. 3.1 The tree associated with the tree matrix W

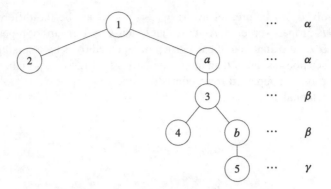

Fig. 3.2 Tree matrix associated with a matrix with repeated rows

As a second example consider a matrix with repeated rows.

$$U = \begin{pmatrix} \alpha & \alpha & \alpha & \alpha & \alpha \\ \alpha & \alpha & \alpha & \alpha & \alpha \\ \alpha & \alpha & \beta & \beta & \beta \\ \alpha & \alpha & \beta & \beta & \beta \\ \alpha & \alpha & \beta & \beta & \gamma \end{pmatrix},$$

where $\alpha < \beta < \gamma$. Then U^ϵ is

$$U^\epsilon = \begin{pmatrix} \alpha & \alpha & \alpha & \alpha & \alpha \\ \alpha & \alpha+\epsilon & \alpha & \alpha & \alpha \\ \alpha & \alpha & \beta & \beta & \beta \\ \alpha & \alpha & \beta & \beta+\epsilon & \beta \\ \alpha & \alpha & \beta & \beta & \gamma \end{pmatrix}.$$

The minimal supporting tree for U is given in Fig. 3.2 (here $I = \{1, 2, 3, 4, 5\}$ and $\tilde{I} = I \cup \{a, b\}$).

We will see later on that the minimal tree extension of an ultrametric matrix U has all the information that is needed to understand the graph associated with U^{-1}.

3.3.2 A Tree Representation of GUM

In order to represent a GUM we need to consider nonsymmetric tree matrices. Let T be a rooted tree, with root r and level function L, and two systems of weights associated with the levels of the tree that we denote w^l and w^r (l stands for left, and r for right). We assume that $w^l \leq w^r$ pointwise. We also require a a total order $\leq$

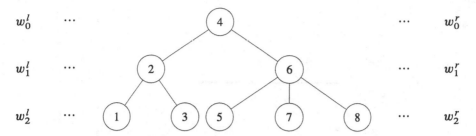

Fig. 3.3 A rooted tree with two weight systems

on the set of vertices of T to distinguish left and right. We call $(T, r, w^l, w^r, \leqslant)$ a
general weighted tree. The matrix associated with this weighted tree is

$$
U_{ij} =
\begin{cases}
w^r_{L(i \wedge j)} & \text{if } j \leqslant i \\
\\
w^l_{L(i \wedge j)} & \text{otherwise.}
\end{cases}
$$

It is straightforward to show that U is a GUM.

Here is an example of a nonsymmetric tree matrix given by the tree on Fig. 3.3.

$$
W =
\begin{pmatrix}
w^r_2 & w^l_1 & w^l_1 & w^l_0 & w^l_0 & w^l_0 & w^l_0 & w^l_0 \\
w^r_1 & w^r_1 & w^l_1 & w^l_0 & w^l_0 & w^l_0 & w^l_0 & w^l_0 \\
w^r_1 & w^r_1 & w^r_2 & w^l_0 & w^l_0 & w^l_0 & w^l_0 & w^l_0 \\
w^r_0 & w^r_0 & w^r_0 & w^r_0 & w^l_0 & w^l_0 & w^l_0 & w^l_0 \\
w^r_0 & w^r_0 & w^r_0 & w^r_0 & w^r_2 & w^l_1 & w^l_1 & w^l_1 \\
w^r_0 & w^r_0 & w^r_0 & w^r_0 & w^r_1 & w^r_1 & w^l_1 & w^l_1 \\
w^r_0 & w^r_0 & w^r_0 & w^r_0 & w^r_1 & w^r_1 & w^r_2 & w^l_1 \\
w^r_0 & w^r_0 & w^r_0 & w^r_0 & w^r_1 & w^r_1 & w^r_1 & w^r_2
\end{pmatrix} .
$$

In this example $\leqslant$ is the standard order in $\{1, 2, 3, 4, 5, 6, 7, 8\}$.

Recall that our intention is to associate a tree to a GUM. In a first thought one is
tempted to believe that the tree matrix associated with a GUM is the tree associated
with the upper or lower part of the matrix. This in general does not work because
these trees can be quite different as shown by the example.

Example 3.18 Consider the NBF matrix

$$
U =
\begin{pmatrix}
4 & 1 & 1 & 1 \\
3 & 4 & 1 & 1 \\
2 & 2 & 4 & 1 \\
2 & 2 & 3 & 4
\end{pmatrix}
$$

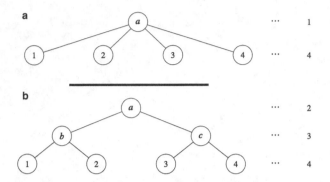

Fig. 3.4 The weighted tree for (**a**) the upper part of U and (**b**) the lower part

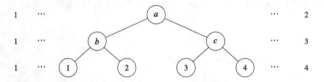

Fig. 3.5 The general weighted tree for the GUM U in Example 3.18

In this example the tree associated with the lower part is an extension of the one associated with the upper part. This is not the general case and both trees can be incomparable. For this reason is better to consider the (symmetric) ultrametric matrix $U + U'$, whose tree is an extension of both (Fig. 3.4).

Theorem 3.19 *U defined on I is a GUM if and only if there exists $(\hat{T}, \hat{r}, w^l, w^r, \leqslant)$ a general weighted tree defined on some $\hat{I}$, which contains I, such that its associated matrix $\hat{U}$ is an extension of U, namely $\hat{U}_{II} = U$.*

Moreover, there exits one minimal weighted tree $(\tilde{T}, \tilde{r}, \tilde{w}^l, \tilde{w}^r, \tilde{\leqslant})$ that satisfies this property. That is, $(\tilde{T}, \tilde{r}, \tilde{w}^l, \tilde{w}^r, \tilde{\leqslant})$ is embedded in any other general weighted tree extension of U.

Proof Without loss of generality we can assume that U is an $n \times n$ NBF matrix. Consider the ultrametric matrix $V = U + U'$ and $(\tilde{T}, \tilde{r}, \tilde{w})$ its minimal weighted tree representation on $\tilde{I}$ (see Theorem 3.16). Now, decompose $\tilde{w} = \tilde{w}^l + \tilde{w}^r$ according to the upper and lower triangular parts of U (such decomposition is unique because the monotone properties of the blocks in a NBF see Definition 3.3). The only thing left is to define the order $\tilde{\leqslant}$ on $\tilde{I}$. Since U is in NBF this order is the standard one on $I = \{1, \cdots, n\}$ which we extend to any total order on $\tilde{I}$ (here there is possibility for more than one selection). It is not difficult to see that $(\tilde{T}, \tilde{r}, \tilde{w}^l, \tilde{w}^r, \tilde{\leqslant})$ is a possible solution.

The general weighted tree associated with U of Example 3.18 is given in Fig. 3.5.

3.3.3   Gomory-Hu Theorem: The Maximal Flow Problem

The main goal of this section is to show the relation between the maximal flow problem in non oriented graphs and ultrametric matrices. In addition, given an ultrametric matrix we construct a linear graph whose maximal flow matrix is the initial matrix.

Consider a non oriented and connected graph (V, E) where V is the set of vertices and E the set of edges. To each edge e we associate a nonnegative number $c(e)$ called the **capacity** of this edge, which represents the maximal flow allowed to be sent trough this edge. Given two vertices $s \neq t$ we denote f_{st} the value of a maximal flow that can be sent from s to t trough the graph satisfying the restriction that on each edge e the flow cannot be larger than $c(e)$ (see [32] for definitions).

The theorem of Gomory-Hu (see [35]) relates this problem and a flow on a associated tree. It is clear that the maximal flow satisfies the ultrametric inequality

$$f_{st} \geq \min\{f_{sq}, f_{qt}\},$$

for any $q \notin \{s, t\}$.

The diagonal of the matrix f should represent the flow that can be generated at each vertex. This diagonal should satisfies

$$\forall s \in V \quad f_{ss} \geq \max\{f_{st} : t \neq s\}.$$

The diagonal can be obtained by considering loops at every vertex. Notice that f is an ultrametric matrix.

Example 3.20 In Fig. 3.6 we consider a symmetric graph where we have put on each edge its capacity.

We have in this example that $f_{12} = a \vee (b \wedge d \wedge c)$, $\quad f_{14} = (a \wedge c) \vee (b \wedge d)$, $\quad f_{15} = f_{14} \wedge e$.

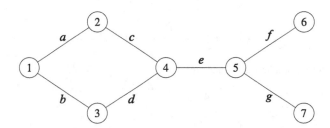

Fig. 3.6 The flow on a graph

We now associate with each ultrametric matrix U a maximum flow problem on a linear tree, for which the maximum flow matrix (except for the diagonal) is U. For that purpose after a permutation we can assume that the matrix U is in nested block form

$$U = \begin{pmatrix} A & \alpha \mathbb{1}_p \mathbb{1}'_q \\ \alpha \mathbb{1}_q \mathbb{1}'_p & B \end{pmatrix},$$

where $\alpha = \min\{U\}$, A, B are ultrametric matrices of sizes $p \times p$ and $q \times q$ respectively. Using an argument based on induction, we can construct two linear rooted trees T_A, T_B associated with A and B. Consider the linear tree obtained by connecting with an edge the leaf of T_A with the root of T_B. The capacity of this edge is α. Is quite simple to prove that the maximum flow matrix is U.

Example 3.21 Consider the following ultrametric matrix

$$U = \begin{pmatrix} 3 & 2 & 1 & 1 \\ 2 & 4 & 1 & 1 \\ 1 & 1 & 1 & 1 \\ 1 & 1 & 1 & 2 \end{pmatrix}$$

and the linear tree associated with U according to the previous algorithm (Fig. 3.7).
   We include loops at every vertex to account for the diagonal as shown in Fig. 3.8.

We notice that the graph associated with U is a linear tree with loops at every vertex. According to Definition 3.11 this graph, which is obviously connected, contains no cycles.
   This example illustrates that an ultrametric matrix is determined by its two main diagonals, and the rest of the matrix is obtained by the ultrametric inequality. The main diagonal elements correspond to loops, and the elements in the secondary diagonal corresponds to the flow between neighbors nodes in a linear tree. Here is a 4×4 example

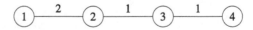

Fig. 3.7 A path flow problem associated with U

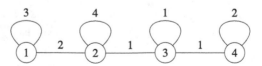

Fig. 3.8 The addition of loops to recover the diagonal of U

$$U = \begin{pmatrix} \gamma_1 & \alpha_1 & \alpha_1 \wedge \alpha_2 & \alpha_1 \wedge \alpha_2 \wedge \alpha_3 \\ \alpha_1 & \gamma_2 & \alpha_2 & \alpha_2 \wedge \alpha_3 \\ \alpha_1 \wedge \alpha_2 & \alpha_2 & \gamma_3 & \alpha_3 \\ \alpha_1 \wedge \alpha_2 \wedge \alpha_3 & \alpha_2 \wedge \alpha_3 & \alpha_3 & \gamma_4 \end{pmatrix}.$$

When $\alpha_1 < \alpha_2 \cdots < \alpha_{n-1} < \alpha_n$ and the diagonal elements $\gamma_i = \alpha_i$, one obtains the type-D matrices considered in [43].

A similar discussion can be made for a GUM in nested block form. Here we need three sets of nonnegative parameters $\alpha_1, \cdots, \alpha_{n-1}, \beta_1, \cdots, \beta_{n-1}$ and $\gamma_1, \cdots, \gamma_n$. Of course we require

$$\forall i \;\; \alpha_i \leq \beta_i$$

$$\forall i \;\; \beta_i \vee \beta_{i-1} \leq \gamma_i,$$

where $\alpha_0 = \beta_0 = 0$ and $\alpha_n = \alpha_{n-1}, \beta_n = \beta_{n-1}$. This is not enough and certain compatibility among α's and β's is needed. There has to be a permutation π that orders both sequence (α_i) and (β_i) in an increasing way

$$\alpha_{\pi(i)} \leq \alpha_{\pi(2)} \leq \cdots \leq \alpha_{\pi(n-1)},$$
$$\beta_{\pi(i)} \leq \beta_{\pi(2)} \leq \cdots \leq \beta_{\pi(n-1)}.$$

A GUM matrix is determined by its three main diagonals. The rest of the matrix is obtained by the ultrametric inequality.

Here is a 4×4 example (Fig. 3.9)

$$U = \begin{pmatrix} \gamma_1 & \alpha_1 & \alpha_1 \wedge \alpha_2 & \alpha_1 \wedge \alpha_2 \wedge \alpha_3 \\ \beta_1 & \gamma_2 & \alpha_2 & \alpha_2 \wedge \alpha_3 \\ \beta_1 \wedge \beta_2 & \beta_2 & \gamma_3 & \alpha_3 \\ \beta_1 \wedge \beta_2 \wedge \beta_3 & \beta_2 \wedge \beta_3 & \beta_3 & \gamma_4 \end{pmatrix}.$$

Due to the needed compatibility between α's and β's, not all maximal flow matrices in oriented graphs (even linear graphs) are GUM. This is a big difference with the symmetric case.

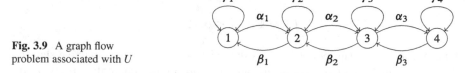

Fig. 3.9 A graph flow problem associated with U

3.3.4   Potential Matrices on Graphs with Cycles of Length Smaller Than 3

As we have seen every ultrametric matrix is the potential of a Markov process on a finite set. In other words the inverse of an ultrametric matrix U is proportional to $\mathbb{I} - P$ for some symmetric substochastic matrix P. The question we wish to answer is what type of graphs are associated with P. This question will be answered in Sect. 4.1. Here we will show that graphs with cycles less than 3 are naturally associated with ultrametricity.

The main result in this section is the following one.

Theorem 3.22 *Consider a connected non oriented graph $\mathscr{G}$ with vertex set I. Fix a node $r \in I$ and call it the root. Assume that P is a symmetric substochastic matrix carried by $\mathscr{G}$ (except maybe at the diagonal), that is*

$$i \neq j \quad and \quad P_{ij} > 0 \Rightarrow (i, j) \in \mathscr{G}.$$

We also assume that P is irreducible and P is only defective at r, that is $\sum_{j \in I} P_{ij} = 1$ for all $i \neq r$ and $\sum_{j \in I} P_{rj} < 1$. Consider $U = (\mathbb{I} - P)^{-1}$ then

  (i) if the cycles of $\mathscr{G}$ have length at most 3 then U is ultrametric;
  (ii) if $\mathscr{G}$ has a cycle of length at least 4 then it is possible to construct P such that U is not ultrametric.

Notice that this result implies that a symmetric random walk on a tree that is defective only at one node (the root) has a potential that is ultrametric.

Proof We recall that U_{ij} represents the expected number of visits to j starting from i for the Markov chain $X = (X_n)$ associated with the kernel P (see Sect. 2.3, where we add an extra absorbing state $\partial \notin I$). That is

$$U_{ij} = \mathbb{E}_i \left(\sum_{n=0}^{\infty} \mathbb{1}_{X_n = j} \right).$$

We recall that $\mathscr{T}_j = \inf\{n \geq 0 : X_n = j\}$ is the (random) hitting time to j for the first time (with the convention $\inf \emptyset = \infty$). Then the strong Markov property implies (see Proposition 2.8)

$$U_{ij} = \mathbb{P}_i(\mathscr{T}_j < \infty) U_{jj}.$$

One of the main assumptions on P is the existence of a unique defective state r. This means that the chain reaches the auxiliary absorbing state only through r. In other words for all $i \in I$ we have $\mathbb{P}_i(\mathscr{T}_r < \infty) = 1$.

Fig. 3.10 An example that
$s = i \wedge j$ is not in $\text{Geod}(i, j)$

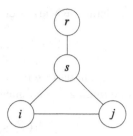

(i). The hypothesis made on the cycles of $\mathscr{G}$ implies that for every pair of vertices there is a unique geodesic connecting them. Assume the contrary and take any pair i_0, j_0 for which there are two different geodesic joining them with the extremal property that these points are also the closest (with respect to $d_{\mathscr{G}}$) among the pairs with multiple geodesics. It is clear that the distance between them is at least 2. The optimality for this couple implies that the geodesics between them can only touch at the extremes: i_0 and j_0. We conclude that the graph has a cycle of length at least 4, which is a contradiction.

Then, as we have done for trees we can define $i \wedge j$ as the farthest point from the root in the nonempty set $\text{Geod}(i, r) \cap \text{Geod}(j, r)$. Also, we denote by $l \preccurlyeq m$ if $l \subset \text{Geod}(m, r)$. The main difference with a tree is that $i \wedge j$ is not necessarily in $\text{Geod}(i, j)$ as shown by the example in Fig. 3.10.

An important property is that any path γ connecting i and r must contain $\text{Geod}(i, r)$. This is done by induction on the distance from i to r. The property is obvious if the distance is 0 because in this case $i = r$. So assume the property holds for any point at a distance at most n from r. Take any vertex i at a distance $n + 1$ from r, that is $\text{Geod}(i, r) = \{i_0 = r \prec i_1 \cdots i_n \prec i_{n+1} = i\}$ and let γ be any path from i to r. By adding to γ the point i_n, at the beginning, we get another path starting from i_n to r, which by induction hypothesis must contain $\{i_0, \cdots, i_{n-1}\}$. Therefore, we need to prove that γ contains i_n. If not, consider the part of γ that connects i and i_{n-1} and transform it on a simple path (with no repetition of points) called ζ. This path has length at least 2, otherwise $(i, i_{n-1}) \in \mathscr{G}$ and we would have two geodesics from i to r. By adding the two points i_n, i to ζ we get a cycle of length at least 4. Thus γ contains $\text{Geod}(i, r)$. Similarly, any path from i to j must contain $\text{Geod}(i, j)$.

Now given 3 different vertices i, j, k the vertices $i \wedge j$ and $i \wedge k$ are elements of $\text{Geod}(i, r)$ and they are comparable in the sense that one of them must belong to the geodesic from the other one to the root. For the same reason $i \wedge j$, $i \wedge k$ and $j \wedge k$ are all comparable and we take t to be the closest node to r.

Now we prove that U is ultrametric. As a first case let us assume that $t = i \wedge k \prec s = i \wedge j$. Then every path starting from i or j to k passes trough s. Indeed, if $s \in \text{Geod}(k, r)$ we would get $s \preccurlyeq t$, because $s \in \text{Geod}(i, r)$, which is a contradiction. Thus $s \notin \text{Geod}(k, r)$ and if there is a path γ from i to k that avoids s we get a contradiction because we construct a path from i to r which does not contains $s \in \text{Geod}(i, r)$. The proof for j is similar.

Using the Markov chain associated with P we have

$$\mathbb{P}_i(\mathscr{T}_s < \mathscr{T}_k, \mathscr{T}_s < \infty) = \mathbb{P}_i(\mathscr{T}_s < \infty).$$

The chain starting from i must pass through s before reaching r, and the probability of attaining r in a finite time is one, because r is the only point losing mass. Hence

$$\mathbb{P}_i(\mathscr{T}_s < \infty) = \mathbb{P}_i(\mathscr{T}_s < \mathscr{T}_r) = \mathbb{P}_i(\mathscr{T}_r < \infty) = 1.$$

Then we have

$$U_{ik} = \mathbb{E}_i\left(\sum_{n=\mathscr{T}_s}^{\infty} \mathbb{1}_{X_n=k}\right) = \mathbb{P}_i(\mathscr{T}_s < \infty)\mathbb{E}_s\left(\sum_{n=0}^{\infty} \mathbb{1}_{X_n=k}\right) = U_{sk}.$$

Similarly $U_{jk} = U_{sk}$ and $U_{sk} = U_{ks} = \mathbb{P}_k(\mathscr{T}_s < \infty)U_{ss} \leq U_{ss}$. On the other hand $U_{ij} \geq \mathbb{E}_i(\sum_{n=\mathscr{T}_s}^{\infty} \mathbb{1}_{X_n=j}) = \mathbb{P}_i(\mathscr{T}_s < \infty)U_{sj} = U_{sj}$ and this last quantity satisfies

$$U_{sj} = U_{js} = \mathbb{P}_j(\mathscr{T}_s < \infty)U_{ss} = U_{ss}.$$

That is $U_{ij} \geq U_{ik} = U_{jk}$ proving that the following three inequalities hold

$$U_{ij} \geq \min\{U_{ik}, U_{kj}\}; \quad U_{ik} \geq \min\{U_{ij}, U_{jk}\}; \quad U_{jk} \geq \min\{U_{ji}, U_{ik}\}. \tag{3.5}$$

So, the case left to analyze is $s = i \wedge j = i \wedge k = j \wedge k$. As a subcase we assume one of the three nodes, say k, is equal to s. We have as before that every path connecting i and k passes by s and

$$U_{ij} \geq U_{ik} = \mathbb{P}_i(\mathscr{T}_k < \infty)U_{kk} = U_{kk} = U_{jk},$$

and the ultrametric inequalities (3.5) hold again.

Finally, we assume $s = i \wedge j = i \wedge k = j \wedge k$ and s is different from i, j, k. Consider $\hat{i}$ the unique point on $\mathrm{Geod}(i, s)$ that satisfies $(\hat{i}, s) \in \mathscr{G}$. Similarly we consider $\hat{j}$ and $\hat{k}$. Assume there is a path γ from $\hat{i}$ to $\hat{j}$ avoiding s. We can also assume that γ contains no repeated points. If the length of γ is at least two then we get a cycle of length at least 4 by adding s and $\hat{i}$ to γ which is a contradiction. Since $\hat{i} \neq \hat{j}$ we conclude that $(\hat{i}, \hat{j}) \in \mathscr{G}$. On the other hand if we assume the same holds between $\hat{i}$ and $\hat{k}$, that is, there exists a path connecting both points avoiding s, we obtain $(\hat{i}, \hat{k}) \in \mathscr{G}$ and we arrive again to a contradiction because then $\hat{j}, \hat{i}, \hat{k}, s$ form a cycle of length 4.

In conclusion if there is a path connecting i and j avoiding s then every path from i or j to k must visit s and then as before we have the set of inequalities (3.5). This means that U is ultrametric and we have finished the proof of (i).

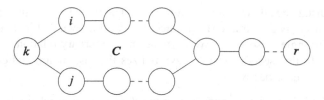

Fig. 3.11 A graph $\hat{\mathscr{G}}$ with a cycle greater than 3

(ii). Consider a graph $\mathscr{G}$ which contains a cycle C of length at least 4. We start by defining a subgraph $\hat{\mathscr{G}}$ which is formed by this cycle and a simple path from a vertex in this cycle to r.

Consider i, k, j as in the Fig. 3.11 and any P symmetric substochastic matrix defective only at r and carried by $\hat{\mathscr{G}}$ (except maybe at the diagonal). Notice first that

$$U_{ji} = \mathbb{P}_j(\mathscr{T}_i < \infty)U_{ii} < U_{ii}$$

because $\mathbb{P}_j(\mathscr{T}_i < \infty) < 1$ due to the fact that there is a path, with positive probability, starting at j ending at r, which does not pass through i. Then, there is a positive probability to reach ∂ (the absorbing state) without visiting i. Similarly, we have $U_{ij} < U_{ii}$. On the other hand $\mathbb{P}_k(\mathscr{T}_i < \mathscr{T}_j) + \mathbb{P}_k(\mathscr{T}_j < \mathscr{T}_i) = 1$ and therefore

$$U_{kj} = \mathbb{P}_k(\mathscr{T}_i < \mathscr{T}_j)U_{ij} + \mathbb{P}_k(\mathscr{T}_j < \mathscr{T}_i)U_{jj} > (\mathbb{P}_k(\mathscr{T}_i < \mathscr{T}_j) + \mathbb{P}_k(\mathscr{T}_j < \mathscr{T}_i))U_{ij}$$
$$= U_{ij}.$$

A similar argument yields

$$U_{ki} > U_{ij},$$

and therefore $U_{ij} < \min\{U_{ik}, U_{kj}\}$ which violates the ultrametric inequality. Thus, U is not ultrametric. An argument based on a perturbation shows that there exists a matrix $\hat{P}$ carried by $\mathscr{G}$ such that its potential $\hat{U}$ satisfies

$$\hat{U}_{ij} < \min\{\hat{U}_{ik}, \hat{U}_{kj}\}$$

and the result is proved.

3.3.5 Ultrametricity and Supermetric Geometry

In this section we shall describe ultrametricity and its generalization to supermetric matrices on purely geometric basis. The idea is to study more closely the underline ultrametric distance and the geometry it induces. The main tool we shall use are the triangles induced by an ultrametric distance. It is well known that in an ultrametric space every triangle is isosceles. We introduce now a ternary relation that captures this concept. In this section $\vee$ and $\wedge$ symbolizes the disjunction and conjunction logical connectors respectively.

Definition 3.23 Consider a nonempty finite set I. A ternary relation $G(\bullet, \bullet, \bullet)$ is said to be an **ultrametric geometry** if

   (i) for all $i \in I$ the relation $G(i, \bullet, \bullet)$ is an equivalence relation;
  (ii) for all $i, j \in I$ we have $G(i, i, j) \Rightarrow i = j$;
 (iii) for all $i, j, k \in I$ we have $G(i, j, k) \vee G(j, i, k) \vee G(k, i, j)$.

We say that the triangle i, j, k is **isosceles** at i if $G(i, j, k)$ is satisfied. We will say that a triangle is **equilateral** if it is not a singleton and it is isosceles at every vertex.

 (iv) If the triangle i, j, k is isosceles at i and the triangle j, k, l is isosceles at j and at least one of them is not equilateral then the triangle i, j, l is isosceles at i.

Given an ultrametric distance d on I it is a simple matter to check that the ternary relation defined by $G(i, j, k) = [d_{ij} = d_{ik}]$ is an ultrametric geometry. The converse is true as is shown by equality (3.6) below. The main ingredient of our proof is the following lemma.

Lemma 3.24 *Assume G is an ultrametric geometry, then there exists two different points j, k such that for all $i \in I \setminus \{j, k\}$ we have $G(i, j, k)$.*

Proof The proof is done by induction on n the size of I. When $n = 3$ the result is trivial from *(iii)* in the definition of an ultrametric geometry. We now assume that the result is true when the cardinal of I is smaller or equal to n and we need to prove it when the cardinal of I is $n + 1$.

Take any $l \in I$ and apply the induction hypothesis to $J = I \setminus \{l\}$. Then there exists $j \neq k \in J$ such that for all $i \in J \setminus \{j, k\}$ we have $G(i, j, k)$. If $G(l, j, k)$ is satisfied, then the induction is finished. If not, due to *(iii)* in Definition 3.23 we can suppose that $G(j, l, k)$ is satisfied. Given that j, k, l is not an equilateral triangle we deduce from *(iv)*, in the same definition, that $G(i, l, k)$ is satisfied for all $i \setminus \{j, k, l\}$. Since $G(j, l, k)$ also holds we deduce that the couple l, k satisfies the desired property.

The construction of an ultrametric distance that induces G is done by the following backward algorithm that supplies a chain of comparable partitions associated with G. We start with the finest partition $\mathfrak{F} = \{\{i\} : i \in I\}$ and now we show how to give the first backward step. Consider the couple j, k given in the previous lemma. The key properties to be proven are

$$(a) \quad G(i, l, k) \iff G(i, l, j)$$

$$(b) \quad G(k, i, l) \iff G(j, i, l)$$

To show (a), notice that $G(i, l, k) \wedge G(i, k, j)$ implies $G(i, l, j)$, because $G(i, \bullet, \bullet)$ is an equivalence relation.

(b) is more involved to prove. Due to the symmetric role between k, j it is enough to show $G(k, i, l)$ implies $G(j, i, l)$ for different i, k, l. By using (iii) in Definition 3.23, we can assume $G(i, j, l)$ holds (which is equivalent to $G(i, l, j)$). From part (a), we obtain that $G(i, k, l)$ holds and then $G(k, i, l) \wedge G(i, k, l)$. This shows i, k, l is an equilateral triangle (by (iv)) that is $G(l, k, i)$ and again $G(l, j, i)$. So, $G(i, j, l) \wedge G(l, j, i)$ and then the triangle i, j, l is also equilateral showing that $G(j, i, l)$. This finishes the proof of (b).

We now consider all points that form an equilateral triangle with k, j, that is $A = \{l \in I : G(k, l, j)\} \cup \{k, j\}$. We identify all of them in a single point and consider the partition $\mathfrak{R} = \{\{i\} : i \in I \setminus A\} \cup \{A\}$. We define

$$G_{\mathfrak{R}}(\mathfrak{i}, \mathfrak{h}, \mathfrak{l}) = G(i, h, l),$$

where $i \in \mathfrak{i}, h \in \mathfrak{h}, l \in \mathfrak{l}$ and $\mathfrak{i}, \mathfrak{h}, \mathfrak{l} \in \mathfrak{R}$. It is tedious but simple to check that $G_{\mathfrak{R}}$ defines an ultrametric geometry on $\mathfrak{R}$. In this way we construct a sequence of partitions $\mathfrak{N} = \mathfrak{R}_1 = \{I\} \prec \mathfrak{R}_2 \prec \cdots \prec \mathfrak{R}_m = \mathfrak{F} = \{\{i\} : i \in I\}$. We define

$$t(i, j) = \max\{s : i, j \text{ belongs to the same atom of } \mathfrak{R}_s\}.$$

Clearly $1 \le t(i, j) \le m - 1$. We now define an ultrametric distance by the formula

$$d_{ij} = \begin{cases} 0 & \text{if } i = j \\ \frac{1}{t(i,j)} & \text{if } i \ne j \end{cases}.$$

Since d is an ultrametric distance we conclude that $d_{ij} = d_{ik}$ implies $d_{ij} = d_{ik} \ge d_{jk}$, which is translated to: if the last time i, j are together is the same as the one for i, k then j, k must be in the same atom at that instant. This fact and an argument based on induction shows the following fundamental relation

$$\forall i, j, k \quad d_{ij} = d_{ik} \Leftrightarrow G(i, j, k). \tag{3.6}$$

Thus, any ultrametric geometry is induced by an ultrametric distance.

We introduce a new class of matrices related to this construction. Recall that a symmetric nonnegative matrix U is ultrametric then there exists an ultrametric distance d such that $d_{ij} \le d_{ik}$ implies $U_{ij} \ge U_{ik}$. If in addition U is strictly ultrametric then $d_{ij} \le d_{ik}$ is equivalent to $U_{ij} \ge U_{ik}$. This remark is the basis of the following definition.

Definition 3.25 A symmetric matrix U is said to be a **supermetric** matrix if there exists an ultrametric distance d such that $d_{ij} = d_{ik}$ implies $U_{ij} = U_{ik}$.

In what follows, given a symmetric matrix U we denote by G_U the following ternary relation

$$G_U(i, j, k) \Leftrightarrow [U_{ij} = U_{ik} \text{ or } (i = j \Rightarrow i = k)].$$

Consider a nonnegative symmetric matrix U. It is straightforward from the definition that U is supermetric if and only if there exists an ultrametric geometry G such that $G \Rightarrow G_U$.

Supermetric matrices are permutation of CBF, which are the constant by blocks matrices (see (3.2)).

Theorem 3.26 *Assume U is a symmetric matrix. The following are equivalent*

 (i) *U is a supermetric matrix;*
(ii) *U is a permutation of a CBF matrix.*

Proof ((i) $\Rightarrow$ (ii)) Assume that U is a supermetric matrix over I and consider d an ultrametric distance such that for all $i, j, k \in I$ it holds $d_{ij} = d_{ik}$ implies $U_{ij} = U_{ik}$. Take $a = \max\{d(k, l) : k \neq l\}$ and two different nodes i, j realizing this maximum. Consider the sets $L = \{l \in I : d(i, l) < a\}$ and $J = L^c$. They form a partition of I and for every $l \in L, k \in J$ we have $d(l, k) = a$. Indeed, the ultrametric inequality implies

$$a = d(i, j) \leq \max\{d(i, l), d(l, j)\}.$$

Since $d(i, l) < a$, the optimality of a implies that $d(l, j) = a$ and similarly we conclude $d(i, k) = a$. Interchanging the roles of i and l we conclude that $d(l, k) = a$.

The supermetric property implies that $U_{lk} = U_{ij}$, thus the block U_{LL^c} is constant. The property that U is a permutation CBF follows now by induction.

((ii) $\Rightarrow$ (i)) Without loss of generality we can assume that U is a CBF matrix. Therefore there is a partition of I in L, L^c and a constant $\alpha \in \mathbb{R}$ such that

$$U = \begin{pmatrix} A & \alpha \\ \alpha & B \end{pmatrix},$$

where $A = U_{LL}, B = U_{L^c L^c}$ are CBF matrices. We can assume inductively the existence of two ultrametric distances d_1 on L and d_2 on L^c that make A, B supermetric matrices. Take any $a \in \mathbb{R}$ such that $\max\{\max\{d_1(i, j) : i, j \in L\}, \max\{d_2(i, j) : i, j \in L^c\}\} < a$ and define d on I by

$$d(i, j) = \begin{cases} d_1(i, j) & \text{if } i, j \in L \\ d_2(i, j) & \text{if } i, j \in L^c \\ a & \text{otherwise} \end{cases}.$$

It is straightforward to prove that d is an ultrametric distance and it is obvious that $d_{ij} = d_{ik}$ implies $U_{ij} = U_{ik}$. Therefore U is supermetric.

Chapter 4
Graph of Ultrametric Type Matrices

Ultrametric and GUM matrices can be seen as the potential matrices of Markov chains on finite state spaces. In this chapter we study the connections of these chains and emphasize the characterization of roots, which are those points where there is a loss of mass. This is equivalent to studying the incidence graph for the inverse matrix. The main notions and results of this chapter are based on [20] for the ultrametric case and [22] for the GUM case.

A key notion in the ultrametric case is the concept of tree matrix. This class of matrices is defined in terms of a supporting tree with the remarkable property that the incidence graph of their inverses coincides with their supporting tree. This is shown in Theorem 4.6. We then show that every ultrametric matrix is embedded into a tree matrix and use this fact to study the incidence graph of their inverses, see Theorem 4.7. In [46] the reader can find a characterization of the GUM matrices whose graph is a tree.

In the GUM case we construct an algorithm to find the roots, see Theorem 4.14, and then describe the graph of their inverses in Theorem 4.16. In these algorithms we make strong use of the properties of equilibrium potentials.

4.1 Graph of an Ultrametric Matrix

Let U be a nonsingular ultrametric matrix defined over the set I. Recall that U is symmetric and $U^{-1} = \kappa(\mathbb{I} - P)$ for some constant $\kappa > 0$ and a doubly substochastic matrix P. In this way there is a Markov process based on an extension $\overline{I} = I \cup \{\partial\}$ of I, where $\partial \notin I$ is an absorbing state for the process. Notice that κ and P are not unique and therefore there are several Markov processes associated with U as discussed in the introduction. Nevertheless, they share the same skeleton and their connections are studied in this section.

© Springer International Publishing Switzerland 2014
C. Dellacherie et al., *Inverse M-Matrices and Ultrametric Matrices*, Lecture Notes in Mathematics 2118, DOI 10.1007/978-3-319-10298-6_4

Definition 4.1 Consider an inverse M-matrix W. The graph $\mathscr{G}^W$ associated with W is the incidence graph of W^{-1} that is

$$(i, j) \in \mathscr{G}^W \Leftrightarrow (W^{-1})_{ij} \neq 0.$$

We remark that in general $\mathscr{G}^W$ is an oriented graph. We shall say that i and j are **neighbors** if $(i, j) \in \mathscr{G}^W$ and $(j, i) \in \mathscr{G}^W$.

Assume that $U^{-1} = \kappa(\mathbb{I} - P)$ is a potential. For different nodes i, j we have the equivalence $(i, j) \in \mathscr{G}^U$ if and only if $P_{ij} > 0$ and this property does not depends on the different decompositions of U^{-1}. Since U^{-1} is row diagonally dominant we have $(U^{-1})_{ii} > 0$ for all i, which means that the graph $\mathscr{G}^U$ has a loop at every i. Nevertheless, for a given i the property $P_{ii} > 0$ depends on the decomposition we choose for U^{-1} and therefore there posses a problem in interpreting this diagonal elements. The information of $\mathscr{G}^U$ is clear from a probabilistic point of view and represents the structure of possible one step transitions of the skeleton associated with U.

We recall the definition of roots for a matrix. This concept must not be confused with the root of a tree. Nevertheless, as we shall see, they are related. A point $i \in I$ is a root of $\mathscr{G}^U$ (or simply of U) if

$$\sum_{j \in I} (U^{-1})_{ij} > 0,$$

that is U^{-1} is strictly row diagonally dominant at row i. We denote by $\mathscr{R}(U)$ the set of roots of U.

Notice that U^{-1} must be strictly diagonally dominant at some row, otherwise we get $U^{-1}\mathbb{1} = 0$, which is not possible. It is not difficult to see that i is a root if and only if for some decomposition (and then for all decompositions)

$$\sum_{j \in I} P_{ij} < 1.$$

This is equivalent to the fact that i belongs to the support of the right equilibrium potential for U, which is $\lambda^U = U^{-1}\mathbb{1}$. This means the Markov process whose potential is U is defective at every root and these nodes are the unique points on I where the process can jump to the absorbing state ∂.

Let us start with a general result for the inverse M-matrix case.

Proposition 4.2 *Assume U is the inverse of an M-matrix. If $U_{ij} = 0$ then $(U^{-1})_{ij} = 0$. Moreover $U_{ki} > 0$ and $U_{ij} > 0$ imply that $U_{kj} > 0$.*

Proof The hypothesis on U is that $U^{-1} = a\mathbb{I} - L$ where L is a nonnegative matrix and a is bigger than the spectral radius of L . Then we have

$$U = \frac{1}{a} \sum_{n \geq 0} \left(\frac{L}{a}\right)^n.$$

From this equality we have $U_{ij} = 0$ implies $L_{ij} = 0$ and $(U^{-1})_{ij} = 0$.

For the second part of the proposition we notice that for some $n, m \in \mathbb{N}$ we have $L_{ij}^n > 0$ and $L_{ki}^m > 0$. Since $L_{kj}^{n+m} \geq L_{ki}^m L_{ij}^n > 0$ we obtain the result.

This proposition, when U is symmetric and nonsingular shows that the relation $i \sim j \Leftrightarrow U_{ij} > 0$ is an equivalence relation. We can decompose U according to the equivalence classes to study the graph $\mathscr{G}^U$. A consequence of this observation is that in most of what follows we can assume that U is a positive matrix.

We next describe the structure of roots and their connections for an ultrametric matrix U in terms of the associated NBF. For that purpose, we require some previous lemmas. Then, in the main result, Theorem 4.7, we characterize the connections and roots of U using its minimal tree extension (see Theorem 3.16).

For a nonsingular ultrametric matrix we recall the existence (and uniqueness) of a right equilibrium potential which is a nonnegative vector that satisfies $U\lambda^U = \mathbb{1}$. Its total mass will be denoted by $\rho(U) = \mathbb{1}'\lambda^U$. We also denote by $\alpha(U) = \min\{U\}$. With these concepts we can state the following:

Lemma 4.3 *Let U be a positive and nonsingular ultrametric matrix. We have $\rho(U) \leq (\alpha(U))^{-1}$ and the equality holds if and only if there exists a constant column of value $\alpha(U)$, i.e. $\exists i \in I$ such that $U_{\bullet i} = \alpha(U)\mathbb{1}$.*

Proof We first notice that $U - \alpha(U)\mathbb{1}\mathbb{1}'$ is a nonnegative ultrametric matrix and therefore it is positive semi-definite. Thus

$$\alpha(U)(\rho(U))^2 = \alpha(U) \sum_{i,j} (\lambda^U)_i (\lambda^U)_j \leq \sum_{i,j} U_{ij} (\lambda^U)_i (\lambda^U)_j \qquad (4.1)$$
$$= (U\lambda^U)'\lambda^U = \mathbb{1}'\lambda^U = \rho(U).$$

Then the inequality holds.

If the i-th column $U_{\bullet i} = \alpha(U)\mathbb{1}$ then by uniqueness of the equilibrium potential we get $\lambda^U = (\alpha(U))^{-1}e$ where e is the ith vector of the canonical basis and then $\rho(U) = (\alpha(U))^{-1}$.

Conversely, if the equality $\rho(U) = (\alpha(U))^{-1}$ holds, we necessarily have that λ^U is concentrated at a unique point. In fact, if $(\lambda^U)_i > 0$, $(\lambda^U)_j > 0$ for $i \neq j$ we get from (4.1) that $U_{ii} = U_{ij} = U_{jj} = \alpha(U)$. From (1.2) and since $\alpha(U)$ is the minimal value of U, we obtain $U_{jk} = U_{ik}$, for all k, which is a contradiction because U is invertible. Therefore if i is in the support of λ^U, that is $(\lambda^U)_i > 0$, then necessarily $U_{ji} = \alpha(U)$ for every $j \in I$. From which the result follows.

Notice that for an ultrametric matrix U the only possible constant columns are the columns of minimum value $\alpha(U)$. Indeed, assume the column k is constant and take any couple (i, j) such that $U_{ij} = \alpha(U)$. Then the ultrametric inequality gives that

$$\alpha(U) = U_{ij} \geq \min\{U_{ik}, U_{kj}\} \geq \alpha(U),$$

which implies that either $U_{ik} = \alpha(U)$ or $U_{jk} = \alpha(U)$. This proves the claim.

In some of the formulas that follow, given $J, K \subset I$ it will be more convenient to use $J \times K$ as the index set for the submatrix U_{JK}, instead of relabeling it to $\{1, \cdots, \#J\} \times \{1, \cdots, \#K\}$, because otherwise results will be very difficult to read.

Recall that any principal submatrix of an inverse M-matrix is an inverse M-matrix and so every principal submatrix of a nonsingular ultrametric matrix is also nonsingular. The next result is important in studying the graph of an ultrametric matrix.

Lemma 4.4 *Let U be a positive and nonsingular ultrametric matrix. Consider $I = J \cup K$ a non trivial partition of I such that $U_{J \times K}$ is the constant block $\alpha(U)$ and define the nonsingular ultrametric matrices $A = U_{JJ}$ and $B = U_{KK}$. Then, the following conditions hold:*

(i) $(U^{-1})_{ij} = (A^{-1})_{ij}$ for $(i, j) \in J \times J \setminus \mathcal{R}(A) \times \mathcal{R}(A)$;
(ii) For $(i, j) \in J \times K$ we have $(U^{-1})_{ij} < 0$ if and only if $(i, j) \in \mathcal{R}(A) \times \mathcal{R}(B)$;
(iii) If $U_{\bullet i} \neq \alpha \mathbb{1}$, for every $i \in I$, then $\mathcal{R}(U) = \mathcal{R}(A) \cup \mathcal{R}(B)$. On the other hand, if $U_{\bullet i} = \alpha \mathbb{1}$ then

$$\mathcal{R}(U) = \{i\} = \begin{cases} \mathcal{R}(A) & \text{if } i \in J \\ \mathcal{R}(B) & \text{if } i \in K \end{cases}.$$

Proof Denote $p = |J|, q = |K|$ and $\alpha = \alpha(U)$. Observe that after a permutation we can assume that

$$U = \begin{pmatrix} A & \alpha \mathbb{1}_p \mathbb{1}'_q \\ \alpha \mathbb{1}_q \mathbb{1}'_p & B \end{pmatrix}.$$

From Schur's decomposition we obtain

$$U^{-1} = \begin{pmatrix} V^{-1} & T \\ T' & Z^{-1} \end{pmatrix}.$$

where $V = A - \alpha^2 \mathbb{1}_p \mathbb{1}'_q B^{-1} \mathbb{1}_q \mathbb{1}'_p$ and $T = -\alpha V^{-1} B^{-1}$. Consider λ^A, λ^B the right equilibrium potentials of A and B, that is $A\lambda^A = \mathbb{1}_p, B\lambda^B = \mathbb{1}_q$. We put $\rho(A) = \mathbb{1}'_p \lambda^A, \rho(B) = \mathbb{1}'_q \lambda^B$. From Lemma (4.3) these quantities are smaller than α^{-1}.

Then $V = A - \alpha^2 \rho(B) \mathbb{1}_p \mathbb{1}'_p$ and

$$V\lambda^A = (A - \alpha^2 \rho(B) \mathbb{1}_p \mathbb{1}'_p)\lambda^A = (1 - \alpha^2 \rho(B)\rho(A))\mathbb{1}_p.$$

U is positive-definite and therefore V is too, which yields

$$0 < (\lambda^A)'V\lambda^A = (1 - \alpha^2 \rho(B)\rho(A))\rho(A).$$

Thus $\alpha^2 \rho(B)\rho(A) < 1$.

Let us now compute V^{-1}. We start from

$$V = A - \alpha^2 \rho(B) \mathbb{1}_p \mathbb{1}_p' = A(\mathbb{I} - \alpha^2 \rho(B) A^{-1} \mathbb{1}_p \mathbb{1}_p')$$

and we get

$$V^{-1} = (\mathbb{I} - \alpha^2 \rho(B) \lambda^A \mathbb{1}_p')^{-1} A^{-1}.$$

Denote by $Q = \lambda^A \mathbb{1}_p'$ then the powers of this matrix are easy to compute since $Q^2 = \lambda^A \mathbb{1}_p' \lambda^A \mathbb{1}_p' = \rho(A) Q$ and so $Q^n = (\rho(A))^{n-1} Q$. Since $\alpha^2 \rho(B) \rho(A) < 1$ we have

$$(\mathbb{I} - \alpha^2 \rho(B) Q)^{-1} = \sum_{n \geq 0} (\alpha^2 \rho(B))^n Q^n = \mathbb{I} + \alpha^2 \rho(B) \sum_{n \geq 0} (\lambda^2 \rho(B) \rho(A))^n Q$$

$$= \mathbb{I} + \frac{\alpha^2 \rho(B)}{1 - \alpha^2 \rho(B) \rho(A)} \lambda^A \mathbb{1}_p'.$$

Therefore

$$V^{-1} = A^{-1} + \frac{\alpha^2 \rho(B)}{1 - \alpha^2 \rho(B) \rho(A)} \lambda^A (\lambda^A)'.$$

and if i, j belong to J

$$(V^{-1})_{ij} = (U^{-1})_{ij} = (A^{-1})_{ij} + \frac{\alpha^2 \rho(B)}{1 - \alpha^2 \rho^B \rho(A)} \lambda_i^A \lambda_j^A. \tag{4.2}$$

We deduce that if $(i, j) \in J \times J \setminus \mathscr{R}(A) \times \mathscr{R}(A)$ then $(U^{-1})_{ij} = (A^{-1})_{ij}$ and (i) holds.

On the other hand

$$T = -\alpha V^{-1} \mathbb{1}_p \mathbb{1}_q' B^{-1} = -\alpha (1 - \alpha^2 \rho(B) \rho(A))^{-1} \lambda^A (\lambda^B)'. \tag{4.3}$$

Then we deduce that, for $i \in J, j \in K$,

$$(U^{-1})_{ij} < 0 \Leftrightarrow \lambda_i^A > 0, \lambda_j^B > 0 \Leftrightarrow (i, j) \in \mathscr{R}(A) \times \mathscr{R}(B),$$

and (ii) holds.

Now observe that $\rho(B) = \alpha^{-1}$ if and only if B has a column, say k, equal to $\alpha \mathbb{1}_q$. This k is unique because B is nonsingular and in this case λ^B is concentrated on k, so $\mathscr{R}(B) = \{k\}$. Since $U_{ik} = \alpha$ for $i \in J$ we deduce that the k column of U is constant α. Hence, $\mathscr{R}(U) = \mathscr{R}(B) = \{k\}$. Similarly, if $\rho(A) = \alpha^{-1}$ then $\mathscr{R}(U) = \mathscr{R}(A)$ is reduced to a unique point.

Now assume $\rho(B)$ and $\rho(A)$ are strictly smaller than α^{-1}. Take $i \in J$, from above equalities (4.2) and (4.3) we deduce

$$\sum_{j \in I}(U^{-1})_{ij} = \frac{\lambda_i^A(1 - \alpha\rho(B))}{1 - \alpha^2\rho(A)\rho(B)}.$$

Because $\alpha\rho(B) < 1$ we obtain for $i \in J$ that $i \in \mathcal{R}^U$ if and only if $i \in \mathcal{R}(A)$. Similarly since $\alpha\rho(A) < 1$ we conclude that $\mathcal{R}(B) = \mathcal{R}(U) \cap K$. Hence we conclude $\mathcal{R}(U) = \mathcal{R}(A) \cup \mathcal{R}(B)$.

4.1.1   Graph of a Tree Matrix

We first study the graph of a tree matrix and use it to describe the graph of a general ultrametric matrix.

Assume that (T, r) is a rooted tree with level function L. For any $s \in T$ we consider T_s the subtree that hangs from s. Formally, this subtree is defined on the set $D_s = \{i \in I : s \preccurlyeq i\}$ and $T_s = T \cap D_s$. This is considered as a rooted tree with root s and level function $L_s(i) = L(i) - L(s)$. If w is a weight system in T, then the induced weight system in T_s will be denoted by w^s, where $w_p^s = w_{L(s)+p}$.

Let U be a tree matrix associated with the weighted tree (T, r, w) (see Definition 3.12). Recall that U is nonsingular if w is strictly positive and strictly increasing. Also observe that $L(j \wedge r) = 0$ for any $j \in I$ which gives a constant column in U with value $\alpha(U) = w_0$. Then λ^U is concentrated on r and the tree matrix U has only one root which coincides with the root of the tree that is $\mathcal{R}(U) = \{r\}$.

Lemma 4.5 *Let U be a tree matrix associated with the weighted tree (T, r, w) over the set I. For $s \in I$ the ultrametric matrix $U_{D_s D_s}$ is a tree matrix associated with the weighted tree (T_s, s, w_s). Moreover,*

$$\forall (i, j) \in D_s \times D_s \setminus \{(s, s)\} \quad ((U_{D_s D_s})^{-1})_{ij} = (U^{-1})_{ij}.$$

Proof The fact that $U_{D_s D_s}$ is a tree matrix associated with the weighted tree (T_s, s, w^s) is obvious. The rest of the proof is based on induction, and it suffices to prove the result when $(s, r) \in T$, $s \neq r$. Following the notation of Lemma 4.4 take $J = D_s$, $K = I \setminus D_s$, $A = U_{JJ}$ and $B = U_{KK}$. We remark that U_{JK} is constant $w_0 = \alpha(U)$.

From part (i) of the cited lemma we get

$$(U^{-1})_{ij} = (A^{-1})_{ij} \text{ for } (i, j) \in J \times J \setminus \mathcal{R}(A) \times \mathcal{R}(A).$$

Notice that A is a tree matrix and therefore $\mathcal{R}(A) = \{s\}$, showing the result.

The following result asserts that for nonsingular tree matrices the graph $\mathcal{G}^U$ coincides with T (except at the diagonal $\Delta \subset I \times I$). We recall that in a rooted tree T with root r, every node $i \neq r$ possesses a unique immediate predecessor i^- which is characterized as $(i^-, i) \in T$ and $i^- \prec i$. On the other hand if i is not a leaf then $\mathcal{S}(i) = \{j : (i, j) \in T, \ i \prec j\}$ is the set of immediate successors of i.

Theorem 4.6 *If U is a nonsingular tree matrix associated with the weighted tree (T, r, w), then U^{-1} is supported by T, that is for all $i \neq j$*

$$(U^{-1})_{ij} < 0 \Leftrightarrow (i, j) \in T.$$

Moreover we have a formula for U^{-1} given by: For $i \neq r$ with level $L(i) = \ell$

$$(U^{-1})_{ij} = \begin{cases} \frac{1}{U_{i^- - U_{ii}}} = \frac{1}{w_{\ell-1}-w_\ell} & \text{for } j = i^- \\[2mm] \frac{1}{U_{ji}-U_{jj}} = \frac{1}{w_\ell-w_{\ell+1}} & \text{for } j \in \mathcal{S}(i) ; \\[2mm] \frac{1}{w_\ell-w_{\ell-1}} + \frac{\#\mathcal{S}(i)}{w_{\ell+1}-w_\ell} & \text{for } j = i \end{cases}$$

and for the root r

$$(U^{-1})_{rj} = \begin{cases} \frac{1}{w_0-w_1} & \text{for } j \in \mathcal{S}(r) \\[2mm] \frac{1}{w_0} + \frac{\#\mathcal{S}(r)}{w_1-w_0} & \text{for } j = r \end{cases}.$$

U^{-1} *is only defective at r, that is for $i \neq r$ we have $\sum_j U_{ij}^{-1} = 0$ and $\sum_j U_{rj}^{-1} = 1/w_0 > 0$.*

Proof We know that $U^{-1} = \kappa(\mathbb{I} - P)$ for some constant $\kappa > 0$ and a substochastic matrix P. There is no loss of generality in assuming that $\kappa = 1$. This is done by considering the matrix κU which is a tree matrix associated with the weighted tree $(T, r, \kappa w)$.

Observe that $U_{ri} = w_0$ for any $i \in I$ so $\mathcal{R}(U) = \{r\}$ and λ^U is concentrated on $\{r\}$ that is

$$(\lambda^U)_i = \begin{cases} w_0^{-1} & \text{if } i = r \\ 0 & \text{otherwise} \end{cases}.$$

This shows the last assertion of the theorem.

As usual we denote by ∂ the absorbing state added to I, and the natural extension of P to $\bar{I} = I \cup \{\partial\}$ given by

$$P_{r\partial} := 1 - \sum_{j \in I} P_{rj} > 0;$$
$$P_{i\partial} := 1 - \sum_{j \in I} P_{ij} = 0 \quad \text{if} \quad i \in I, i \neq r;$$
$$P_{\partial\partial} = 1.$$

Consider two different points $i, j \in I \setminus \{r\}$, such that $i \wedge j = r$. We shall prove that $P_{ij} = 0$. Observe that $i \wedge j = r$ is equivalent to $U_{ij} = w_0$. On the other hand $j \neq r$ implies $U_{jj} \geq w_1$. Since $i \neq r$ we obtain P is not defective at i, that is $\sum_{\ell \in I} P_{i\ell} = 1$. Then the Markov property gives

$$U_{ij} = w_0 = P_{ij} U_{jj} + \sum_{k \neq i, k \in I} P_{ik} U_{kj} \geq P_{ij} w_1 + w_0 \sum_{k \neq i, k \in I} P_{ik} = w_0 + P_{ij}(w_1 - w_0).$$

From $w_1 > w_0 > 0$ we deduce that $P_{ij} = 0$. We have proved

$$[i, j \in I \setminus \{r\}, i \neq j, i \wedge j = r] \Rightarrow P_{ij} = 0. \tag{4.4}$$

Consider now $i \neq j$ such that $s = i \wedge j$ is different from i, j. In particular $(i, j) \notin T$. So, i, j are different points in D_s and both are different from the root s of the tree T_s. Hence, from what we have proved $(U_{D_s D_s})_{ij}^{-1} = 0$ and from Lemma 4.5, we conclude $P_{ij} = 0$. So far we have proved that $P_{ij} > 0$ implies that necessarily $i \wedge j \in \{i, j\}$.

The next step is to prove that for $i \neq r$ it holds

$$P_{ir} > 0 \Leftrightarrow (i, r) \in T. \tag{4.5}$$

If $(i, r) \notin T$ take $j \prec i$ such that $(i, j) \in T$. Then $j = j \wedge i$ and $U_{ij} = U_{jj}$. Recall that $\mathcal{T}_j$ is the time of first visit to j. Then, the strong Markov property yields

$$U_{ij} = \mathbb{P}_i(\mathcal{T}_j < \infty) U_{jj},$$

and we deduce that $\mathbb{P}_i(\mathcal{T}_j < \infty) = 1$, that is, if the chain starts at i, the point j will be visited with probability one. Finally, if we assume $P_{ir} > 0$ we get $\mathbb{P}_i(\mathcal{T}_j = \infty) \geq P_{ir} P_{r\partial} > 0$, which is a contradiction. Thus

$$(i, r) \notin T \Rightarrow P_{ir} = 0. \tag{4.6}$$

This property is propagated throughout the tree, using again Lemma 4.5. Indeed, let i, j be two different points in T, such that $(i, j) \notin T$. There are two different cases. If $s = i \wedge j$ is different from i, j then $P_{ij} = 0$. On the other hand, if $s = j$ then from (4.6) we get $(U_{D_s D_s})_{is}^{-1} = 0$ and finally $P_{ij} = 0$. Thus we have proved

$$(i, j) \notin T \Rightarrow P_{ij} = 0.$$

Conversely, let us show that for $i \neq j$ if $(i,j) \in T$ then $P_{ij} > 0$. Without loss of generality we can assume that $j = i^-$ is the immediate predecessor of i. Since $U_{ij} = \sum_{n \geq 0} P_{ij}^n = U_{jj} > 0$, we deduce that $P_{ij}^m > 0$ for some $m \geq 1$. A fortiori, there exists a path

$$i_0 = i \to i_1 \to \cdots \to i_{m-1} \to i_m = j$$

such that $P_{i_s, i_{s+1}} > 0$ for $s = 0, \cdots, m-1$ and $P_{i_{m-1}, j} > 0$. We can take $i_0, \cdots, i_{m-1}, j$ all different. We shall prove that necessarily $m = 1$ which is exactly $P_{ij} > 0$ as we claimed.

So, assume that $m \geq 2$. Then, necessarily $(i_1, i_0) \in T$ and $i_1 \neq j$, which implies that i_1 must be a successor of i. Since the only allowed transitions are in T, we deduce that $i_1, \cdots, i_{m-1}$ are all successors of i. Being $i_{m-1} \neq i$ and a successor of i, we have $(i_{m-1}, j) \notin T$ contradicting the fact that $P_{i_{m-1}j} > 0$.

Let us prove the last part of the Theorem. Since $UU^{-1} = \mathbb{I}$ we get for $i \neq r$

$$1 = \sum_{j:(i,j) \in T} U_{ij}(U^{-1})_{ji} = U_{ii}(U^{-1})_{ii} + U_{ii^-}(U^{-1})_{i^-i} + U_{ii} \sum_{j \in \mathscr{S}(i)} (U^{-1})_{ji},$$

where i^- is the immediate predecessor of i. For any immediate successor j of i (if it has one, which means that i is not a leaf), we have used the fact that $i \wedge j = i$ and U_{ij} is constant equal to U_{ii}. We point out that the last sum is 0 if i is a leaf. Since U^{-1} is not defective at i we get

$$\sum_{j \in \mathscr{S}(i)} (U^{-1})_{ji} + (U^{-1})_{i^-i} + U_{ii} = 0,$$

and therefore

$$(U^{-1})_{i^-i} = (U^{-1})_{ii^-} = \frac{1}{U_{ii^-} - U_{ii}}.$$

Thus we have the following formula, for $i \neq r$ with level $L(i) = \ell$

$$(U^{-1})_{ij} = \begin{cases} \frac{1}{U_{ii^-} - U_{ii}} = \frac{1}{w_{\ell-1} - w_\ell} & \text{for } j = i^- \\[2ex] \frac{1}{U_{ji} - U_{jj}} = \frac{1}{w_\ell - w_{\ell+1}} & \text{for } j \in \mathscr{S}(i) . \\[2ex] \frac{1}{w_\ell - w_{\ell-1}} + \frac{\#\mathscr{S}(i)}{w_{\ell+1} - w_\ell} & \text{for } j = i \end{cases}$$

Notice that $(U^{-1})_{ii}$ is computed using the fact that U^{-1} is not defective at i. For the root we need only to compute $(U^{-1})_{rr}$ because we already have for $j \in \mathscr{S}(r)$ that

$$U_{rj} = \frac{1}{w_0 - w_1}.$$

The computation is done by using the fact that the r-th row of U is constant. This yields

$$w_0(U^{-1})_{rr} + w_0 \frac{\#\mathscr{S}(r)}{w_0 - w_1} = 1,$$

and then

$$(U^{-1})_{rj} = \begin{cases} \frac{1}{w_0 - w_1} & \text{for } j \in \mathscr{S}(r) \\[2mm] \frac{1}{w_0} + \frac{\#\mathscr{S}(r)}{w_1 - w_0} & \text{for } j = r \end{cases}.$$

Notice that the defect at r is $1/w_0$.

Now, we study the graph of a general ultrametric matrix U. In what follows we denote by $\tilde{U}$ the minimal tree matrix constructed in Theorem 3.16 which gives an extension $\tilde{I}$ of I and $\tilde{U}_{II} = U$. We also denote by $(\tilde{T}, \tilde{r}, \tilde{w})$ the weighted tree associated with $\tilde{U}$, by $\tilde{L}$ the level function of $\tilde{T}$ and by $\widetilde{\text{Geod}}(\bullet, \bullet)$ the geodesic in $\tilde{T}$. If U is strictly positive and nonsingular we get $\tilde{U}$ is also nonsingular and it has only one root $\tilde{r}$, that is, the row sums of $\tilde{U}^{-1}$ are all 0 except the one associated with $\tilde{r}$, which is strictly positive.

Our main result is the following characterization of $\mathscr{G}^U$ for an ultrametric matrix U.

Theorem 4.7 *Let U be a positive nonsingular ultrametric matrix and $i \neq j$ in I. Then, $(i, j) \in \mathscr{G}^U$ if and only if the geodesic in $\tilde{T}$ joining i and j does not contain other points in I than i and j, that is*

$$\widetilde{\text{Geod}}(i, j) \cap I = \{i, j\}.$$

Proof We know that $(\tilde{U})^{-1} = \kappa(\mathbb{I} - \tilde{P})$ for some constant $\kappa > 0$ and a substochastic matrix $\tilde{P}$, which we know is carried by $\tilde{T}$ and it is only defective at $\tilde{r}$ (see Theorem 4.6). As usual add a state $\tilde{\partial} \notin \tilde{I}$ as the absorbing state and set $\tilde{P}_{\tilde{k}, \tilde{\partial}} = 1 - \sum_{\tilde{\ell} \in \tilde{I}} \tilde{P}_{\tilde{k}, \tilde{\ell}}$. Denote by $\{\tilde{X}_n : n \geq 0\}$ the Markov chain on $\tilde{I} \cup \{\tilde{\partial}\}$ with transition probability $\tilde{P}$ and by $\tilde{\mathscr{T}}_{\tilde{\partial}}$ the first time that the chain attains $\tilde{\partial}$. By $\tilde{\mathbb{P}}_{\tilde{k}}$, we denote the law of the chain when starting from $\tilde{k}$ and by $\tilde{\mathbb{E}}_{\tilde{k}}$ the mean expected value associated with this law.

Remind that $I \subseteq \tilde{I}$ and $\tilde{U}_{II} = U$. Starting from $i \in I$, the random times at which the chain visits I is a finite sequence $\tilde{\mathbb{P}}_i$—a.e. and we denote them by $\mathscr{T}_0 = 0 < \mathscr{T}_1 < \ldots < \mathscr{T}_N$. Set

$$P_{ij} = \tilde{\mathbb{P}}_i\{\mathscr{T}_1 < \infty, \, \tilde{X}_{\mathscr{T}_1} = j\} \text{ for } i, j \in I. \tag{4.7}$$

This is the transition kernel associated with the induced Markov chain on I, when starting from I. This Markov chain, when started from $i \in I$ is given by: $X_0 = i, X_1 = \tilde{X}_{\mathscr{T}_1}, \cdots, X_N = \tilde{X}_{\mathscr{T}_N}$ and $X_m = \partial$ for $m \geq N + 1$. Is not difficult to see that

$$(P^m)_{ij} = \mathbb{P}_i(X_m = j) = \tilde{\mathbb{P}}_i(\mathscr{T}_m < \infty, \tilde{X}_{\mathscr{T}_m} = j).$$

Thus, we have the following equalities

$$(\kappa U)_{ij} = (\kappa \tilde{U})_{ij} = \tilde{\mathbb{E}}_i \left(\sum_{n \geq 0} 1_{\tilde{X}_n = j}\right) = \tilde{\mathbb{E}}_i \left(\sum_{m \geq 0} 1_{\tilde{X}_{\mathscr{T}_m} = j}\right) = \sum_{m \geq 0} \tilde{\mathbb{E}}_i \left(1_{\tilde{X}_{\mathscr{T}_m} = j}\right)$$
$$= \sum_{m \geq 0} \tilde{\mathbb{P}}_i(\mathscr{T}_m < \infty, \tilde{X}_{\mathscr{T}_m} = j) = \sum_{n \geq 0} (P^m)_{ij} = (\mathbb{I} - P)_{ij}^{-1}$$

Hence $P = \mathbb{I} - \kappa^{-1} U^{-1}$ is a substochastic matrix associated with U.

The important point is that every trajectory in $\tilde{T}$ that connects i and j must contain $\widetilde{\text{Geod}}(i, j)$. Now, for $i \neq j$ we have $0 < P_{ij} = \tilde{\mathbb{P}}_i\{\mathscr{T}^1 < \infty, \tilde{X}_{\mathscr{T}^1} = j\}$ if and only if there exists a trajectory in $\tilde{T}$ that connects i and j for which the first entrance to I, after time zero, must occur when visiting j. This is possible if and only if $\widetilde{\text{Geod}}(i, j) \cap I = \{i, j\}$ and the result is proved.

As a consequence of this theorem we obtain the next result.

Corollary 4.8 *Let U be a positive nonsingular ultrametric matrix. If $\mathscr{G}^U$ contains a cycle $(i_0, \ldots, i_m, i_0)$ that is $(i_s, i_{s+1}) \in \mathscr{G}^U$ for $i = 0, \ldots, m$ and $i_{m+1} = i_0$, then it contains the complete graph of this cycle that is $(i_s, i_q) \in \mathscr{G}^U$ for all $s, q = 0, \cdots, m$.*

Proof We can assume $m > 0$ and all points (except the extremes) in the cycle are different. The result will be proven by contradiction, so denote by q the first integer for which there exists $s > q$ such that $(i_q, i_s) \notin \mathscr{G}^U$. In particular $i_q \neq i_s$. Then,

$$\pi = \bigcup_{t=q}^{s-1} \widetilde{\text{Geod}}(i_t, i_{t+1}) \text{ and } \pi' = \bigcup_{t=s}^{m} \widetilde{\text{Geod}}(i_t, i_{t+1}) \cup \bigcup_{t=0}^{q-1} \widetilde{\text{Geod}}(i_t, i_{t+1}),$$

are two paths in the tree $\tilde{T}$ joining i_q and i_s. So, both of them contain $\widetilde{\text{Geod}}(i_q, i_s)$. From the characterization obtained in Theorem 4.7 we get $(i_q, i_s) \notin \mathscr{G}^U$ implies the existence of some $k \in (I \setminus \{i_q, i_s\}) \cap \widetilde{\text{Geod}}(i_q, i_s)$. Hence $k \in \pi \cap \pi'$, i.e.

$k = i_t = i_{t'}$ for $t \in [q, s]$ and $t' \in [0, q] \cup [s, m]$. This implies $k = i_q$ or $k = i_s$, which is a contradiction.

As a second corollary of the proof given for Theorem 4.7 we have

Proposition 4.9 *Let U be a positive nonsingular ultrametric matrix and $m \geq 1$. Then, different points i, j are connected in m steps in $\mathscr{G}^U$, that is $P_{ij}^m > 0$, if and only if $\#(\widetilde{\text{Geod}}(i, j) \cap I) \leq m + 1$.*

The following result summarizes the description of the set of roots for an ultrametric matrix, as well as an algorithm which allows to obtain this set. For $j \in I$ we denote by $I_j = \{i \in I : U_{ij} = U_{jj}\}$ the set of points with the property that the chain starting from them will visit for sure the node j.

Theorem 4.10 *If U is a positive nonsingular ultrametric matrix then $\mathscr{R}(U)$ is a singleton $\{r\}$ if and only if $U_{\bullet r} = \alpha(U)\mathbb{1}$ or equivalently if $r = \tilde{r}$.*

On the other hand, if $\mathscr{R}(U)$ is not a singleton then it is given by

$$\mathscr{R}(U) = \{i \in I : \widetilde{\text{Geod}}(i, \tilde{r}) \cap I = \{i\}\}.$$

Moreover, the graph $\mathscr{G}^U \cap \mathscr{R}(U) \times \mathscr{R}(U)$ is a complete graph.

The class of sets $(I_r : r \in \mathscr{R}(U))$ is a partition of I and the set of roots $\mathscr{R}(U)$ is the set of points $\{r_m : m \geq 1\}$ given by the algorithm

$$U_{r_m r_m} = \min\{U_{ii} : i \in I \setminus \bigcup_{0 \leq s < m} I_{r_s}\},$$

where we start with $I_{r_0} = \phi$, the empty set.

If $r \in \mathscr{R}(U)$ and $\ell \notin I_r$ then any path in $\mathscr{G}^U$ from ℓ to I_r contains r. On the other hand, for $\ell \in I \setminus \mathscr{R}(U)$ there exists a unique $r \in \mathscr{R}^U$ such that any path in $\mathscr{G}^U$ from ℓ to $\tilde{\partial}$ contains r.

Proof Recall that the matrix $\tilde{U}$ has only one root which is $\tilde{r}$ because it is a nonsingular tree matrix. By definition of $\tilde{r}$ we deduce that $r = \tilde{r}$ for some point $r \in I$ if and only if $U_{\bullet r} = \alpha(U)$ where $\alpha(U)$ is the minimal value of U. This is equivalent to have $\mathscr{R}(U) = \{r\}$.

From formula (4.7) we have for $i \in I$

$$\sum_{j \in I} P_{ij} = \tilde{\mathbb{P}}_i\{\mathscr{T}_1 < \infty\}.$$

Notice that i is defective or equivalently i is a root for U if and only if this last probability is less than 1, that is, if and only if there is a trajectory joining i and $\tilde{\partial}$, for the Markov chain $(\tilde{X}_n)$, that never visits a point in I (except the initial state i at time 0). This is equivalent to have $\widetilde{\text{Geod}}(i, \tilde{r}) \cap I = \{i\}$, from which we have the desired characterization of the roots of U.

Now, let us show the roots are neighbors. Let $i \neq j$ be in $\mathscr{R}(U)$. Then $\widetilde{\mathrm{Geod}}(i, \tilde{r}) \cap I = \{i\}$ and $\widetilde{\mathrm{Geod}}(j, \tilde{r}) \cap I = \{j\}$. We have

$$\widetilde{\mathrm{Geod}}(i, j) = \widetilde{\mathrm{Geod}}(i, i \wedge j) \cup \widetilde{\mathrm{Geod}}(i \wedge j, j).$$

Since $\widetilde{\mathrm{Geod}}(i, i \wedge j) \subseteq \widetilde{\mathrm{Geod}}(i, \tilde{r})$ and $\widetilde{\mathrm{Geod}}(i \wedge j, j) \subseteq \widetilde{\mathrm{Geod}}(j, \tilde{r})$, we deduce

$$\widetilde{\mathrm{Geod}}(i, j) \cap I = \{i, j\}.$$

From the previous theorem we conclude $(i, j) \in \mathscr{G}^U$.

Now, let us prove the algorithm we proposed provides the roots of U. For that purpose we need to study I_r. If $U_{ir} = U_{rr}$ this means that $\tilde{U}_{ir} = \tilde{U}_{rr}$, which is equivalently to $i \wedge r = r$ and therefore $r \preccurlyeq i$. Thus, we have shown the equivalence

$$I_r = \{i \in I : U_{ir} = U_{rr}\} = \{i \in I : r \preccurlyeq i\} = \{i \in I : r \in \widetilde{\mathrm{Geod}}(i, \tilde{r})\}.$$

The first step in the algorithm is to show that if r_1 satisfies $U_{r_1 r_1} = \min\{U_{ii} : i \in I\}$ then r_1 must be a root for U. Assume the contrary, that is, there exists $k \in \widetilde{\mathrm{Geod}}(r_1, \tilde{r}) \cap (I \setminus \{r_1\})$. This is not possible because $U_{kk} = \tilde{U}_{kk} = \tilde{U}_{kr_1} < \tilde{U}_{r_1 r_1} = U_{r_1 r_1}$. The conclusion is $r_1 \in \mathscr{R}(U)$.

On the other hand $i \in \bigcup_{s<m} I_s$ if and only if $\widetilde{\mathrm{Geod}}(i, \tilde{r})$ contains some $\{r_s : s < m\}$. A similar argument as before gives that if r_m satisfies $U_{r_m r_m} = \min\{U_{ii} : i \in I \setminus \bigcup_{s<m} I_s\}$ then r_m is also a root. The rest of the theorem follows easily.

Example 4.11 Take the matrix U defined by $U_{ii} = a_i$, $U_{ij} = \alpha$ for $i \neq j$, where we assume $0 < \alpha \leq a_1 < \min\{a_2, \ldots, a_n\}$. This is a nonsingular ultrametric matrix. If $\alpha < a_1$ then U is strictly ultrametric, so $\mathscr{R}(U) = I$ and $\mathscr{G}^U$ is the complete graph (see the result below). A change of behavior occurs at $\alpha = a_1$, because U is not strictly ultrametric. In this case U has a constant column and then it has a unique root $r = 1$. Also it is straightforward to show $(i, j) \in \mathscr{G}^U$, for $i \neq j$, only if $i = 1$ or $j = 1$.

The next result summarizes some properties that hold for general strictly ultrametric matrices, not necessarily positive.

Proposition 4.12 *Assume that U is a strictly ultrametric matrix. We have $\mathscr{R}(U) = I$ and*

$$U_{ij} = 0 \Leftrightarrow (U^{-1})_{ij} = 0.$$

If we consider the classes of the equivalence relation

$$k \sim l \Leftrightarrow U_{kl} > 0,$$

then $\mathscr{G}^U$ is the complete graph on each class and no edge is found between these classes. In particular if U is positive then $\mathscr{G}^U$ is the complete graph.

Proof Consider $J \subseteq I$ a class for the equivalence relation $\sim$. Then $V = U_{JJ}$ is a positive strictly ultrametric matrix and so it is nonsingular. It is easy to see that in the minimal tree extension on $\tilde{J}$, the set J is exactly the set of leaves of the weighted tree $(\tilde{T}, \tilde{r}, \tilde{w})$ associated with the extension matrix $\tilde{V}$, because the diagonal elements of V strictly dominate the values out of the diagonal. Then, for $k, l \in J$ we have $\widetilde{\text{Geod}}(k, l) \cap I = \{k, l\}$ and then $(k, l) \in \mathscr{G}^V$. Since always the diagonal of $J \times J$ is a subset of $\mathscr{G}^V$ we conclude the graph $\mathscr{G}^V$ is the complete graph. Also $\mathscr{R}(V) = J$. For two different classes J, K we have $U_{KJ} = 0$ and therefore we have a block structure that implies $(U^{-1})_{KJ} = 0$. This yields that $\mathscr{G}^U$ is the complete graph on each class and there is no connections between different classes. Also we have $\mathscr{R}(U) = I$.

Finally, from the general result for inverse M-matrices in Proposition 4.2 we need only to prove $(U^{-1})_{ij} = 0 \Rightarrow U_{ij} = 0$, or equivalently $U_{ij} > 0 \Rightarrow (U^{-1})_{ij} > 0$ which has already been proved.

4.2 Graph of a Generalized Ultrametric Matrix

In this section we shall study the graph associated with the Markov process induced by a GUM. A main role will be played by **dyadic** trees, which are those trees whose points have two successors, except of course for the leaves.

There is a one-to-one correspondence between dyadic trees and a dyadic chain of partitions. Let (T, r) be a dyadic rooted tree on the set $\tilde{I}$, with level function L and set of leaves $\mathscr{L} = I$. Recall that $H(T)$ is the height of T. It is important to notice that each point $t \in T$ can be identified with the set $\mathscr{L}(t) = \{i \in I : t \preccurlyeq i\}$, which is the subset of leaves that hang from t. Clearly we have $\mathscr{L}(r) = I$ the set of all leaves of T.

For each $0 \leq q \leq H(T)$ the partition $\mathfrak{R}_q$ on I is given by

$$\mathfrak{R}_q = \{\mathscr{L}(t) : L(t) = q\} \bigcup \{\{i\} : i \in I, L(i) \leq q\}.$$

Notice that each non trivial atom of this partition $\mathscr{L}(t)$ is divided in two atoms $\mathscr{L}(t^-), \mathscr{L}(t^+)$, of the next partition, where t^-, t^+ are the two successors of t. For the moment how we choose them to be t^- or t^+ is irrelevant and will be studied latter.

Conversely, to each dyadic chain of partitions **F** on a set I, the dyadic tree associated with it is the tree whose nodes are all the atoms of all partitions of this chain of partitions. Edges connect each nontrivial atom of a partition with the corresponding two atoms of the next generation. We point out that I is the set of leaves of T.

4.2.1   A Dyadic Tree Supporting a GUM

In what follows we identify $t \in T$ and $\mathscr{L}(t)$. Now, we give a construction of GUM close in spirit to the one introduced in Theorem 3.19.

Proposition 4.13 *Let U be a GUM over the set I. Then, there exist a rooted dyadic tree (T, r) with set of leaves I, and two positive real functions $\boldsymbol{\alpha} = (\alpha_t : t \in T)$, $\boldsymbol{\beta} = (\beta_t : t \in T)$ such that $\boldsymbol{\alpha}|_I = \boldsymbol{\beta}|_I$, which also satisfy:*

(i) $\alpha_t \leq \beta_t$ for $t \in T$;
(ii) $\boldsymbol{\alpha}$ and $\boldsymbol{\beta}$ are increasing in $(T, \preceq)$ i.e. $\alpha_t \leq \alpha_s$ and $\beta_t \leq \beta_s$ for $t \preceq s$;
(iii) $U_{ij} = \alpha_t$ if $(i, j) \in t^- \times t^+$ and $U_{ij} = \beta_t$ if $(i, j) \in t^+ \times t^-$, where $t = i \wedge j$;
(iv) $U_{ii} = \alpha_i = \beta_i$ for $i \in I$.

In this case we will say that $(T, r, \boldsymbol{\alpha}, \boldsymbol{\beta})$ is associated with or supports U.

This is an equivalent definition for a GUM, whose proof is essentially given in Theorem 3.3 in [52]. Notice that ultrametric matrices are those GUM with $\boldsymbol{\alpha} = \boldsymbol{\beta}$.

Proof We can assume that U is in NFB

$$U - \begin{pmatrix} A & \alpha \mathbb{1}_p \mathbb{1}_q' \\ \beta \mathbb{1}_q \mathbb{1}_p' & B \end{pmatrix},$$

where $A = U_J$ and $B = U_K$ are NBF. We recall that U_J means U_{JJ}. The parameters $0 \leq \alpha \leq \beta$ also satisfy

(i) $\alpha = \min\{U\}$;
(ii) $\beta \leq \min\{A_{ij} : i \geq j\}$ and $\beta \leq \min\{B_{ij} : i \geq j\}$.

We consider the root $t = \tilde{r}$ which is identified with I and its successors t^-, t^+ which are identified with J and K respectively. We also define $\alpha_t = \alpha$ and $\beta_t = \beta$. The rest of the proof is done by induction considering the trees associated with A and B.

In what follows, we assume that U is a nonsingular GUM and $(T, \tilde{r}, \boldsymbol{\alpha}, \boldsymbol{\beta})$ is its associated dyadic rooted tree. Recall that a remarkable easy criterion for non-singularity of GUM matrices is that it has no row (column) of zeros and all rows (or columns) are different. Like in the previous chapter, a main role to understand the connections in U^{-1} is played by the equilibrium potentials. We recall that U^{-1} is a row and column diagonally dominant M-matrix which implies that U has a left and a right equilibrium potentials, which are denoted by μ^U and λ^U, respectively. These are two nonnegative vectors that satisfy $U' \mu^U = \mathbb{1}$, $U \lambda^U = \mathbb{1}$. Their total masses $\bar{\mu}^U = \mathbb{1}' \mu^U$ and $\bar{\lambda}^U = \mathbb{1}' \lambda^U$ are equal because

$$\mathbb{1}' \lambda^U = (U' \mu^U)' \lambda^U = (\mu^U)' U \lambda^U = (\mu^U)' \mathbb{1}.$$

As in the case of an ultrametric matrix we have the existence of a constant $\kappa > 0$ and a doubly substochastic matrix P such that $U^{-1} = \kappa(\mathbb{I} - P)$. Again we are interested in the graph $\mathcal{G}^U$, which gives the connections of the one step transitions associated with P, completed by the diagonal. That is $(i, j) \in \mathcal{G}^U$ if and only if $(U^{-1})_{ij} \neq 0$, which for $i \neq j$ is equivalent to $P_{ij} > 0$. Of particular interest is the set of roots of U, denoted by $\mathcal{R}(U)$, which consists of points where the Markov chain, constructed from P loses mass. That is $i \in \mathcal{R}(U)$ if and only if $\sum_j P_{ij} < 1$, which is also equivalent to $\sum_j (U^{-1})_{ij} > 0$. These roots will be called exiting roots of U, to distinguish them from the roots of U', which we call entering roots. The set of entering roots is denoted by $\mathcal{R}_{ent}(U) = \mathcal{R}(U')$.

4.2.2 Roots and Connections

In this section we state the main results on the study of $\mathcal{G}^U$. Our first result is an algorithm to obtain the roots of U, which can be seen as a generalization of Theorem 4.10.

Theorem 4.14 *Let U be a GUM. The set of roots $\mathcal{R}(U)$ is given by the following algorithm: Initially we put $I_0 = I$, $\mathcal{R}(U)_{-1} = \phi$ and $k = 0$,*

$$
Step\ k : \begin{cases} i_k \in \operatorname{argmin}\{i \in I_k : \sum_{j \in I} U_{ij}\} \\ H_k = \{j \in I_k : U_{ji_k} = U_{i_k i_k}\} \\ \mathcal{R}(U)_k = \mathcal{R}(U)_{k-1} \cup \{i_k\}, \quad I_{k+1} = I_k \setminus H_k \end{cases}.
$$

If $I_{k+1} = \phi$ then $\mathcal{R}(U) = \mathcal{R}(U)_k$ and we stop. If not we continue with step $k + 1$. Moreover the sets

$$
\mathcal{H}_r = \{j \in I : U_{jr} = U_{rr}\}\ for\ r \in \mathcal{R}(U), \tag{4.8}
$$

form a partition of I, and $\mathcal{H}_r = H_k$ where $r = i_k$. Also we have

$$
[r \in \mathcal{R}(U),\ i \in \mathcal{H}_r \setminus \{r\},\ j \notin \mathcal{H}_r] \Rightarrow (U^{-1})_{ij} = 0. \tag{4.9}
$$

That is, the unique connections between sets $\mathcal{H}_r$, $\mathcal{H}_s$ are through the roots r, s.

Recall that $H \in T$ is seeing as a subset of I, the set of leaves of T. Then the matrix U_H is a GUM and the above algorithm can be applied to obtain its roots.

Consider $(T, \tilde{r}, \boldsymbol{\alpha}, \boldsymbol{\beta})$ associated with U, as in Proposition 4.13. Let us introduce the following subsets (remind $\alpha_i = \beta_i$ for i a leaf):

$$
\mathcal{N}_i^+ = \{t \in T : t \prec i, \alpha_t = \alpha_i\}\ and\ \mathcal{N}_i^- = \{t \in T : t \prec i, \beta_t = \beta_i\}.
$$

Since α and β increase in $(T, \preceq)$ we get that $\mathcal{N}_i^+$ and $\mathcal{N}_i^-$ are the sets of constancy of α and β starting from the leaf i:

$$t \in \mathcal{N}_i^- \text{ (respectively } \mathcal{N}_i^+) \text{ implies } \mathrm{Geod}(i, t) \subseteq \mathcal{N}_i^- \text{ (respectively } \mathcal{N}_i^+).$$

In particular if $t \in \mathcal{N}_i^-$ (respectively $\mathcal{N}_i^+$) then t^- or t^+ belongs to $\mathcal{N}_i^-$ (respectively $\mathcal{N}_i^+$). We will prove as a corollary of Lemma 4.24 the inclusion

$$\forall i \in I : \mathcal{N}_i^+ \subseteq \mathcal{N}_i^-. \tag{4.10}$$

Now we construct the following sets of (forbidden non-oriented) arcs Γ and Γ':

$$(t, s) \in \Gamma \Leftrightarrow [s = t^-, t \in \mathcal{N}_i^+ \text{ for some } i \in t^+] \text{ or } [s = t^+, t \in \mathcal{N}_i^- \text{ for some } i \in t^-]$$

$$(t, s) \in \Gamma' \Leftrightarrow [s = t^-, t \in \mathcal{N}_i^- \text{ for some } i \in t^+] \text{ or } [s = t^+, t \in \mathcal{N}_i^+ \text{ for some } i \in t^-] \tag{4.11}$$

The following result furnishes another algorithm for the characterization of the roots at each one of the levels of the tree, in particular the roots of U are find at the first level I of the tree (which is the last in the algorithm).

Theorem 4.15 *Let U be a nonsingular GUM and $(T, \tilde{r}, \alpha, \beta)$ a dyadic tree supporting it. Let $H \in T$, then for $i \in H$ we have*

(i) $i \in \mathcal{R}(U_H)$ if and only if there exists a trajectory joining i and H in T that avoids Γ, or equivalently $\mathrm{Geod}(i, H) \cap \Gamma = \phi$;
(ii) $i \in \mathcal{R}_{ent}(U_H)$ if and only if there exists a trajectory joining i and H in T that avoids Γ' or equivalently $\mathrm{Geod}(i, H) \cap \Gamma' = \phi$.

Furthermore, assume U is strictly row and column diagonally dominant, that is, for all i it holds $U_{ii} > \sup\{U_{ij}, U_{ji} : j \neq i\}$. Then, Γ, Γ' are empty, $\mathcal{R}(U) = \mathcal{R}_{ent}(U) = I$ and $\mathcal{G}^U$ is the complete graph.

Theorem 4.16 *Let U be a nonsingular GUM. For $i \neq j$ consider $H = i \wedge j$ and $S(H) = \{H', H''\}$, with $i \in H'$ and $j \in H''$. Then*

(i) $(i, j) \in \mathcal{G}^{U_H} \Leftrightarrow i \in \mathcal{R}(U_{H'})$ and $j \in \mathcal{R}_{ent}(U_{H''})$;
(ii) $(i, j) \in \mathcal{G}^U \Leftrightarrow (i, j) \in \mathcal{G}^{U_H}$ and one (and only one) of the following two conditions is satisfied:

(ii.1) $U_{ij} = \beta_{i \wedge j} > \alpha_{i \wedge j}$;
(ii.2) $U_{ij} = \alpha_{i \wedge j}$ and for every $M \prec H$ such that $\alpha_M = \alpha_{i \wedge j}$ it holds

$$[\{i, j\} \subseteq M^- \Rightarrow (M, M^-) \notin \Gamma'] \text{ and } [\{i, j\} \subseteq M^+ \Rightarrow (M, M^+) \notin \Gamma].$$

Part (i) shows that condition $i \in \mathcal{R}(U_{H'})$, $j \in \mathcal{R}_{ent}(U_{H''})$ is equivalent to the fact that $(i, j) \in \mathcal{G}^{U_H}$ which means that i, j are connected in the one step transition for the Markov chain associated with the GUM U_H. Therefore part (ii) characterizes when this connection pursues until the coarsest level.

Example 4.17 Let $\alpha < \beta < \gamma$, $I = \{1, \ldots, 6\}$. Consider the following matrix

$$U = \begin{pmatrix} \beta & \alpha & \alpha & \alpha & \alpha & \alpha \\ \beta & \gamma & \alpha & \alpha & \alpha & \alpha \\ \beta & \beta & \beta & \alpha & \alpha & \alpha \\ \beta & \beta & \beta & \gamma & \beta & \beta \\ \beta & \beta & \beta & \beta & \beta & \beta \\ \beta & \beta & \beta & \beta & \beta & \gamma \end{pmatrix}$$

In Fig. 4.1 we display a dyadic tree supporting the GUM structure of U. For instance $\alpha_I = \alpha, \beta_I = \beta, \alpha_J = \alpha, \beta_J = \alpha, \alpha_K = \beta, \beta_K = \beta$.

In this example $(U_J^{-1})_{31} < 0$ because $J = \{1, 2, 3\} = 3 \wedge 1$, $J^- = \{1, 2\}$, $J^+ = \{3\}$, $3 \in \mathscr{R}(U_{\{3\}})$ and $1 \in \mathscr{R}_{ent}(U_{J^-})$ (according to Theorem 4.15).

On the other hand, since $U_{31} = \beta_{3 \wedge 1} = \beta > \alpha = \alpha_{3 \wedge 1}$ we conclude by Theorem 4.16 part *(ii)* that $U_{31}^{-1} < 0$. Now $(U_J^{-1})_{13} < 0$ by analogous reasons. Again from Theorem 4.16 we obtain $U_{13}^{-1} = 0$ because $U_{13} = \alpha_{1 \wedge 3}$, $I \prec J = 1 \wedge 3$ with $\alpha_I = \alpha_{1 \wedge 3}$, $\{1, 3\} \subset I^-$ and $(I, I^-) \in \Gamma'$ due to the fact that $I \in \mathscr{N}_5^-$ where $5 \in K = I^+$.

In what follows we shall assume, without loss of generality, that U is a nonsingular NBF matrix indexed by I. We also take $(T, \tilde{r}, \boldsymbol{\alpha}, \boldsymbol{\beta})$ a supporting tree for U. In the next results we denote by

$$\alpha = \alpha^U = \min\{U_{ij} : i, j \in I\} \text{ and } \beta = \beta^U = \min\{U_{ij} : i \geq j\}.$$

It is possible that $\alpha = \beta = 0$ but in this case the study of the inverse of U and $\mathscr{G}^U$ can be done simply by blocks. For this reason we shall impose that $\beta > 0$. We partition $I = I^- \cup I^+$ according to the immediate successors of $\tilde{r}$, the root of T,

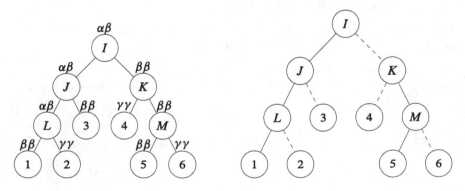

Fig. 4.1 The tree T (*on the left*) and the set of forbidden arcs — — in Γ

which is naturally identified with I. We simply denote by $J = I^-$, $K = I^+$ and $p = \#J$, $q = \#K$. Then $U_{JK} = \alpha \mathbb{1}_p \mathbb{1}'_q$, $U_{KJ} = \beta \mathbb{1}_q \mathbb{1}'_p$ and

$$U = \begin{pmatrix} A & \alpha \mathbb{1}_p \mathbb{1}'_q \\ \beta \mathbb{1}_q \mathbb{1}'_p & B \end{pmatrix}, \tag{4.12}$$

where $A = U_J$, $B = U_K$ are also nonsingular matrices in NBF. Recall that $U_{\bullet j}$ is the j-th column of U, $U_{i\bullet}$ is the i-th row of U and $e(i)$ is the i-th vector of the canonical basis.

Lemma 4.18 *Consider a positive matrix U in NBF as in (4.12) then*

(i) $\alpha \bar{\lambda}^U \leq 1$ and $\beta \bar{\lambda}^U \leq 1$;

(ii) $\alpha \bar{\lambda}^U = 1$ if and only if there exists a unique $j_0 \in I$ such that $U_{\bullet j_0} = \alpha \mathbb{1}$ and $U_{j_0 \bullet} = \alpha \mathbb{1}'$. Moreover $\beta = \alpha$ and $\mu^U = \lambda^U = \alpha^{-1} e(j_0)$. This characterizes the case when $\mathcal{R}(U) = \{j_0\}$;

(iii) $\beta \bar{\lambda}^U = 1$ implies $U_{nj} = \beta$ for all $j \in \mathcal{R}(U)$.

Proof (i) The first inequality holds because α is the minimum of the matrix. The second one because by the NBF property β is the minimum of the n-th row, which implies $1 = \sum_{j \in I} U_{nj}(\lambda^U)_j \geq \beta \bar{\lambda}^U$.

(ii) Since for all i it holds $1 = \sum_{j \in I} U_{ij}(\lambda^U)_j \geq \alpha \bar{\lambda}^U$. The last inequality becomes an equality if and only if

$$\lambda^U_{j_0} > 0 \Rightarrow \forall i \in I \ U_{ij_0} = \alpha.$$

This implies that at least for one j_0 the corresponding column is constant equal to α. It is clear that j_0 is unique since U is nonsingular. In particular, we conclude that $\alpha \leq \beta \leq U_{j_0 j_0} = \alpha$, that is $\beta = \alpha$ and the j_0 row of U is constant α.

(iii) Since U is a NBF matrix we have $U_{ij} \geq \beta$ for every $i \geq j \in I$. Then, for every j we have $U_{nj} \geq \beta$. Since $1 = \sum_{j \in I} U_{nj}(\lambda^U)_j \geq \beta \bar{\lambda}^U$, we deduce that:

$\beta \bar{\lambda}^U = 1$ implies $U_{nj} = \beta$ for every $j \in \mathcal{R}(U)$.

The next result enters into the details of part *(ii)* in the previous lemma.

Lemma 4.19 *We have $\alpha \bar{\lambda}^U = 1$ if and only if one and only one of the two equalities holds: $\alpha \bar{\lambda}^A = 1$ or $\alpha \bar{\lambda}^B = 1$. In the first case $j_0 \in J$ and in the second $j_0 \in K$, where j_0 is the index of previous lemma part (ii).*

This lemma says that U has a unique root if and only if one (and only one) of the two matrices A and B has a unique root, which is preserved to the next level.

Proof The condition is obviously necessary because if $\alpha \bar{\lambda}^U = 1$ then U has a unique constant column with value α and therefore either A or B has the same property, which implies the desired proposition.

On the other hand assume $\alpha \bar{\lambda}^A = 1$. Since $\alpha \leq \alpha^A$ and $\alpha^A \bar{\lambda}^A \leq 1$ we obtain $\alpha^A = \alpha$ and $\alpha^A \bar{\lambda}^A = 1$ which holds if and only if there exists a column $j_0 \in J$ such that $A_{\bullet j_0} = \alpha \mathbb{1}_p$. This implies in particular that $A_{j_0 j_0} = \alpha$ and then $\beta = \alpha$. Since $U_{K j_0} = \alpha \mathbb{1}_q$ we deduce $U_{\bullet j_0} = \alpha \mathbb{1}$ and also $U_{j_0 \bullet} = \alpha \mathbb{1}'$. An analogous reasoning gives the other alternative for the case $\alpha \bar{\lambda}^B = 1$. We notice that it is not possible that both alternatives are true because then U would have two constant columns.

Lemma 4.20 *The quantity $\Delta = 1 - \alpha \beta \bar{\lambda}^A \bar{\lambda}^B$ is strictly positive and the right and left equilibrium potentials for U are given by*

$$\lambda^U = \begin{pmatrix} a\lambda^A \\ b\lambda^B \end{pmatrix}, \ \mu^U = \begin{pmatrix} c\mu^A \\ d\mu^B \end{pmatrix},$$

with

$$a = \Delta^{-1}(1 - \alpha \bar{\lambda}^B), \ b = \Delta^{-1}(1 - \beta \bar{\lambda}^A),$$
$$c = \Delta^{-1}(1 - \beta \bar{\lambda}^B), \ d = \Delta^{-1}(1 - \alpha \bar{\lambda}^A). \tag{4.13}$$

Moreover, we have $\mathscr{R}(U) \subseteq \mathscr{R}(A) \cup \mathscr{R}(B)$ and

$$\mathscr{R}(U) = \begin{cases} \mathscr{R}(A) & \text{if } \beta \bar{\lambda}^A = 1 \\ \mathscr{R}(B) & \text{if } \alpha \bar{\lambda}^B = 1 \ . \\ \mathscr{R}(A) \cup \mathscr{R}(B) & \text{otherwise} \end{cases} \tag{4.14}$$

An analogous statement holds for the roots of U'.

Proof The right and left signed equilibrium potentials of any nonsingular matrix F have the same total mass: $\bar{\lambda}^F = \bar{\mu}^F$. Now, using the special block form of U we get the equations $U\lambda^U = \mathbb{1}$, $U'\mu^U = \mathbb{1}$ are respectively equivalent to

$$a \qquad + b\,\alpha \bar{\lambda}^B = 1$$
$$a\,\beta \bar{\lambda}^A + b \qquad = 1;$$

$$c \qquad + d\,\beta \bar{\lambda}^B = 1$$
$$c\,\alpha \bar{\lambda}^A + d \qquad = 1.$$

The determinant of both equations is Δ. Since $\beta \bar{\lambda}^A \leq 1$ and $\beta \bar{\lambda}^B \leq 1$, we get $\Delta \geq 0$. This determinant is zero only when both $\alpha \bar{\lambda}^A = 1$ and $\alpha \bar{\lambda}^B = 1$. Both equalities cannot hold simultaneously when U is nonsingular, so $\Delta > 0$. Thus a, b, c, d are well defined and unique. The rest of the proof follows at once.

Lemma 4.21 *We have* $\beta\bar\lambda^U = 1$ *if and only if* $\left[\beta\bar\lambda^A = 1 \text{ or } \beta\bar\lambda^B = 1\right]$. *The first case implies* $U_{pj} = \beta$ *for all* $j \in \mathcal{R}(A)$ *and the second case* $U_{nj} = \beta$ *for all* $j \in \mathcal{R}(B)$, *where* $p = \#J$ *and* n *is the size of* U.

Proof We only prove the first equivalence because the other implications follow easily from Lemma 4.18 (c). Using the relation obtained between λ^U and λ^A, λ^B, we get

$$\beta\bar\lambda^U = \beta(a\bar\lambda^A + b\bar\lambda^B) = \frac{\beta}{\Delta}(\bar\lambda^A + \bar\lambda^B - (\alpha + \beta)\bar\lambda^A\bar\lambda^B).$$

A simple computation yields

$$\beta\bar\lambda^U \le 1 \text{ is equivalent to } (1 - \beta\bar\lambda^A)(1 - \beta\bar\lambda^B) \ge 0,$$

with equality satisfied simultaneously in both equations. Then, the desired equivalence is shown.

Let us study the inverse of U. We decompose U^{-1} also in blocks as U is decomposed in (4.12)

$$U^{-1} = \begin{pmatrix} C & D \\ E & F \end{pmatrix}.$$

By using Schur's decomposition we get

$$E = -\beta B^{-1}\mathbb{1}_q\mathbb{1}'_p C = -\beta\lambda^B\mathbb{1}'_p C. \tag{4.15}$$

We also obtain

$$C = (A - \alpha\beta\bar\lambda^B\mathbb{1}_p\mathbb{1}'_p)^{-1} = A^{-1} + \frac{\alpha\beta\bar\lambda^B}{1 - \alpha\beta\bar\lambda^A\bar\lambda^B} = A^{-1} + \frac{\alpha\beta\bar\lambda^B}{\Delta}\lambda^A(\mu^A)'. \tag{4.16}$$

By replacing (4.16) in (4.15) we find

$$E = -\frac{\beta}{\Delta}\lambda^B(\mu^A)'. \tag{4.17}$$

For D, F we find analogous expressions to (4.16), (4.17), namely

$$D = -\frac{\alpha}{\Delta}\lambda^A(\mu^B)',$$

$$F = B^{-1} + \frac{\alpha\beta\bar\lambda^A}{\Delta}\lambda^B(\mu^B)'. \tag{4.18}$$

From (4.17) and (4.18) we deduce that for $(i, j) \in (J \times K) \cup (K \times J)$ it holds

$$(U^{-1})_{ij} < 0 \text{ if and only if } (i, j) \in \mathscr{R}(A) \times \mathscr{R}_{ent}(B) \cup \mathscr{R}(B) \times \mathscr{R}_{ent}(A).$$

Previous to analyze what happens when $i \neq j$ for $(i, j) \in J \times J \cup K \times K$, let us supply some elementary properties. Denote $U(\gamma) = U - \gamma \mathbb{1}\mathbb{1}^t$. Observe that $\alpha \bar{\lambda}^U < 1$ is equivalent to $\alpha < \min\{U_{ii} : i \in I\}$. Under this condition, for any $\gamma \in [0, \alpha]$ we get

$$U(\gamma)^{-1} = U^{-1} + \frac{\gamma}{1 - \gamma \bar{\lambda}^U} \lambda^U (\mu^U)'. \tag{4.19}$$

A straightforward computation yields

$$\lambda^{U(\gamma)} = \frac{1}{1 - \gamma \bar{\lambda}^U} \lambda^U, \quad \mu^{U(\gamma)} = \frac{1}{1 - \gamma \bar{\lambda}^U} \mu^U.$$

For our analysis we will need the following key result.

Lemma 4.22 *Let $\gamma \in [0, \alpha]$ and $\gamma < \min\{U_{ii} : i \in I\}$. Then, for $i \neq j$:*

$$U_{ij} > \gamma \Rightarrow [(U^{-1})_{ij} < 0 \Leftrightarrow (U(\gamma)^{-1})_{ij} < 0];$$
$$U_{ij} = \gamma \Rightarrow (U(\gamma)^{-1})_{ij} = 0. \tag{4.20}$$

Proof From hypothesis $U(\gamma)$ is a NBF and it is nonsingular because the diagonal is strictly positive and no two rows are equal. Denote by $P(\gamma)$ a substochastic matrix such that $U(\gamma)^{-1} = \kappa(\mathbb{I} - P(\gamma))$, for some $\kappa > 0$. For $i \neq j$ we have $(P(\gamma))_{ij} = 0$ if and only if $(U(\gamma)^{-1})_{ij} = 0$. On the other hand we have

$$U(\gamma) = \kappa^{-1} \sum_{m \geq 0} (P(\gamma))^m.$$

Then, $(U(\gamma))_{ij} = 0$ implies $(P(\gamma))_{ij} = 0$ and then the second relation in (4.20) is satisfied.

Let us prove the first relation in (4.20). From (4.19) and $\gamma \geq 0$, we obtain that if $(U(\gamma)^{-1})_{ij} < 0$ then necessarily $(U^{-1})_{ij} < 0$. So, we need to prove that

$$[U_{ij} > \gamma \text{ and } (U^{-1})_{ij} < 0] \Rightarrow (U(\gamma)^{-1})_{ij} < 0.$$

From equation (4.19) this relation obviously holds when $\lambda_i^U \mu_j^U = 0$. So in the rest of the proof we assume that

$$\lambda_i^U \mu_j^U > 0,$$

which implies that $U(\gamma)_{ij}^{-1}$ is a strictly increasing function of $\gamma \in [0, \alpha]$.

To continue with the proof, we first study the case $\gamma < \alpha$. Assume $(U(\gamma)^{-1})_{ij} = 0$. Since $(U(\gamma)^{-1})_{ij}$ is strictly increasing, we are able to find some $\hat{\gamma} < \alpha$ such that $(U(\hat{\gamma})^{-1})_{ij} > 0$ which contradicts the fact that $U(\hat{\gamma})^{-1}$ is an M-matrix (recall that $U(\gamma)$ is a NBF and then its inverse is an M-matrix). Therefore, if $\gamma < \alpha$ we have $(U(\gamma)^{-1})_{ij} < 0$.

For the rest of the proof we assume $\gamma = \alpha$. By hypothesis of the lemma we have $\alpha \bar{\lambda}^U < 1$, then $\alpha \bar{\lambda}^A < 1$ and $\alpha \bar{\lambda}^B < 1$. Hence

$$U(\alpha) = \begin{pmatrix} A(\alpha) & 0 \\ (\beta - \alpha)\mathbb{1}_q \mathbb{1}'_p & B(\alpha) \end{pmatrix} \text{ and}$$

$$U(\alpha)^{-1} = \begin{pmatrix} A(\alpha)^{-1} & 0 \\ -(\beta - \alpha)\lambda^{B(\alpha)}(\mu^{A(\alpha)})' & B(\alpha)^{-1} \end{pmatrix}.$$

Let $i \neq j$ be such that $U_{ij} > \alpha$ or equivalently $(U(\alpha))_{ij} > 0$. For $(i, j) \in J \times J$ (analogously if $(i, j) \in K \times K$) we get $(U(\alpha)^{-1})_{ij} = (A(\alpha)^{-1})_{ij}$. From (4.16) $(A^{-1})_{ij} \leq (U^{-1})_{ij} < 0$, so we obtain the result by induction on the dimension of the matrix.

Therefore, we can assume $i \in K$, $j \in J$. Since $(U(\alpha))_{ij} > 0$ we have $\beta > \alpha$. The relations

$$(U^{-1})_{ij} = -\tfrac{\beta}{\Delta} \lambda_i^B \mu_j^A, \quad (U(\alpha)^{-1})_{ij} = -(\beta - \alpha)\lambda_i^{B(\alpha)} \mu_j^{A(\alpha)},$$

$$\lambda^{B(\alpha)} = \tfrac{1}{1 - \alpha \bar{\lambda}^B} \lambda^B, \quad \mu^{A(\alpha)} = \tfrac{1}{1 - \alpha \bar{\lambda}^A} \mu^A,$$

give $(U(\alpha)^{-1})_{ij} = \tfrac{\Delta(\beta - \alpha)}{\beta}(U^{-1})_{ij}$, and the result is proven.

Lemma 4.23 *For $i \neq j$ in J we have*

$$(U^{-1})_{ij} = 0 \Leftrightarrow (A^{-1})_{ij} = 0 \text{ or } [U_{ij} = \alpha \text{ and } \beta \bar{\lambda}^B = 1]. \tag{4.21}$$

Furthermore $\alpha \bar{\lambda}^A = 1$ implies $(U^{-1})_{ij} = (A^{-1})_{ij}$.

Proof From (4.16) we have

$$(U^{-1})_J = A^{-1} + \frac{\gamma}{1 - \gamma \bar{\lambda}^A} \lambda^A (\mu^A)' = A(\gamma)^{-1}$$

with $\gamma = \alpha \beta \bar{\lambda}^B \leq \alpha \leq \min\{A_{ij} : i, j \in J\}$.

Let us prove the last assertion of the lemma. For that reason assume that $\alpha \bar{\lambda}^A = 1$. Recall again that $\alpha^A \geq \alpha$ and $\alpha^A \bar{\lambda}^A \leq 1$. Thus, we have $\alpha = \alpha^A$ and λ^A is supported at a unique point j_0, or equivalently A has a constant column of value α. This implies that $\alpha = \beta = A_{j_0 j_0}$ and the j_0 row of U is also constant. In particular we have $\lambda^A = \mu^A = \alpha^{-1} e(j_0)$. Hence we conclude that $(U^{-1})_{ij} = (A^{-1})_{ij}$, as required.

We now turn to the proof of equivalence (4.21). From $(A^{-1})_{ij} \leq (U^{-1})_{ij} \leq 0$ we get $(A^{-1})_{ij} = 0 \Rightarrow (U^{-1})_{ij} = 0$.

Thus, for the rest of the proof we may assume $(A^{-1})_{ij} < 0$ and we must show the following equivalence

$$(U^{-1})_{ij} = 0 \Leftrightarrow [U_{ij} = \alpha \text{ and } \beta\bar{\lambda}^B = 1].$$

Consider first the case $\alpha\bar{\lambda}^A = 1$. As has been already proved we have $(U^{-1})_{ij} = (A^{-1})_{ij} < 0$. Since $\beta\bar{\lambda}^B < 1$ (recall that $\Delta = 1 - \alpha\beta\bar{\lambda}^A\bar{\lambda}^B > 0$) we get the desired equivalence.

Now, we assume that $\alpha\bar{\lambda}^A < 1$. This implies that $\alpha < \min\{A_{ii} : i \in J\}$ and we obtain $\gamma \leq \alpha < \min\{A_{ii} : i \in J\}$. So, we can use Lemma 4.22 for the matrix A. Here, there are two possible cases.

If $U_{ij} = \alpha$ and $\beta\bar{\lambda}^B = 1$, we have $A_{ij} = U_{ij} = \alpha$. We are in the situation where $A_{ij} = \gamma$, and therefore $(A(\gamma)^{-1})_{ij} = 0$. Since $(U^{-1})_{ij} = (A(\gamma)^{-1})_{ij} = 0$ we conclude the equivalence in this case.

Finally, we assume that $U_{ij} > \alpha$ or $\beta\bar{\lambda}^B < 1$. In both cases we have $A_{ij} > \gamma$, which implies $(A(\gamma)^{-1})_{ij} < 0 \Leftrightarrow (A^{-1})_{ij} < 0$ (see Lemma 4.22). Since we have assumed $(A^{-1})_{ij} < 0$ we deduce $(U^{-1})_{ij} = A(\gamma)^{-1})_{ij} < 0$ and the result holds.

The results already obtained in these lemmas, even if they are formulated for the first level of the tree $I = I^- \cup I^+$ can be applied to anyone of them. More precisely, any element $J \in T$, which is not a leaf, is partitioned as $J = J^- \cup J^+$. In this case we redefine $\alpha = \alpha_J$, $\beta = \beta_J$, $n = |J|$ (recall we consider J as an element of T and as a subset of I where it corresponds), and all the results obtained before apply to this situation. For instance, by induction and the last part of Lemma 4.23 it follows

$$\text{if } \tilde{r} \in \bigcup_{i \in I} \mathcal{N}_i^+ \text{ then for } i \neq j, \ (U^{-1})_{ij} = (U_J^{-1})_{ij}, \text{ where } J = i \wedge j.$$

Lemma 4.24 *Assume $J \in T$ is not a leaf then*

(i) $\alpha_J\bar{\lambda}^{U_J} = 1$ *if and only if $J \in \mathcal{N}_i^+$ for some leaf $i \in J$. In this case $\beta_J = \alpha_J = \alpha_i$.*
(ii) $\beta_J\bar{\lambda}^{U_J} = 1$ *if and only if $J \in \mathcal{N}_i^-$ for some leaf $i \in J$.*

Proof (i) From Lemma 4.18 part (ii) the relation $\alpha_J\bar{\lambda}^{U_J} = 1$ holds if and only if there exists a column $i \in J$ such that $(U_J)_{\bullet i} = \alpha_J \mathbb{1}_J$ and $U_{ii} = \alpha_i = \alpha_J$. Thus $J \in \mathcal{N}_i^+$ for the leaf i. Conversely assume $\alpha_i = \alpha_J$ for $i \in J$. Since $U_{ii} = \alpha_i = \alpha_J$ we conclude the result. The equality $\beta_J = \alpha_J = \alpha_i$ follows from the same cited lemma.

Now we turn to the proof of (ii). Assume that $\beta_J\bar{\lambda}_{U_J} = 1$. We must show that there exists a leaf $i \in J$ such that $\beta_J = \beta_i$. From Lemma 4.21 consider $M \succ J$ such that $\beta_M = \beta_J$ and $\beta_M\lambda^{U_M} = 1$. By recurrence we show the existence of a leaf $i \in J$ such that $\beta_i = \beta_J$ and therefore $J \in \mathcal{N}_i^-$.

Conversely, the fact $\beta_J = \beta_i$ for $i \in J$ implies $\beta_M = \beta_J$ for all elements $M \in \text{Geod}(i, J)$. Then, using Lemma 4.21 and recurrence, we show $\beta_M \lambda^{U_M} = 1$ for all $M \in \text{Geod}(i, J)$.

Corollary 4.25 *For every $i \in I$ we have $\mathcal{N}_i^+ \subseteq \mathcal{N}_i^-$.*

Proof Observe that $\alpha_J \leq \beta_J \leq \beta_i = \alpha_i = \alpha_J$ for any $J \in \mathcal{N}_i^+$. This implies the desired result.

Now we give the proofs of the main theorems of the section.

Proof (Theorem 4.14) We use recurrence on the dimension of U and assume it is a NBF matrix. For this reason we partition U as in (4.12) (we use the notation therein introduced). Also we fix the function *argmin* by choosing the smallest optimal index i in case there are several ones.

Notice that if $U_{i^*i^*} = \alpha$ for some $i^* \in I$, which is necessarily unique, then in the algorithm $i_0 = i^*$, $I_0 = I$, $I_1 = \phi$. Hence $\mathcal{R}(U) = \{i_0\}$, and the result is shown. In the sequel we assume $U_{ii} > \alpha$ for every $i \in I$.

We remind the notation $J = I^-$, $K = I^+$ as the immediate successors of the root $\tilde{r}$ of T which was identified with I.

Case $U_{ii} > \beta$ for all $i \in J$. In this situation for every step k in the algorithm we have

$$[i_k \in J \Rightarrow H_k \subset J] \text{ and } [i_k \in K \Rightarrow H_k \subset K]. \tag{4.22}$$

In fact, if $i_k \in J$ and $j \in K$ we have $U_{ji_k} = \beta < U_{i_k i_k}$, and if $i_k \in K$ and $j \in J$ we have $U_{ji_k} = \alpha < U_{i_k i_k}$. From Lemma 4.24 we have $\alpha_J \bar{\lambda}^{U_J} < 1$ and $\beta_J \bar{\lambda}^{U_J} < 1$. Thus according to Lemma 4.20 we deduce $\mathcal{R}(U) = \mathcal{R}(A) \cup \mathcal{R}(B)$, and we must prove that our algorithm supplies the desired roots.

We assume by an inductive argument, that our algorithm gives the right answer for both matrices A and B: $\mathcal{R}(A) = \{i_0^A, \ldots, i_u^A\}$, $\mathcal{R}(B) = \{i_0^B, \ldots, i_v^B\}$. Also we use H_k^A, J_k, H_k^B, K_k to denote the sets obtained when applying the algorithm to A and B. Suppose that when using the algorithm to the matrix U at steps $k_0 < \cdots < k_l$ the corresponding points $i_{k_0}, \ldots, i_{k_l}$ belong to J.

We now prove that $l = u$ and $\{i_{k_0}, \ldots, i_{k_l}\} = \{i_0^A, \ldots, i_u^A\}$. Using (4.22) we obtain that $J \subseteq I_{k_0}$. Since $\sum_{\ell \in J} A_{i\ell} + \alpha \#K = \sum_{\ell \in I} U_{i\ell}$ for any $i \in J$, we obtain that $i_0^A = i_{k_0}$, so $H_0^A = H_{k_0}$, $J_1 = I_{k_0+1} \cap J$, $I_{k_0+1} \cap K = I_{k_0} \cap K$. Repeating this procedure we get the desired relation. We argue similarly for the matrix B and conclude $\mathcal{R}(U) = \mathcal{R}(A) \cup \mathcal{R}(B)$.

Case $U_{i^*i^*} = \beta$ for Some $i^* \in J$. In this situation we have $\beta_J = \beta_{i^*}$, which is exactly the fact that $J \in \mathcal{N}_{i^*}^-$. Thus, Lemma 4.24 (*ii*) and then Lemma 4.20, allow us to conclude $\mathcal{R}(U) = \mathcal{R}(A)$. Hence, we need to show our algorithm supplies this result. Notice that we have assumed implicitly $\alpha < \beta$ in this case, then

$$\text{for every } j \in K \text{ we have } \sum_{\ell \in I} U_{i^*\ell} \leq \beta|J| + \alpha|K| < \sum_{\ell \in I} U_{j\ell}.$$

This implies that if m denotes the first step in the algorithm such that $i^* \in H_m$, we necessarily have $i_0, \ldots, i_m \in J$ and $H_k \cap K = \phi$ for every $k < m$. Now

$$\beta = U_{i^* i^*} \geq U_{i^* i_m} = U_{i_m i_m} \geq \beta,$$

then $U_{i_m i_m} = \beta = U_{j i_m}$ for any $j \in K$. We deduce $H_m = H_m^A \cup K$ and conclude the algorithm supplies the equality $\mathcal{R}(U) = \mathcal{R}(A)$.

We now must prove that $\mathcal{H}_r = H_k$ for any root r, where $r = i_k$. As an intermediate result we will show relation (4.9), that is,

$$[r \in \mathcal{R}(U), \; i \in \mathcal{H}_r \setminus \{r\}, \; j \notin \mathcal{H}_r] \Rightarrow (U^{-1})_{ij} = 0.$$

By construction it is clear that $H_k \subseteq \mathcal{H}_r$. In order to prove the equality it is enough to show that for two different roots r, r' we have $\mathcal{H}_r \cap \mathcal{H}_{r'} = \phi$, because by construction the collection $(H_q : q)$ is a partition of I.

Let us take $i \in \mathcal{H}_r$, $i \neq r$, so $\sum_{s \in I} (U^{-1})_{is} U_{sr} = 0$. From this equality, we get

$$U_{rr} \sum_{s \in \mathcal{H}_r} (U^{-1})_{is} = - \sum_{s \in I \setminus \mathcal{H}_r} (U^{-1})_{is} U_{sr}.$$

On the other hand $-\sum_{s \in I \setminus \mathcal{H}_r} (U^{-1})_{is} \geq 0$. If this last sum were strictly positive we would arrive to

$$U_{rr} \sum_{s \in \mathcal{H}_r} (U^{-1})_{is} < -U_{rr} \sum_{s \in I \setminus \mathcal{H}_r} (U^{-1})_{is},$$

because for $s \in I \setminus \mathcal{H}_r$ we have $U_{sr} < U_{rr}$. Then, $U_{rr} \sum_{s \in I} (U^{-1})_{is} < 0$, which contradicts the fact that U^{-1} is diagonally dominant. Hence, we conclude that

$$(U^{-1})_{ij} = 0 \text{ for any } j \notin \mathcal{H}_r \text{ and } \sum_{s \in \mathcal{H}_r} (U^{-1})_{is} = 0,$$

so $i \notin \mathcal{R}(U)$. In particular we have proven (4.9).

Assume now $\mathcal{H}_r \cap \mathcal{H}_{r'}$ is not empty, and take i any element in this intersection. From the previous discussion we obtain $r \notin \mathcal{H}_{r'}, r' \notin \mathcal{H}_r$, and $(U^{-1})_{is} < 0$ implies $s \in \mathcal{H}_r \cap \mathcal{H}_{r'}$. Now, we use that $U = \kappa^{-1} \sum_{m \geq 0} P^m$ or equivalently $U^{-1} = \kappa(\mathbb{I} - P)$, for some substochastic kernel P. Since $U_{ir} = U_{rr} > 0$ we deduce there exists $m \geq 1$ satisfying $P_{ir}^m > 0$. Consider

$$m_0 = \min \{m \geq 1 : \; P_{\ell r}^m > 0 \text{ and } \ell \in \mathcal{H}_r \cap \mathcal{H}_{r'}\}.$$

We denote by $i_0 \in \mathcal{H}_r \cap \mathcal{H}_{r'}$ any optimal site for the above minimization problem. If $m_0 \geq 2$ we obtain

$$0 < P_{i_0 r}^{m_0} = \sum_{s \in I} P_{i_0 s} P_{sr}^{m_0 - 1},$$

but this last sum is zero because $P_{i_0s} > 0$ only if $s \in \mathcal{H}_r \cap \mathcal{H}_{r'}$ and then by definition of m_0 we get $P_{sr}^{m_0-1} = 0$, which is a contradiction. Therefore $m_0 = 1$, which also implies that $P_{jor} > 0$ or equivalently $(U^{-1})_{i_0r} < 0$, and then $r \in \mathcal{H}_r \cap \mathcal{H}_{r'}$ given another contradiction. Thus, the result is proven.

Proof (Theorem 4.15) We only prove part (i), the proof of (ii) being analogous. We proceed by induction on the geodesic distance between i and H. When $H = i$ the result is obvious (recall that we identify i with $\{i\}$). Assume the result holds up to $X \in T$ and we would like to prove it for the immediate ancestor H of X. Of course we assume that $i \in X \subset H$. The other immediate successor of H is denoted by Y. Then from formulas (4.13) and (4.14) in Lemma 4.20 we get

$$i \in \mathcal{R}(U_H) \Leftrightarrow i \in \mathcal{R}(U_X) \text{ and } \left[[X = H^-, \alpha_H \bar{\lambda}^{U_Y} < 1] \text{ or } [X = H^+, \beta_H \bar{\lambda}^{U_Y} < 1] \right].$$

Now, we prove that the proposition inside the big brackets is equivalent to the fact that $(X, H) \notin \Gamma$. First assume that $X = H^-$. Notice that $\alpha_H \bar{\lambda}^{U_Y} = 1$ is equivalent to $\alpha_H = \alpha_Y$ and $\alpha_Y \bar{\lambda}^{U_Y} = 1$ (this is by using $\alpha_H \leq \alpha_Y$ and $\alpha_Y \bar{\lambda}^{U_Y} \leq 1$). Now, part (i) in Lemma 4.24 shows this is equivalent to

$$\alpha_H = \alpha_Y \text{ and } Y \in \mathcal{N}_k^+ \text{ for some leaf } k \in Y,$$

and by definition we obtain the equivalence with: $H \in \mathcal{N}_k^+$ for some leaf $k \in Y$. From (4.11) we conclude this is equivalent to $(H, X) \in \Gamma$.

Now, in the case $X = H^+$ the equivalence to be shown, follows from the fact that $\beta_H \bar{\lambda}^{U_Y} = 1$ is equivalent to $\beta_H = \beta_Y$ and $\beta_Y \bar{\lambda}^{U_Y} = 1$ which by part (ii) in Lemma 4.24 turns to be equivalent to

$$H \in \mathcal{N}_k^- \text{ for some leaf } k \in Y,$$

and the result is proven, because this holds if and only if $(H, X) \in \Gamma$.

Proof (Theorem 4.16) Recall that $i \wedge j = H$. From (4.17) and (4.18) (that is Schur's decomposition) applied to the matrix U_H we get that every $(i, j) \in H' \times H''$ satisfies

$$(U_H^{-1})_{ij} < 0 \text{ if and only if } (i, j) \in \mathcal{R}(U_{H'}) \times \mathcal{R}_{ent}(U_{H''})$$

Then part (i) follows.

Let us prove part (ii). If $H = I$ there is nothing to prove, so in what follows we assume $H \succ I$ and denote by J the successor of I that contains i, j. We notice that $(ii.1)$ and $(ii.2)$ cannot be satisfied simultaneously.

Clearly $(U^{-1})_{ij} < 0 \Rightarrow (U_H^{-1})_{ij} < 0$ (use induction and formula (4.16)), then we are reduced to prove that under the condition $(U_H^{-1})_{ij} < 0$ we have

$$(U^{-1})_{ij} < 0 \Leftrightarrow [(ii.1) \text{ or } (ii.2)].$$

We first assume (*ii*.1), that is $U_{ij} = \beta_{i \wedge j} > \alpha_{i \wedge j}$. We suppose that $I \preccurlyeq M \prec H$ and prove by induction that $(U_M^{-1})_{ij} < 0$. When M^* is the immediate ancestor of H we use relation (4.21) in Lemma 4.23. Notice that $U_{ij} = (U_{M^*})_{ij} > \alpha_{i \wedge j} \geq \alpha^{U_{M^*}}$ and then $(U_{M^*}^{-1})_{ij} < 0$. An inductive argument shows the desired property $(U^{-1})_{ij} < 0$. An important point is that for every M such that $I \preccurlyeq M \prec H$ we have $(U_M)_{ij} > \alpha_{i \wedge j} \geq \alpha^M$.

The next step is to prove that (*ii*.1) holds under the assumptions $(U^{-1})_{ij} < 0$ and the negation of (*ii*.2). A detailed analysis shows that the only interesting case is the existence of a point $\hat{M}$ satisfying $I \preccurlyeq \hat{M} \prec H$, $\alpha_{\hat{M}} = \alpha_H$ and

$$\{i, j\} \subseteq \hat{M}^- \text{ and } (\hat{M}, \hat{M}^-) \in \Gamma', \text{ or}$$
$$\{i, j\} \subseteq \hat{M}^+ \text{ and } (\hat{M}, \hat{M}^+) \in \Gamma.$$

From the definition of Γ and Γ', we obtain

$$\{i, j\} \subseteq \hat{M}^- \text{ and } \hat{M} \in \mathscr{N}_k^-, \text{ for some } k \in \hat{M}^+ \tag{4.23}$$
$$\{i, j\} \subseteq \hat{M}^+ \text{ and } \hat{M} \in \mathscr{N}_k^-, \text{ for some } k \in \hat{M}^-.$$

In both cases we get $\hat{M} \in \mathscr{N}_k^-$, which means that $\beta_{\hat{M}} = \beta_k$ and therefore $\beta_\bullet$ is constant in $\text{Geod}(k, M)$. According to Lemma 4.24, we conclude

$$\beta_X \bar{\lambda}^{U_X} = 1, \tag{4.24}$$

for all $X \in \text{Geod}(\hat{M}, k)$.

On the other hand the assumption $(U_H^{-1})_{ij} < 0, (U^{-1})_{ij} < 0$ implies that for all $I \preccurlyeq M \preccurlyeq H$ one has $(U_M^{-1})_{ij} < 0$. Denote the immediate successors of M as $S(M) = \{M', M''\}$ and we take M' such that $i, j \in M'$. According to Lemma 4.23 we necessarily have

$$U_{ij} > \alpha_M \text{ or } \beta_M \bar{\lambda}^{U_{M''}} < 1. \tag{4.25}$$

In particular, if we take $M = \hat{M}$, we conclude from (4.23) and (4.24) that $\beta_{M''} \bar{\lambda}^{U_{M''}} = 1$, and since $\beta_{M''} = \beta_M$ we have

$$\beta_M \bar{\lambda}^{U_{M''}} = 1.$$

Thanks to (4.25), we get

$$U_{ij} > \alpha_M = \alpha_H$$

and therefore $U_{ij} = \beta_{i \wedge j} > \alpha_{i \wedge j}$ proving that (*ii*.1) holds.

The only case left to be analyzed is that if (*ii*.1) is not satisfied then $(U^{-1})_{ij} < 0 \Leftrightarrow$ (*ii*.2) holds. When (*ii*.1) is not satisfied we have $U_{ij} = \alpha_H$. Consider any

$M \prec H$ such that $\alpha_M = \alpha_H$ and denote $S(M) = \{M', M''\}$ with $\{i, j\} \subseteq M'$. Since $(U_M)_{ij} = \alpha_H = \alpha_M = \alpha^{U_M}$, we get from Lemma (4.23)

$$(U_M^{-1})_{ij} < 0 \Leftrightarrow [(U_{M'}^{-1})_{ij} < 0 \text{ and } \beta_M \bar{\lambda}^{U_{M''}} < 1].$$

Using formulas in (4.13) in Lemma 4.20, we obtain that $\beta_M \bar{\lambda}^{M''} < 1$ is equivalent to

$$[M' = M^- \Rightarrow \mathscr{R}_{ent}(U_{M'}) \subseteq \mathscr{R}_{ent}(U_M)] \text{ and } [M' = M^+ \Rightarrow \mathscr{R}(U_{M'}) \subseteq \mathscr{R}(U_M)].$$

Now, Theorem 4.15 allows us to conclude this statement is equivalent to

$$[M' = M^- \Rightarrow (M, M^-) \notin \Gamma'] \text{ and } [M' = M^+ \Rightarrow (M, M^+) \notin \Gamma].$$

So, an argument based on induction permits us to deduce $(U^{-1})_{ij} < 0 \Leftrightarrow (ii.2)$.

4.3   Permutations That Preserve a NFB

Recall that if U is a GUM then there is a permutation Π such that $\Pi' U \Pi$ is in NBF (see Definition 3.7). We shall make this permutation explicit in terms of the following order relation $\leq_T$ in I as

$$\text{for } i \neq j \text{ we put } i <_T j \text{ if } i \in t^-, j \in t^+ \text{ with } t = i \wedge j.$$

It is easy to see that $\leq_T$ is a total order in I. The only non trivial fact is transitivity, which follows from the tree structure. We say that a set $Y \subseteq I$ is a **interval** if $i, k \in Y$ and $i \leq_T j \leq_T k$ imply $j \in Y$.

As before, we consider $I = \{1, \cdots, n\}$. Since $(I, \leq_T)$ is totally ordered there exists a minimal element of I which for the moment we denote by 1_T. Inductively, we define $(i + 1)_T$ to be the successor of i_T with respect to the total order $\leq_T$. Observe that $i \leq_T j \leq_T k$ implies

$$U_{ik} \leq U_{jk} \leq U_{kk} \text{ and } U_{ii} \geq U_{ij} \geq U_{ik}.$$

Consider the permutation $\varphi_T : I \to I$ defined by $\varphi_T(i) = j$ if $i = j_T$, which is the inverse of the function $i \to i_T$. This permutation will put U in NBF by defining

$$V_{ij} = U_{\varphi(i)\varphi(j)}.$$

We shall study later on the combinatorics of permutations that put U in NBF.

For the moment by permuting I with φ_T, we can assume $\leq_T$ is the usual order relation $\leq$ on I. If this is the case we prove that U is in NBF. Indeed, we have

$$U_{i,i+1} = \alpha_{i \wedge i+1}, \quad U_{i+1,i} = \beta_{i \wedge i+1},$$

and the increasing property of α, β with respect to $\preceq$ implies that in general

$$
U_{ij} = \begin{cases}
\min\{\alpha_{i \wedge i+1}, \ldots, \alpha_{j-1 \wedge j}\} & \text{if } i < j \\
\\
\min\{\beta_{j \wedge j+1}, \ldots, \beta_{i-1 \wedge i}\} & \text{if } i > j
\end{cases},
$$

because $i \wedge j = i \wedge (i+1) \wedge \ldots \wedge j$ if $i < j$ and similarly in the other case.

Observe there exists i_0 with $i_0 \wedge i_0 + 1 = \tilde{r}$ the root of T, so

$$
\alpha_{i_0 \wedge i_0+1} = \alpha_{\tilde{r}} = \min\{\alpha_{i \wedge i+1} : i = 1, \ldots, n-1\} = \min\{U\}
$$

and

$$
\beta_{i_0 \wedge i_0+1} = \beta_{\tilde{r}} = \min\{\beta_{i \wedge i+1} : i = 1, \ldots, n-1\}.
$$

In fact, take $i_0 = \max\{i : i \in (\tilde{r})^-\}$ and $j_0 = \min\{j : j \in (\tilde{r})^+\}$. It is clear that $i_0 < j_0$ and there cannot be an element between them, which implies that $j_0 = i_0 + 1$. This situation takes place at all levels of the tree which gives the NBF property of U. We denote by $\tilde{\alpha}_i = U_{i,i+1}$ and $\tilde{\beta}_i = U_{i+1,i}$ for $i = 1, \cdots, n-1$.

The following algorithm gives also a dyadic chain of partitions associated with the NBF matrix U. Start from the trivial partition $\mathfrak{N}$ at step $k = 0$ and stop at the discrete partition $\mathfrak{F}$. At an intermediate step k, for each $J \in \mathfrak{R}_k$ which is not a singleton we define $\tilde{J} = J \setminus \{\bar{i}\}$, where $\bar{i} = \max\{i : i \in J\}$.

From the above arguments there exists $i_0 = i(J) \in J$ satisfying

$$
\begin{aligned}
&i_0 + 1 \in J \\
&\tilde{\alpha}_{i_0} = \inf\{\tilde{\alpha}_j : j \in \tilde{J}\} \\
&\tilde{\beta}_{i_0} = \inf\{\tilde{\beta}_j : j \in \tilde{J}\}
\end{aligned}
$$

We put $J^- = \{i \in J : i \le i_0\}$ and $J^+ = \{i \in J : i > i_0\}$ and define the new partition $\mathfrak{R}_{k+1}$ accordingly. Notice that $U_{ij} = \tilde{\alpha}_{i(J)}$ for $(i, j) \in J^- \times J^+$ and $U_{ij} = \tilde{\beta}_{i(J)}$ for $(i, j) \in J^+ \times J^-$. The dyadic chain of partitions $\mathbf{F}$ constructed by the algorithm is said to be associated with U. If T is the tree associated with this chain of partitions and we put $\alpha_J = \tilde{\alpha}_{i(J)}, \beta_J = \tilde{\beta}_{i(J)}$, then (T, α, β) supports U.

If U is in NBF the chain of partitions $\mathbf{F}$ is not necessarily unique. In fact, at some step k there could exist several $i(J)$ satisfying the same set of conditions. On the other hand, an argument based on induction on the dimension of the matrix proves that the fact that the algorithm gives a suitable chain of partitions does not depend on the choice of $i(J)$. This means that all possible selections at every level are admissible. In Theorem 4.27 we explicit all the chain of partitions associated with a fixed NBF.

We come back to the framework U a GUM supported by the dyadic tree (T, α, β) (for its definition see Proposition 4.13). It follows from the definition of the total order $\leq_T$, that any $J \in T$ is a $\leq_T$-interval. We introduce now a special class of partitions of J and also a special class of permutations.

Definition 4.26 A partition $\mathfrak{R}_J$ of $J \in T$ is said to be a T-**partition** if any atom $H \in \mathfrak{R}_J$ is a node of T. A permutation $\varphi_J : J \rightarrow J$ is said to be a $\mathfrak{R}_J$-**interval exchange** if for any $K \in \mathfrak{R}_J$ the image set $\varphi_J(K)$ is an interval of J and the restriction $\varphi_J|_K$ is an increasing function with respect to the order $\leq_T$. We will also denote by φ_J the extension of this permutation to I, where we put $\varphi_J(i) = i$ if $i \notin J$.

In what follows for any $L \in T$, $L \neq I$ denote by $\mathscr{A}(L)$ the immediate ancestor of L in the tree. Now, fix $J \in T$, and introduce the following T-partition $\mathfrak{R}_J^{\alpha,\beta}$ of J. If J is a leaf we put $\mathfrak{R}_J^{\alpha,\beta} = \{J\}$. When J is not a leaf we consider the partition of J whose atoms, all elements of T, are those $K \succ J$ satisfying

$$[(\alpha_{\mathscr{A}(K)}, \beta_{\mathscr{A}(K)}) = (\alpha_J, \beta_J)] \text{ and } [K \text{ is a leaf or } (\alpha_K, \beta_K) \neq (\alpha_J, \beta_J)].$$

It is simple to prove that this is a T-partition of J.

We say a chain of partitions $\tilde{\mathbf{F}}_J = \{\tilde{\mathfrak{R}}_0 \prec \ldots \prec \tilde{\mathfrak{R}}_q\}$, on J, is **compatible** with the partition $\mathfrak{R}_J^{\alpha,\beta}$ if $\tilde{\mathfrak{R}}_0 = \{J\}$ is the trivial one, $\tilde{\mathfrak{R}}_q = \mathfrak{R}_J^{\alpha,\beta}$ and the atoms of any of the partitions $\tilde{\mathfrak{R}}_k$ are union of consecutive atoms of $\mathfrak{R}_J^{\alpha,\beta}$ (consecutive with respect to the order $\leq_T$ among intervals, in particular they are also $\leq_T$-intervals). We also assume that $\tilde{\mathbf{F}}_J$ is dyadic in the sense that every non trivial atom of one partition is divided in two atoms of the next partition. Here the trivial atoms are the ones of the finer partition $\mathfrak{R}_J^{\alpha,\beta}$, which need not to be singletons (see the example and figure below).

Consider the subtree $T_J^{\alpha,\beta}$ of T whose vertex set is given by

$$\{K \succeq J : (\alpha_K, \beta_K) = (\alpha_J, \beta_J)\} \cup \mathfrak{R}_J^{\alpha,\beta}.$$

$T_J^{\alpha,\beta}$ is a dyadic tree whose root is J and $\mathfrak{R}_J^{\alpha,\beta}$ is its set of leaves. We denote by $\mathbf{F}_J^{\alpha,\beta}$ the associated chain of partitions, which is clearly compatible with $\mathfrak{R}_J^{\alpha,\beta}$.

Continuation of Example 4.17 In Fig. 4.2 below, we display the dyadic chain of partitions $\mathscr{F}_I^{\alpha,\beta}$ for the matrix considered in Example 4.17. The chain of partitions $\mathscr{F}_I^{\alpha,\beta}$ is given by the sequence of refining partitions

$$\{I\}, \{J, K\}, \{L, \{3\}, K\}, \{\{1\}, \{2\}, \{3\}, K\}.$$

Recall that K is identified with $\{4, 5, 6\}$, which is the set of leaves that hang from K. Similar considerations are made for I, J, L. In this example we have $\mathfrak{R}_I^{\alpha,\beta} = \{\{1\}, \{2\}, \{3\}, K\}$ and $\mathscr{F}_I^{\alpha,\beta}$ is compatible with $\mathfrak{R}_I^{\alpha,\beta}$. In the same figure we show another chain of partitions $\tilde{\mathscr{F}}$ which is $\mathfrak{R}_I^{\alpha,\beta}$-compatible. The sequence of partitions for this new chain of partitions is $\{I\}, \{L, \{3\} \cup K\}, \{\{1\}, \{2\}, \{3\}, K\}$.

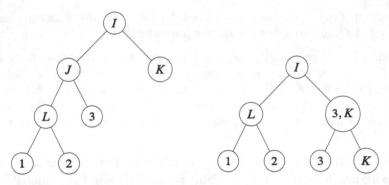

Fig. 4.2 On the left a chain of partitions $_I^{\alpha,\beta}$ and a compatible chain of partitions $\tilde{\mathscr{F}}$ on the right

An application of the Theorem 4.27 (below) to this example shows that the set of permutations which preserve the NBF are those which fix the points $\{1, 2, 3\}$.

The following result studies the permutations compatible with the NBF.

Theorem 4.27 *Let* U *be a NBF matrix on the set* I. *Consider* $\varphi : I \to I$ *a permutation and define* $U^\varphi = (U_{\varphi(i)\varphi(j)} : i, j \in I)$. *Then*

(i) U^φ *is in NBF if and only if* φ *is a composition of permutations* φ_J, *where* φ_J *is a* $\mathfrak{R}_J^{\alpha,\beta}$*-exchange of intervals and* J *satisfies* $\alpha_J = \beta_J$;

(ii) *Let* $\mathbf{F}$ *be a fixed dyadic chain of partitions associated with* U. *Then the class of total dyadic chains of partitions associated with the NBF, is constructed by making for* $J \in T$ *all possible replacements of* $\mathbf{F}_J^{\alpha,\beta}$ *with a chain of partitions* $\tilde{\mathbf{F}}_J$ *compatible with* $\mathfrak{R}_J^{\alpha,\beta}$.

Proof The proof is straightforward. Notice that in statements $(i), (ii)$ of the Theorem it suffices to consider $J = I$ and those J satisfying $(\alpha_{\mathscr{A}(J)}, \beta_{\mathscr{A}(J)}) \neq (\alpha_J, \beta_J)$, because they generate dyadic maximal subtrees $T_J^{\alpha,\beta}$.

We finish this section by proving that the sets $\mathscr{H}_r : r \in \mathscr{R}(U)$, defined in (4.8), are intervals. For that purpose take $i \neq j \in \mathscr{H}_r$ such that $i \leq_T j$ and consider any k different from i, j such that $i \leq_T k \leq_T j$. We consider two possible cases.

Case: $k \leq_T r$. In this situation we have $i \leq_T k \leq_T r$ and therefore we obtain for some $s, t, u \in T: i \in t^-, k \in t^+, i \in s^-, r \in s^+$ and $k \in u^-, r \in u^+$. Necessarily we have $i \wedge r \preccurlyeq i \wedge k$ and $i \wedge r \preccurlyeq k \wedge r$ and then

$$U_{rr} = U_{ir} \leq U_{kr} \leq U_{rr},$$

proving that $U_{kr} = U_{rr}$.

Case: $r \leq_T k$. In this situation consider $r \leq_T k \leq_T j$ and conclude as before because $r \wedge j \preccurlyeq k \wedge r$ and then

$$U_{rr} = U_{jr} \leq U_{kr} \leq U_{rr}.$$

We notice that the sets $\mathscr{H}_r$ are not necessarily elements of T. For instance take $I = \{1, 2, 3\}$, and consider the GUM

$$U = \begin{pmatrix} \delta & \gamma & \alpha \\ \beta & \beta & \alpha \\ \beta & \beta & \delta \end{pmatrix},$$

where $\delta > \beta > \gamma > \alpha > 0$. It is easy to see that $T = \{\{1, 2, 3\}, \{1, 2\}, \{1\}, \{2\}, \{3\}\}$ and that $\leq_T$ is the usual order in I. We have $\mathscr{R}(U) = \{1, 2\}$ and $\mathscr{H}_2 = \{2, 3\} \notin T$.

Chapter 5
Filtered Matrices

In this chapter we introduce certain notions that are frequent in probability theory and stochastic analysis, which prove fruitful in describing inverse M-matrices. A main concept is mean conditional expectation matrix. This type matrices is the linear version of partitions and they satisfy the complete maximum principle, so they are natural objects in theory of inverse M-matrices. Additionally, they are also projections that preserve positivity and the constant vectors.

We study a broad class of inverse M-matrices that can be decomposed as a sum of a nondecreasing family of conditional expectations, much in the spirit of a spectral decomposition. This class of matrices is called weakly filtered. The coefficients of these decompositions may be diagonal matrices, instead of just real numbers. Unlike the spectral case, these decompositions are not unique, which introduces an additional difficulty. In Propositions 5.20 and 5.21, we provide canonical versions of these decompositions.

Ultrametric matrices, as well as GUM are weakly filtered matrices (see Proposition 5.17). A main objective of this chapter is to characterize when a weakly filtered matrix is a potential. This is completely answered with a backward algorithm that determines when a weakly filtered matrix is an inverse M-matrix and when this is the case it also provides a decomposition of the inverse (see Sect. 5.3.1). It is difficult to provide explicit treatable conditions on the coefficients of a weakly filtered matrix that approves this algorithm. Nevertheless, we are able to supply two sets of conditions leading to inverse M-matrices (see Theorem 5.27) containing as a special case the GUM class.

We conclude the chapter by reviewing the class of matrices whose spectral decompositions are given in terms of a filtration and relate this concept to Dirichlet-Markov matrices, see Theorem 5.35.

© Springer International Publishing Switzerland 2014
C. Dellacherie et al., *Inverse M-Matrices and Ultrametric Matrices*, Lecture Notes in Mathematics 2118, DOI 10.1007/978-3-319-10298-6_5

5.1 Conditional Expectations

Consider $I = \{1, \cdots, n\}$ and the usual Euclidean structure on $\mathbb{R}^n$. In particular, we denote $\langle\, ,\, \rangle$ the standard inner product on $\mathbb{R}^n$. Recall that a projection is an idempotent symmetric matrix and given two projections P, Q, it is said that Q is finer than P, which is denoted by $P \leq Q$, if they commute and

$$PQ = QP = P.$$

We now introduce a special class of projections intimately related to CBF matrices and use them to extend the notion of GUM.

Definition 5.1 A **conditional expectation** $\mathbb{E}$ is a stochastic projection matrix, that is

$$\mathbb{E} \geq 0, \ \mathbb{E}\mathbb{1} = \mathbb{1}, \ \mathbb{E}' = \mathbb{E}, \ \mathbb{E}^2 = \mathbb{E}.$$

The first example of a conditional expectation is the identity matrix $\mathbb{I}$. The following is another natural example: the *mean value matrix* of size $n \times n$, denoted by $\mathbb{M}_n$, is

$$\mathbb{M}_n = \frac{1}{n}\mathbb{1}\mathbb{1}' = \frac{1}{n}\begin{pmatrix} 1 & 1 & \cdots & 1 \\ 1 & 1 & \cdots & 1 \\ \vdots & \vdots & \ddots & \vdots \\ 1 & 1 & \cdots & 1 \end{pmatrix}.$$

Here is another example

$$\mathbb{E} = \begin{pmatrix} \mathbb{M}_3 & 0 \\ 0 & \mathbb{M}_2 \end{pmatrix} = \begin{pmatrix} 1/3 & 1/3 & 1/3 & 0 & 0 \\ 1/3 & 1/3 & 1/3 & 0 & 0 \\ 1/3 & 1/3 & 1/3 & 0 & 0 \\ 0 & 0 & 0 & 1/2 & 1/2 \\ 0 & 0 & 0 & 1/2 & 1/2 \end{pmatrix}. \tag{5.1}$$

This is not a simple example. Indeed, in Lemma 5.6 we will show that after a permutation of rows and columns every conditional expectation $\mathbb{E}$ has the block form

$$\mathbb{E} = \begin{pmatrix} \mathbb{M}_{p_1} & 0 & \cdots & 0 & 0 \\ 0 & \mathbb{M}_{p_2} & \cdots & 0 & 0 \\ \vdots & \vdots & \ddots & \vdots & \vdots \\ 0 & 0 & \cdots & \mathbb{M}_{p_{m-1}} & 0 \\ 0 & 0 & \cdots & 0 & \mathbb{M}_{p_m} \end{pmatrix} \tag{5.2}$$

Remark 5.2 The following observation is elementary, nevertheless it is one of the cornerstones of our development. Consider a conditional expectation $\mathbb{E}$ and $t > 0$, then

$$(\mathbb{E} + t\mathbb{I})^{-1} = \frac{1}{t(1+t)} \left((1+t)\mathbb{I} - \mathbb{E} \right), \tag{5.3}$$

is an M-matrix. This computation will be the basis of our presentation and will show that conditional expectations provide a good and rich framework to develop part of the theory of inverse M-matrices. We also observe that every conditional expectation is a potential, in the large sense of Chap. 2, that is they satisfy the complete maximum principle. However, except for the identity, they are all singular. This explains the need to use a perturbation like $\mathbb{E} + t\mathbb{I}$.

It is natural to see conditional expectations as operators. For that reason it is useful to identify $\mathbb{R}^n$ with $L^2(I, \mathscr{P}(I), \#)$ where $\#$ is the counting measure on I. Here a vector x is interpreted as a function on I and all standard algebraic operations with functions can be used for vectors. En particular $f(x)$ is the vector whose components are the ones transformed by f. Of particular interest are $x^2, |x|, (x)_+$. For $a \in \mathbb{R}$ we interpret $x + a$ as $x + a\mathbb{1}$. Also $x \vee y$ is the pointwise maximum for the vectors x, y. The usual product of functions corresponds to component multiplication of vectors or Hadamard product denoted by $\odot$. We will avoid using this notation unless it is strictly necessary and will simply put xy instead of $x \odot y$. When y has no zero components we denote by x/y instead of $x \odot \frac{1}{y}$. The vector space $\mathbb{R}^n$ with the Hadamard product is an algebra, and with standard order is a lattice.

From the context there will be no possible confusion of this product xy with $x'y = \langle x, y \rangle$ or with the matrix xy'. Given a vector $x \in \mathbb{R}^n$ we denote by D_x the diagonal matrix naturally associated with it and given a square matrix A we associate the vector $\mathrm{diag}(A)$ that corresponds to the diagonal of A. Then the following relations are simple to prove

$$D_x y = D_y x = xy;$$
$$D_x \mathbb{1} = x;$$
$$D_x D_y = D_y D_x = D_{xy};$$
$$(\forall i \ x_i \neq 0) \Rightarrow (D_x)^{-1} = D_{x^{-1}}, \quad \text{where } (x^{-1})_i = 1/x_i.$$

Remark 5.3 We shall write $\mathbb{E}(x)$ instead of $\mathbb{E}x$ which is the standard notation in matrix theory. This is important when we consider the following more general setting.

A measure μ on $(I, \mathscr{P}(I))$ is identified with the vector of its weights $\mu_i = \mu(\{i\})$, which we assume are positive. We define the inner product $\langle x, y \rangle_\mu = \sum_i x_i y_i \mu_i$ and identify $(\mathbb{R}^n, \langle \, , \, \rangle_\mu)$ with the space $L^2(I, \mathscr{P}(I), \mu)$. When μ is the counting measure, the associated inner product is the standard one. In the enlarged context, a conditional expectation $\mathbb{E}^\mu : \mathbb{R}^n \to \mathbb{R}^n$ is a linear operator such that $\mathbb{E}^\mu(\mathbb{1}) = \mathbb{1}$ and

$$\forall x \geq 0 \;\; \mathbb{E}^\mu(x) \geq 0;$$
$$\forall x, y \;\;\; \langle \mathbb{E}^\mu(x), y \rangle_\mu = \langle x, \mathbb{E}^\mu(y) \rangle_\mu;$$
$$\forall x \;\;\;\;\;\; \mathbb{E}^\mu(\mathbb{E}^\mu(x)) = \mathbb{E}^\mu(x).$$

It can be proved that $\mathbb{E}^\mu = \Lambda\, D_\mu$, for some symmetric matrix Λ. As an example we have the general mean value matrix

$$\frac{1}{\bar{\mu}} \begin{pmatrix} \mu_1 & \mu_2 & \cdots & \mu_n \\ \mu_1 & \mu_2 & \cdots & \mu_n \\ \vdots & \vdots & \cdots & \vdots \\ \mu_1 & \mu_2 & \cdots & \mu_n \end{pmatrix} = \frac{n}{\bar{\mu}}\, \mathbb{M}_n\, D_\mu.$$

where $\bar{\mu} = \mathbb{1}'\mu$ is the total mass of μ. Here is a 5×5 example

$$\mathbb{E}^\mu = \begin{pmatrix} \frac{\mu_1}{\mu_1+\mu_2+\mu_3} & \frac{\mu_2}{\mu_1+\mu_2+\mu_3} & \frac{\mu_3}{\mu_1+\mu_2+\mu_3} & 0 & 0 \\ \frac{\mu_1}{\mu_1+\mu_2+\mu_3} & \frac{\mu_2}{\mu_1+\mu_2+\mu_3} & \frac{\mu_3}{\mu_1+\mu_2+\mu_3} & 0 & 0 \\ \frac{\mu_1}{\mu_1+\mu_2+\mu_3} & \frac{\mu_2}{\mu_1+\mu_2+\mu_3} & \frac{\mu_3}{\mu_1+\mu_2+\mu_3} & 0 & 0 \\ 0 & 0 & 0 & \frac{\mu_4}{\mu_4+\mu_5} & \frac{\mu_5}{\mu_4+\mu_5} \\ 0 & 0 & 0 & \frac{\mu_4}{\mu_4+\mu_5} & \frac{\mu_5}{\mu_4+\mu_5} \end{pmatrix}.$$

Let us continue the presentation in the standard setting. To each partition $\mathfrak{R} = \{R_1, \cdots, R_m\}$ we associate, in a natural way, the conditional expectation $\mathbb{E}$ given by

$$\mathbb{E}_{ij} = \begin{cases} (\#(R_p))^{-1} & \text{if } i, j \in R_p \\ 0 & \text{otherwise} \end{cases}. \tag{5.4}$$

This formula defines a unique conditional expectation operator. It can be shown that after a suitable permutation of rows and columns, $\mathbb{E}$ has the block form given in (5.2) where $p_s = \#(R_s)$.

We point out that $\mathbb{M}_n$ is the conditional expectation associated with the trivial partition $\mathfrak{N}$ and that the identity $\mathbb{I}$ is the one associated with the finest partition $\mathfrak{F}$. In example (5.1) the partition is $\mathfrak{R} = \{\{1, 2, 3\}, \{4, 5\}\}$.

It is well known that partitions and equivalence relations on I are the same concepts. Indeed, given $i, j \in I$ the relation

$$i \overset{\mathfrak{R}}{\sim} j \;\Leftrightarrow\; i, j \text{ belong to the same atom in } \mathfrak{R},$$

is the equivalence relation associated with the partition $\mathfrak{R}$.

Lemma 5.4 *Let $\mathbb{E}$ be a conditional expectation then*

(i) *if $x \leq \mathbb{E}(x)$ then $x = \mathbb{E}(x)$. Similarly, if $x \geq \mathbb{E}(x)$ then $x = \mathbb{E}(x)$;*

(ii) *if $x \geq 0$ and $\mathbb{E}x = 0$ then $x = 0$. Furthermore, if $x > 0$ then $\mathbb{E}x > 0$;*

(iii) *if ϕ is a convex function then $\mathbb{E}(\phi(x)) \geq \phi(\mathbb{E}x)$. In particular $\mathbb{E}(x^2) \geq (\mathbb{E}(x))^2$;*

(iv) *the image $\mathrm{Im}(\mathbb{E}) = \{\mathbb{E}(x) : x \in \mathbb{R}^n\}$ of $\mathbb{E}$ is a lattice algebra; moreover if $f : \mathbb{R} \to \mathbb{R}$ is any function then for all x we consider $f(\mathbb{E}(x)) = \mathbb{E}(f(\mathbb{E}(x)))$, this means that if $v \in \mathrm{Im}(\mathbb{E})$ then $f(v) \in \mathrm{Im}(\mathbb{E})$.*

(v) *$\forall x, y \; \mathbb{E}(x\mathbb{E}(y)) = \mathbb{E}(x)\mathbb{E}(y)$;*

(vi) *let $\mathbb{E} \leq \tilde{\mathbb{E}}$ be two conditional expectations and $a \in \mathbb{R}^n$. Then,*

$$\tilde{\mathbb{E}} D_a \mathbb{E} = D_{\tilde{\mathbb{E}}(a)} \mathbb{E}$$

Remark 5.5 Properties $(iv), (v), (vi)$ are very special of conditional expectations and will play a fundamental role in some computations we will do below.

Proof (i) From $\langle \mathbb{E}(x), \mathbb{1} \rangle = \langle x, \mathbb{E}(\mathbb{1}) \rangle = \langle x, \mathbb{1} \rangle$, we deduce that if $x \leq \mathbb{E}(x)$ then $x = \mathbb{E}(x)$. The case $x \leq \mathbb{E}(x)$ follows similarly.

(ii) In particular if $x \geq 0 = \mathbb{E}(x)$ we deduce that $x = \mathbb{E}(x) = 0$. The rest follows immediately by taking $x \geq \epsilon \mathbb{1}$ for some positive ϵ. Indeed, since $\mathbb{E} \geq 0$ it is monotone, that is $\mathbb{E}(x) \geq \epsilon \mathbb{E}(\mathbb{1}) = \epsilon \mathbb{1} > 0$

(iii) Since $\mathbb{E}$ is monotone, we have

$$\mathbb{E}(x) \vee \mathbb{E}(y) \leq \mathbb{E}(x \vee y).$$

Then for all $a, b, c, d \in \mathbb{R}$ the following inequality holds

$$(a\mathbb{E}(x) + b) \vee (c\mathbb{E}(x) + d) \leq \mathbb{E}((ax + b) \vee (cx + d)).$$

In particular if $a = 1, b = c = d = 0$ we get that $(\mathbb{E}(x))_+ \leq \mathbb{E}((x)_+)$, which is the desired property for the convex function $\phi(\bullet) = (\bullet)_+$. For a general convex function $\phi : \mathbb{R} \to \mathbb{R}$ we use the representation

$$\phi(r) = \sup_{n \in \mathbb{N}} (a_n r + b_n),$$

for a suitable countable family of real numbers $\{a_m, b_m : m \in \mathbb{N}\}$. Then, we have

$$a_r \mathbb{E}(x) + b_r \leq \mathbb{E}\left(\sup_{m \in \mathbb{N}} (a_m x + b_m) \right),$$

and deduce

$$\phi(\mathbb{E}(x)) \leq \mathbb{E}(\phi(x)).$$

(iv) We use the inequality $(\mathbb{E}(u))^2 \leq \mathbb{E}(u^2)$ for $u = \mathbb{E}(x)$ to deduce

$$(\mathbb{E}(x))^2 = (\mathbb{E}(\mathbb{E}(x)))^2 \leq \mathbb{E}((\mathbb{E}(x))^2),$$

which by (i) implies

$$(\mathbb{E}(x))^2 = \mathbb{E}((\mathbb{E}(x))^2).$$

Replacing x by $x + y$ and canceling out the quadratic terms we get

$$\mathbb{E}(x)\mathbb{E}(y) = \mathbb{E}((\mathbb{E}(x)\mathbb{E}(y))).$$

This proves that $\mathrm{Im}(\mathbb{E})$ is an algebra. An inductive argument shows that for all $s \in \mathbb{N}$ it holds

$$(\mathbb{E}(x))^s = \mathbb{E}((\mathbb{E}(x))^s).$$

Consider now a function f and an interpolation polynomial p, of degree at most n (the size of the matrix $\mathbb{E}$), such that

$$p(\mathbb{E}(x)_i) = f(\mathbb{E}(x)_i)$$

for $i = 1, \cdots, n$. From $p(\mathbb{E}(x)) = \mathbb{E}(p(\mathbb{E}(x)))$ we conclude $f(\mathbb{E}(x)) = \mathbb{E}(f(\mathbb{E}(x)))$, which proves that $\mathrm{Im}(\mathbb{E})$ is invariant under pointwise evaluation of general functions. In particular

$$(\mathbb{E}(x))_+ = \mathbb{E}((\mathbb{E}(x))_+),$$

which allows us to conclude $\mathrm{Im}(\mathbb{E})$ is a lattice.

(v) The next equalities are a simple consequence of previous relations, for any u,

$$\langle \mathbb{E}(x\mathbb{E}(y)), u \rangle = \langle x\mathbb{E}(y)), \mathbb{E}(u) \rangle = \langle x, \mathbb{E}(u)\mathbb{E}(y) \rangle = \langle x, \mathbb{E}(\mathbb{E}(u)\mathbb{E}(y)) \rangle$$
$$= \langle \mathbb{E}(x), \mathbb{E}(u)\mathbb{E}(y) \rangle = \langle \mathbb{E}(x)\mathbb{E}(y), \mathbb{E}(u) \rangle = \langle \mathbb{E}(\mathbb{E}(x)\mathbb{E}(y)), u \rangle$$
$$= \langle \mathbb{E}(x)\mathbb{E}(y), u \rangle,$$

Then, we conclude the equality

$$\mathbb{E}(x\mathbb{E}(y)) = \mathbb{E}(x)\mathbb{E}(y).$$

(vi) For every $x \in \mathbb{R}^n$ we have

$$\tilde{\mathbb{E}}(D_a\mathbb{E}(x)) = \tilde{\mathbb{E}}(a\mathbb{E}(x)) = \mathbb{E}(x)\tilde{\mathbb{E}}(a).$$

The last equality follows from (v) and the fact that $\mathbb{E} \leq \tilde{\mathbb{E}}$. So, we obtain

$$\tilde{\mathbb{E}}(D_a \mathbb{E}(x)) = D_{\tilde{\mathbb{E}}(a)} \mathbb{E}(x),$$

and the result is proven.

In the following result we prove that every conditional expectation comes from a partition.

Lemma 5.6 *Associated with a $n \times n$ conditional expectation $\mathbb{E}$ there exists a unique partition $\mathfrak{R} = \{R_1, \cdots, R_p\}$ of I such that (5.4) holds. Moreover, the invariant subspace $\mathrm{Im}(\mathbb{E}) = \mathbb{E}(\mathbb{R}^n)$ is characterized as*

$$\mathrm{Im}(\mathbb{E}) = \{x \in \mathbb{R}^n : \forall i \overset{\mathfrak{R}}{\sim} j \ x_i = x_j\}, \tag{5.5}$$

the subspace of vectors that are constant in each atom of $\mathfrak{R}$. Given two conditional expectations $\mathbb{E}_1$ and $\mathbb{E}_2$ with associated partitions $\mathfrak{R}_1$ and $\mathfrak{R}_2$ then

$$\mathbb{E}_1 \leq \mathbb{E}_2 \Leftrightarrow \mathfrak{R}_1 \preccurlyeq \mathfrak{R}_2,$$

where on the left $\leq$ is the standard partial order on projections and on the right $\preccurlyeq$ is the standard partial order on partitions (see Definition 3.1)

Proof The following relation is associate with $\mathbb{E}$

$$i \overset{\mathfrak{R}}{\sim} j \Leftrightarrow \forall x \in \mathbb{R}^n : (\mathbb{E}x)_i = (\mathbb{E}x)_j.$$

We now prove $\mathrm{Im}(\mathbb{E}) = \{x \in \mathbb{R}^n : \forall i \overset{\mathfrak{R}}{\sim} j \ x_i = x_j\}$. It is clear $\overset{\mathfrak{R}}{\sim}$ is an equivalence relation characterized by

$$i \overset{\mathfrak{R}}{\sim} j \Leftrightarrow \forall \ell = 1, \cdots, n \ (\mathbb{E}(e(\ell) + 1))_i = (\mathbb{E}(e(\ell) + 1))_j,$$

where $e(1), \cdots, e(n)$ is the canonical basis of $\mathbb{R}^n$. Since $\mathbb{E}(e(\ell) + 1) \geq 1$ the following vector is well defined for every $i \in I$

$$v(i) = H \left(\left| \frac{\mathbb{E}(e(1) + 1)}{(\mathbb{E}(e(1) + 1))_i} - 1 \right| + \cdots + \left| \frac{\mathbb{E}(e(n) + 1)}{(\mathbb{E}(e(n) + 1))_i} - 1 \right| \right), \tag{5.6}$$

where H is the function $H(0) = 1$ and $H(r) = 0$, for all $r > 0$.

From Lemma 5.4 this vector $v(i) \in \mathrm{Im}(\mathbb{E})$, that is $\mathbb{E}(v(i)) = v(i)$. Also it is clear that $v(i)_i = 1$ and

$$v(i)_j = \begin{cases} 1 & \text{if } j \overset{\mathfrak{R}}{\sim} i \\ 0 & \text{otherwise} \end{cases},$$

moreover $j \overset{\mathfrak{R}}{\sim} i$ iff $v(i) = v(j)$. We denote by $R_1, \cdots, R_m$ the equivalence classes of $\overset{\mathfrak{R}}{\sim}$. Choose $i_1 \in R_1, \cdots, i_m \in R_m$ and consider $\mathscr{A} = \{v(i_1), \cdots, v(i_m)\}$. It is straightforward to prove that $\mathscr{A} \subset \text{Im}(\mathbb{E})$ is a basis of $\{x \in \mathbb{R}^n : \forall i \overset{\mathfrak{R}}{\sim} j \ \ x_i = x_j\}$ proving this vector subspace is contained in $\text{Im}(\mathbb{E})$.

For the opposite inclusion, consider $z = \mathbb{E}(x)$, then it is clear that $i \overset{\mathfrak{R}}{\sim} j$ implies $z(i) = z(j)$ and then

$$\text{Im}(\mathbb{E}) \subseteq \{z \in \mathbb{R}^n : \forall i \overset{\mathfrak{R}}{\sim} j \ \ z_i = z_j\}.$$

Let now $\mathbb{E}_1 \leq \mathbb{E}_2$. Then, for all i we have $\mathbb{E}_2(v(i)) = v(i)$, where $v(i)$ is defined in (5.6) for $\mathbb{E} = \mathbb{E}_1$. Thus, if $i \overset{\mathfrak{R}_2}{\sim} j \Rightarrow i \overset{\mathfrak{R}_1}{\sim} j$, proving one implication.

For the opposite implication, we notice that if $\mathfrak{R}_1 \preccurlyeq \mathfrak{R}_2$ we get from (5.5) that $\text{Im}(\mathbb{E}_1) \subseteq \text{Im}(\mathbb{E}_2)$ and the lemma is proved.

Remark 5.7 According to the previous result, a vector x belongs to the $\text{Im}(\mathbb{E})$ if and only if x is constant on the atoms of $\mathbb{E}$. We shall say that x is $\mathbb{E}$-**measurable**.

Example 5.8 Consider the conditional expectation on $I = \{1, \cdots, 6\}$ given by

$$\mathbb{E} = \begin{pmatrix} M_3 & 0 & 0 \\ 0 & M_2 & 0 \\ 0 & 0 & M_1 \end{pmatrix}.$$

The partition associated with $\mathbb{E}$ is given by $\mathfrak{R} = \{\{1, 2, 3\}, \{4, 5\}, \{6\}\}$. A vector $x \in \mathbb{R}^6$ is $\mathbb{E}$-measurable if and only if it is of the form $x = (a, a, a, b, b, c)'$ for some $a, b, c \in \mathbb{R}$.

The following definition is important in probability theory and it is the natural counterpart of chain of partitions.

Definition 5.9 A **filtration** $\mathbf{F} = (\mathbb{E}_0, \mathbb{E}_1, \cdots, \mathbb{E}_m)$ is an increasing sequence of conditional expectations $\mathbb{E}_0 \leq \mathbb{E}_1 \leq \cdots \leq \mathbb{E}_m$. As in the case of chains of partitions, there could be repetitions of conditional expectations. For that reason a filtration is said to be a **strict filtration** if it is a strictly increasing sequence of conditional expectations, so $\mathbb{E}_0 < \mathbb{E}_1 < \cdots < \mathbb{E}_m$.

A filtration $\mathbf{F} = (\mathbb{E}_0, \cdots, \mathbb{E}_m)$ is **maximal** if it is strict and maximal with respect to the natural order of projections on the set of conditional expectations.

A filtration $\mathbf{F} = (\mathbb{E}_0, \cdots, \mathbb{E}_m)$ is **dyadic** if it strict and all the non trivial atoms in $\mathbb{E}_\ell$ split into two atoms of $\mathbb{E}_{\ell+1}$ at every level $\ell = 0, \cdots, m - 1$.

Notice that chain of partitions, as defined in Definition 3.1, Sect. 3.1, and a filtration of conditional expectations are equivalent concepts. From now on we will use them as synonymous.

A maximal filtration can be described in terms of atoms in the following way. The passage from $\mathbb{E}_\ell$ to $\mathbb{E}_{\ell+1}$ is obtained by splitting one and only one nontrivial atom of $\mathbb{E}_\ell$ in two atoms of $\mathbb{E}_{\ell+1}$.

The following example is the simplest dyadic filtration on $I = \{1, \cdots, n\}$

$$\mathfrak{N} = \{1, \cdots, n\} \cdots \prec \{\{1\}, \{2\}, \cdots, \{i\}, \{i+1, \cdots, n\}\} \prec \cdots \prec \mathfrak{F}.$$

This filtration is also maximal. Here is another example of dyadic filtration on $I = \{1, \cdots, 4\}$, that is not maximal

$$\mathfrak{N} \prec \{\{1, 2\}, \{3, 4\}\} \prec \mathfrak{F},$$

and a maximal on the same set, that is not dyadic

$$\mathfrak{N} \prec \{\{1, 2\}, \{3, 4\}\} \prec \{\{1, 2\}, \{3\}, \{4\}\} \prec \mathfrak{F}.$$

It is not hard to see that a maximal filtration on a set of n elements is always composed of n conditional expectations.

Given a filtration $\mathbf{F} = (\mathbb{E}_0, \mathbb{E}_1, \cdots, \mathbb{E}_m)$, the basic algebraic tool is the commutation rule

$$\mathbb{E}_s \mathbb{E}_t = \mathbb{E}_t \mathbb{E}_s = \mathbb{E}_{s \wedge t},$$

as it happens with any spectral family of projections. Nevertheless, a filtration satisfies very specific commutation properties among which we emphasize the following, for $s \leq t$ and any $a \in \mathbb{R}^n$

$$\mathbb{E}_t D_a \mathbb{E}_s = D_{\mathbb{E}_t(a)} \mathbb{E}_s.$$

This is a consequence of Lemma 5.4 (*vi*).

We show in the next result that a filtration, after a common permutation, can be put as a sequence of compatible block form matrices.

Lemma 5.10 *Assume that* $\mathbf{F} = (\mathbb{E}_0, \mathbb{E}_1, \cdots, \mathbb{E}_m)$ *is a filtration. Then, there exists a fixed permutation* Π *such that for all* $\ell = 0, \cdots, m$

$$\Pi \mathbb{E}_\ell \Pi' = \begin{pmatrix} \mathrm{M}_{p_1^\ell} & 0 & 0 & \cdots & 0 \\ 0 & \mathrm{M}_{p_2^\ell} & 0 & \cdots & 0 \\ \vdots & \vdots & \ddots & \cdots & \vdots \\ 0 & 0 & \cdots & 0 & \mathrm{M}_{p_{k_\ell}^\ell} \end{pmatrix}.$$

Proof The proof is done by induction on m. The case $m = 0$ is evident (see (5.4)) and we will give only some details for the case $m = 1$. So, take Π_0 a permutation that puts $\mathbb{E}_0$ in block form

$$\Pi_0 \mathbb{E}_0 \Pi_0' = \begin{pmatrix} \mathrm{M}_{p_1^0} & 0 & 0 & \cdots & 0 \\ 0 & \mathrm{M}_{p_2^0} & 0 & \cdots & 0 \\ \vdots & \vdots & \ddots & \cdots & \vdots \\ 0 & 0 & \cdots & 0 & \mathrm{M}_{p_{k_0}^0} \end{pmatrix}.$$

We denote by $\mathfrak{R}_0 = \{R_1, \cdots, R_k\}$ the partition associated with $\mathbb{E}_0$, where the atom R_1 corresponds to M_{p_1} and so on. Notice that Π_0 is not unique, in particular Π_0 is arbitrary inside each atom of $\mathfrak{R}_0$. Since $\mathbb{E}_1$ is finer than $\mathbb{E}_0$ we obtain that

$$\Pi_0 \mathbb{E}_0 \Pi_0' \Pi_0 \mathbb{E}_1 \Pi_0' = \Pi_0 \mathbb{E}_0 \Pi_0'.$$

Using the fact that $\Pi_0 \mathbb{E}_0 \Pi_0'$ and $\Pi_0 \mathbb{E}_1 \Pi_0'$ are nonnegative we deduce $\Pi_0 \mathbb{E}_1 \Pi_0'$ has block form

$$\Pi_0 \mathbb{E}_1 \Pi_0' = \begin{pmatrix} A_{p_1^0} & 0 & 0 & \cdots & 0 \\ 0 & A_{p_2^0} & 0 & \cdots & 0 \\ \vdots & \vdots & \ddots & \cdots & \vdots \\ 0 & 0 & \cdots & 0 & A_{p_{k_0}^0} \end{pmatrix}.$$

It is straightforward to show that each A_{p_t} is a conditional expectation on $\mathbb{R}^{p_t}$ and therefore it can be put on a block form, after a suitable permutation σ_t that acts on R_t. If we put together these permutations we can construct a permutation Π_1 such that

$$\Pi_1 \mathbb{E}_0 \Pi_1' = \begin{pmatrix} \mathrm{M}_{p_1^0} & 0 & 0 & \cdots & 0 \\ 0 & \mathrm{M}_{p_2^0} & 0 & \cdots & 0 \\ \vdots & \vdots & \ddots & \cdots & \vdots \\ 0 & 0 & \cdots & 0 & \mathrm{M}_{p_{k_0}^0} \end{pmatrix}.$$

and

$$\Pi_1 \mathbb{E}_1 \Pi_1' = \begin{pmatrix} \mathrm{M}_{p_1^1} & 0 & 0 & \cdots & 0 \\ 0 & \mathrm{M}_{p_2^1} & 0 & \cdots & 0 \\ \vdots & \vdots & \ddots & \cdots & \vdots \\ 0 & 0 & \cdots & 0 & \mathrm{M}_{p_{k_1}^1} \end{pmatrix}.$$

5.2   Filtered Matrices

In this section we relate ultrametricity and some special spectral decompositions based on a filtration. For that purpose the following concept is important.

Definition 5.11 A nonnegative matrix is said to be **strongly filtered** with respect to the filtration $\mathbf{F} = (\mathbb{E}_0, \mathbb{E}_1, \cdots, \mathbb{E}_m)$ if there exists a sequence of nonnegative real numbers $a_0, a_1, \cdots, a_m$ such that

$$U = \sum_{\ell=0}^{m} a_\ell \mathbb{E}_\ell.$$

Notice than we can always assume $\mathbf{F}$ is a maximal filtration.

The class of strongly filtered matrices with respect to a fixed filtration, is a cone closed under matrix multiplication, in particular the powers of a strongly filtered matrix are also strongly filtered.

In the sequel each conditional expectation is identified with the partition associated with it, and every filtration is identified with the chain of respective partitions.

Definition 5.12 Consider a conditional expectation $\mathbb{E}$ with associated equivalence relation $\mathfrak{R}$. The **incidence matrix** of $\mathbb{E}$, or $\mathfrak{R}$, is the matrix $\mathbb{F}$ defined as

$$\mathbb{F}_{ij} = \begin{cases} 1 & \text{if } i \overset{\mathfrak{R}}{\sim} j \\ 0 & \text{otherwise} \end{cases}.$$

Notice that $\mathbb{F}$ is the incidence matrix associated with the graph induced by $\mathfrak{R}$ and $\mathbb{F}_{ij} = 1$ if and only if $\mathbb{E}_{ij} > 0$.

We also consider N, the **counting vector** associated with $\mathbb{E}$ defined as

$$N = \mathbb{F}\mathbb{1},$$

that is, $N_i = \#\{j : i \overset{\mathfrak{R}}{\sim} j\}$. Since N is strictly positive, the vector $1/N$ is well defined and it corresponds to the diagonal of $\mathbb{E}$.

Let us introduce a product between vectors and matrices, which will be useful in our description. Given a vector $x \in \mathbb{R}^n$ and a matrix A of size $n \times m$, we denote by $x \cdot A$ the matrix of size $n \times m$ obtained as

$$x \cdot A = D_x A,$$

which is the standard product of D_x, the diagonal matrix form from x and A. This operation is different from $x'A$, which is the product of the row vector x' and the matrix A. We will also need the matrix product $A \cdot y = AD_y$, which is different from the action of the matrix A on the vector y given by Ay.

With this notation the relation between $\mathbb{E}, \mathbb{F}$ and N is given by

$$\mathbb{E} = D_{1/N}\mathbb{F} = \frac{1}{N} \cdot \mathbb{F}.$$

Example 5.13 Consider $\mathbb{E}$ given in (5.1). Hence

$$\mathbb{F} = \begin{pmatrix} 1\,1\,1\,0\,0 \\ 1\,1\,1\,0\,0 \\ 1\,1\,1\,0\,0 \\ 0\,0\,0\,1\,1 \\ 0\,0\,0\,1\,1 \end{pmatrix}, \quad N = \begin{pmatrix} 3 \\ 3 \\ 3 \\ 2 \\ 2 \end{pmatrix}, \quad \text{and } \mathbb{E} = \frac{1}{N} \cdot \mathbb{F} = \begin{pmatrix} 1/3\,1/3\,1/3\,\,0\,\,\,\,0 \\ 1/3\,1/3\,1/3\,\,0\,\,\,\,0 \\ 1/3\,1/3\,1/3\,\,0\,\,\,\,0 \\ 0\,\,\,\,0\,\,\,\,0\,\,1/2\,1/2 \\ 0\,\,\,\,0\,\,\,\,0\,\,1/2\,1/2 \end{pmatrix}.$$

The next definition plays an important role in extending the class of inverse M-matrices. The basic idea is to replace the constants $a_0, \cdots, a_m$, in the previous definition, by functions (that is, diagonal matrices) that satisfies the following condition.

Definition 5.14 A sequence of vectors $\mathfrak{a}_1, \cdots, \mathfrak{a}_m$ is said to be **adapted** to the filtration $\mathbf{F} = (\mathbb{E}_0, \mathbb{E}_1, \cdots, \mathbb{E}_m)$ if

$$\mathfrak{a}_i \in \mathrm{Im}(\mathbb{E}_i), \quad \text{or equivalently } \mathbb{E}_i \mathfrak{a}_i = \mathfrak{a}_i.$$

With the idea of adaptedness we can introduce the concept of filtered operator.

Definition 5.15 A nonnegative matrix U is said to be **filtered** with respect to the filtration $\mathbf{F} = (\mathbb{E}_0, \mathbb{E}_1, \cdots, \mathbb{E}_m)$ if there exists a sequence of nonnegative and adapted vectors $\mathfrak{a}_0, \cdots, \mathfrak{a}_m$ such that

$$U = \sum_{\ell=0}^{m} \mathfrak{a}_\ell \cdot \mathbb{E}_\ell. \tag{5.7}$$

We can assume without loss of generality that $\mathbf{F}$ is maximal. In particular $\mathbb{E}_0 = \mathbb{M}_n$ (n is the size of U) and $\mathbb{E}_m = \mathbb{I}$.

Example 5.16 Assume that $n = 3$, and consider the dyadic filtration, whose chain of partitions is

$$\mathfrak{N} = \{\{1, 2, 3\}\} \prec \{\{1, 2\}, \{3\}\} \prec \{\{1\}, \{2\}, \{3\}\}.$$

The filtration of conditional expectations is

$$\mathbb{E}_0 = \frac{1}{3} \begin{pmatrix} 1\,1\,1 \\ 1\,1\,1 \\ 1\,1\,1 \end{pmatrix} \prec \mathbb{E}_1 = \begin{pmatrix} \frac{1}{2}\,\frac{1}{2}\,0 \\ \frac{1}{2}\,\frac{1}{2}\,0 \\ 0\,\,0\,\,1 \end{pmatrix} \prec \mathbb{E}_2 = \begin{pmatrix} 1\,0\,0 \\ 0\,1\,0 \\ 0\,0\,1 \end{pmatrix}$$

The most general filtered matrix, using this filtration is the combination $U = \mathfrak{a}_0 \cdot \mathbb{E}_0 + \mathfrak{a}_1 \cdot \mathbb{E}_1 + \mathfrak{a}_2 \mathbb{E}_2$. Adaptedness of $(\mathfrak{a}_k)_k$ is equivalent to the following structure

$$\mathfrak{a}_0 = \begin{pmatrix} \alpha \\ \alpha \\ \alpha \end{pmatrix}, \quad \mathfrak{a}_1 = \begin{pmatrix} \beta \\ \beta \\ \chi \end{pmatrix}, \quad \mathfrak{a}_2 = \begin{pmatrix} \delta \\ \epsilon \\ \eta \end{pmatrix}.$$

Hence we get

$$U = \frac{1}{3} \begin{pmatrix} \alpha & 0 & 0 \\ 0 & \alpha & 0 \\ 0 & 0 & \alpha \end{pmatrix} \begin{pmatrix} 1 & 1 & 1 \\ 1 & 1 & 1 \\ 1 & 1 & 1 \end{pmatrix} + \begin{pmatrix} \beta & 0 & 0 \\ 0 & \beta & 0 \\ 0 & 0 & \chi \end{pmatrix} \begin{pmatrix} \frac{1}{2} & \frac{1}{2} & 0 \\ \frac{1}{2} & \frac{1}{2} & 0 \\ 0 & 0 & 1 \end{pmatrix} + \begin{pmatrix} \delta & 0 & 0 \\ 0 & \epsilon & 0 \\ 0 & 0 & \eta \end{pmatrix} \begin{pmatrix} 1 & 0 & 0 \\ 0 & 1 & 0 \\ 0 & 0 & 1 \end{pmatrix}$$

$$= \begin{pmatrix} \delta + \beta/2 + \alpha/3 & \beta/2 + \alpha/3 & \alpha/3 \\ \beta/2 + \alpha/3 & \epsilon + \beta/2 + \alpha/3 & \alpha/3 \\ \alpha/3 & \alpha/3 & \eta + \beta/2 + \alpha/3 \end{pmatrix}$$

If all the parameters $\alpha, \beta, \chi, \delta, \epsilon, \eta$ are nonnegative, we obtain that U is an ultrametric matrix.

Notice again that in this definition we can always assume $\mathbf{F}$ is a strict filtration. Also it is not difficult to show $\mathfrak{a} \cdot \mathbb{E} = \mathbb{E} \cdot \mathfrak{a}$, that is $D_{\mathfrak{a}} \mathbb{E} = \mathbb{E} D_{\mathfrak{a}}$, whenever $\mathfrak{a}$ is an element of $\mathrm{Im}(\mathbb{E})$. This shows that every filtered matrix is symmetric. However, the class of filtered matrices with respect to a fixed filtration, is not in general closed under matrix multiplication.

Proposition 5.17 *Assume that U is a matrix of size $n \geq 2$.*

(I) The following two conditions are equivalent

 (i) U is ultrametric;
(ii) U is filtered.

Under any of these equivalent conditions it holds

(iii) there exist a maximal filtration $\mathbf{F} = (\mathbb{E}_0 = \mathbb{M}_n, \cdots, \mathbb{E}_{n-1} = \mathbb{I})$ and a nondecreasing sequence of nonnegative numbers $0 \leq b_0 \leq \cdots \leq b_{n-2}$ such that

$$U = \sum_{\ell=0}^{n-2} b_\ell (\mathbb{E}_\ell - \mathbb{E}_{\ell+1}) + D_S,$$

where $S = U \mathbb{1}$ and $D_S = S \cdot \mathbb{I}$ is the diagonal matrix associated with S.

(II) The following are equivalent

 (iv) U is ultrametric and there exists a constant s such that $U \mathbb{1} = s \mathbb{1}$, that is, the row sums of U are constant;
 (v) U is strongly filtered.

Proof We begin with item (I). Let us prove by induction on n that $(i) \Rightarrow (ii)$. When $n = 2$, we have

$$U = \begin{pmatrix} a & \alpha \\ \alpha & b \end{pmatrix}.$$

We notice that U is ultrametric iff $\alpha \le a \wedge b$ and in this case we have

$$U = 2\alpha M_2 + \begin{pmatrix} a - \alpha & 0 \\ 0 & b - \alpha \end{pmatrix} \mathbb{I} = \mathfrak{a}_0 \cdot M_2 + \mathfrak{a}_1 \cdot \mathbb{I},$$

where $\mathfrak{a}_0 = 2\alpha \mathbb{1} \in \text{Im}(M_2)$, $\mathfrak{a}_1 = (a - \alpha, b - \alpha)' \in \text{Im}(\mathbb{I})$, thus U is filtered. Conversely, for $n = 2$, every filtered matrix can be written as

$$U = rM_2 + \begin{pmatrix} s \\ t \end{pmatrix} \cdot \mathbb{I},$$

with r, s, t nonnegative, which implies $U_{12} \le U_{11} \wedge U_{22}$ and U is ultrametric.

Consider now an ultrametric matrix U of size $n + 1$ and take $\alpha = \min\{U\}$. According to Proposition 3.4 there exists a permutation Π and integers $p \ge 1, q \ge 1, p + q = n + 1$ such that

$$\Pi U \Pi' = \begin{pmatrix} V_1 & \alpha \mathbb{1}_p \mathbb{1}'_q \\ \alpha \mathbb{1}_q \mathbb{1}'_p & V_2 \end{pmatrix} = \alpha \mathbb{E}_0 + \begin{pmatrix} V_1 - \alpha & 0 \\ 0 & V_2 - \alpha \end{pmatrix}$$

where $V_1 - \alpha$, $V_2 - \alpha$ are ultrametric matrices of their respective sizes. The rest of the proof follows by using a decomposition of these two ultrametric matrices as filtered matrices.

To show that $(ii) \Rightarrow (i)$ we also use induction. Assume that U is of size $n + 1$ and it is filtered

$$U = \mathfrak{a}_0 \cdot \mathbb{E}_0 + \sum_{\ell=1}^{m} \mathfrak{a}_\ell \cdot \mathbb{E}_\ell,$$

where $\mathbf{F} = (\mathbb{E}_0, \cdots, \mathbb{E}_m)$ can be assumed strict and $\mathbb{E}_0 = M_{n+1}$. Since $\mathfrak{a}_0, \cdots, \mathfrak{a}_m$ are nonnegative we have

$$\alpha = \frac{\mathfrak{a}_0}{n + 1},$$

is the minimum of U. Consider the equivalence relation $\mathfrak{R}$ associated with $\mathbb{E}_1$. Since the filtration is strict we have $\mathbb{E}_0 \ne \mathbb{E}_1$ and therefore for every $\neg(i \overset{\mathfrak{R}}{\sim} j)$ we have $U_{ij} = \alpha$. Take any atom R_1 of this relation, then

$$U_{R_1 R_1^c} = \alpha \mathbb{1}_p \mathbb{1}'_q$$

where $1 \leq p = \#(R_1) \leq n$ and $q = n+1-p$. Then, again there exists a permutation $\tilde{\Pi}$ such that

$$\tilde{\Pi} U \tilde{\Pi}' = \begin{pmatrix} \tilde{V}_1 & \alpha \mathbb{1}_p \mathbb{1}'_q \\ \alpha \mathbb{1}_q \mathbb{1}'_p & \tilde{V}_2 \end{pmatrix}$$

where $\tilde{V}_1, \tilde{V}_2$ are filtered of the respective sizes. The result follows as before by induction.

To finish item (I) it is enough to prove, for example, that $(i) \Rightarrow (iii)$. As usual let U be indexed by $I = \{1, \cdots, n\}$. We show how to construct a filtration such that (iii) holds. As initial partition we consider $\mathfrak{R}^0 = \{I\}$ with only one atom $A_1^0 = I$, whose size is $p_1^0 = n$. We also denote by $U(0) = U$.

From Proposition 3.4 there is a partition of I in two sets K, L such that $U_{KL} = \alpha = \min\{U\}$. We consider $p = \#K, q = n-p = \#L, \mathbb{E}_0 = \mathbb{M}_n$ and the conditional expectation $\mathbb{E}_1$ given by the blocks

$$(\mathbb{E}_1)_{KK} = \mathbb{M}_p, (\mathbb{E}_1)_{LL} = \mathbb{M}_q, (\mathbb{E}_1)_{KL} = 0, (\mathbb{E}_1)_{LK} = 0.$$

Take $b_0 = n\alpha$ and define $U(1) = U - b_0(\mathbb{E}_0 - \mathbb{E}_1)$, which is an ultrametric matrix that satisfies

$$U(1)_{KK} = U_{KK}+\alpha(n/p-1), U(1)_{LL} = U_{LL}+\alpha(n/q-1), U(1)_{KL} = 0, U(1)_{LK} = 0.$$

This defines a partition $\mathfrak{R}^1 = \{K, L\}$ with two atoms $A_1^1 = K, A_2^1 = L$ and its respective sizes $p_1^1 = p, p_2^1 = q$. Now, we explain how to continue with a decomposition of U. Assume we have already constructed $\mathbb{E}_0 < \mathbb{E}_1 < \cdots < \mathbb{E}_k$, associated with a sequence of refining partitions $\mathfrak{R}^m = \{A_1^m, \cdots, A_{m+1}^m\}$, for $m = 0, \cdots, k$, whose atoms have sizes $1 \leq \#A_1^m = p_1^m, \cdots, 1 \leq \#A_{m+1}^m = p_{m+1}^m$ such that for $1 \leq s, t \leq m + 1$

$$(\mathbb{E}_m)_{A_s^m A_t^m} = \begin{cases} \mathbb{M}_{p_s^m} & \text{if } s = t \\ 0 & \text{otherwise} \end{cases}.$$

We also assume constructed a sequence of constants $0 \leq b_0 \leq \cdots \leq b_{k-1}$ such that

$$U(m) = U - \sum_{\ell=0}^{m-1} b_\ell(\mathbb{E}_\ell - \mathbb{E}_{\ell+1}) \geq 0$$

satisfies for $1 \leq s \neq t \leq m + 1$

$$U(m)_{A_s^m A_t^m} = 0.$$

Each block $U(k)_{A_s^k A_s^k}$ is ultrametric for $s = 1, \cdots, k + 1$. For those blocks of size greater than 1 we can find, according to Proposition 3.4, a partition into K_s, L_s such that

$$K_s \cup L_s = A_s^k, \ U(k)_{K_s L_s} = \alpha_s^k \mathbb{1}_{K_s} \mathbb{1}'_{L_s}, \ \text{ with } \ \alpha_s^k = \min\{U(k)_{A_s^k A_s^k}\}.$$

We also include in the induction hypothesis that (which is valid for $k = 1$)

$$b_{k-1} \le \min\{\alpha_s^k p_s^k : p_s^k > 1\}. \tag{5.8}$$

Now, we show how to choose $\mathbb{E}_{k+1}, \mathfrak{R}^{k+1}, b_k$ satisfying the previous requirements. Take for that purpose any s^* realizing the minimum in (5.8) and define the partition $\mathfrak{R}^{k+1}$ obtained from $\mathfrak{R}^k$ by breaking the atom $A_{s^*}^k$ into the disjoint sets K_{s^*}, L_{s^*}, whose cardinals we denote by p^*, q^*. The corresponding conditional expectation $\mathbb{E}_{k+1}$ has the same blocks as $\mathbb{E}_k$ except that the block $(\mathbb{E}_k)_{A_{s^*}^k A_{s^*}^k}$ is now given by

$$(\mathbb{E}_{k+1})_{K_{s^*} K_{s^*}} = \mathbb{M}_{p^*}, (\mathbb{E}_{k+1})_{L_{s^*} L_{s^*}} = \mathbb{M}_{q^*}, (\mathbb{E}_{k+1})_{K_{s^*} L_{s^*}} = 0, (\mathbb{E}_{k+1})_{L_{s^*} K_{s^*}} = 0.$$

We also define $b_k = \alpha_{s^*}^k p_{s^*}^k$, which by definition is at least b_{k-1}. The ultrametric matrix $U(k + 1) = U(k) - b_k(\mathbb{E}_k - \mathbb{E}_{k+1})$ differs from $U(k)$ only at the block $A_{s^*}^k$ and it is given by

$$\begin{aligned}
U(k + 1)_{K_{s^*} K_{s^*}} &= (U(k))_{K_{s^*} K_{s^*}} + \alpha_{s^*}^k (p_{s^*}^k / p^* - 1) \\
U(k + 1)_{L_{s^*} L_{s^*}} &= (U(k))_{L_{s^*} L_{s^*}} + \alpha_{s^*}^k (p_{s^*}^k / q^* - 1) \\
U(k + 1)_{K_{s^*} L_{s^*}} &= 0 \\
U(k + 1)_{L_{s^*} K_{s^*}} &= 0.
\end{aligned}$$

We have that

$$\begin{aligned}
\min\{U(k + 1)_{K_{s^*} K_{s^*}}\} &= \min\{U(k)_{K_{s^*} K_{s^*}}\} + \alpha_{s^*}^k (p_{s^*}^k / p^* - 1) \\
&\ge \alpha_{s^*}^k + \alpha_{s^*}^k (p_{s^*}^k / p^* - 1) \ge \alpha_{s^*}^k p_{s^*}^k / p^* = b_k / p^*.
\end{aligned}$$

Similarly $\min\{U(k + 1)_{L_{s^*} L_{s^*}}\} \ge b_k / q^*$. This fact together with the fact that $U(k)$ and $U(k + 1)$ are equal outside $K_{s^*} \cup L_{s^*}$, implies that $U(k + 1) \ge 0$ and that (5.8) holds for $k + 1$.

The algorithm stops after $n - 1$ steps by constructing a maximal filtration $\mathbb{E}_0 = \mathbb{M}_n < \cdots < \mathbb{E}_{n-1} = \mathbb{I}$ and an increasing sequence of constants $0 \le b_0 \le \cdots \le b_{n-2}$ such that

$$D = U(n - 1) = U - \sum_{\ell=0}^{n-2} b_\ell(\mathbb{E}_\ell - \mathbb{E}_{\ell+1}),$$

is a nonnegative diagonal matrix. We notice that $\mathbb{E}_\ell \mathbb{1} = \mathbb{1}$ which implies that $D = D_S$ and $S = U\mathbb{1}$.

We now prove item (II). We first prove that $(v) \Rightarrow (iv)$, so we assume that U is strongly filtered, that is,

$$U = \sum_{\ell=0}^{m} a_\ell \mathbb{E}_\ell,$$

where $a_0, \cdots, a_m$ are nonnegative constants and $\mathbf{F} = (\mathbb{E}_0, \cdots, \mathbb{E}_m)$ is a filtration. In particular U is filtered and according to what we have proved in (I), U is ultrametric. On the other hand

$$S = U\mathbb{1} = \sum_{\ell=0}^{m} a_\ell \mathbb{E}_\ell \mathbb{1} = \sum_{\ell=0}^{m} a_\ell \mathbb{1},$$

which implies that the row sums are the constant $s = \sum_{\ell=0}^{m} a_\ell$.

To finish the proof of the proposition let us show that $(iv) \Rightarrow (v)$. Since U is ultrametric we know that

$$U = \sum_{\ell=0}^{n-2} b_\ell (\mathbb{E}_\ell - \mathbb{E}_{\ell+1}) + s\mathbb{I},$$

where s is the row sum of U (which is assumed to be constant) and $(b_\ell : \ell = 0, \cdots, n-2)$ is increasing sequence of nonnegative numbers. Reviewing the way $b_0, \cdots, b_{n-2}$ were constructed, it is straightforward to show (by induction) that $b_\ell \leq \max(U(\ell)\mathbb{1})$ for $\ell = 0, \cdots, n-2$. Since $U(\ell)\mathbb{1} = U\mathbb{1}$, we conclude that $b_\ell \leq s$ for all $\ell = 0, \cdots, n-2$. Therefore, we get

$$U = \sum_{\ell=0}^{n-2} b_\ell (\mathbb{E}_\ell - \mathbb{E}_{\ell+1}) + s\mathbb{I} = \sum_{\ell=0}^{n-2} (b_\ell - b_{\ell-1})\mathbb{E}_\ell + (s - b_{n-2})\mathbb{I} = \sum_{\ell=0}^{n-1} a_\ell \mathbb{E}_\ell,$$

where $0 \leq a_\ell = b_\ell - b_{\ell-1}$ for $\ell = 0, \cdots, n-2$, $0 \leq a_{n-1} = s - b_{n-2}$ and as usual $b_{-1} = 0$. This shows that U is strongly filtered and the result is proven.

5.3 Weakly Filtered Matrices

The larger class of matrices for which we can prove that are inverse M-matrices, is obtained by the following generalization of filtered matrices (or equivalently ultrametric matrices). This new class is obtained by allowing some of the coefficients in (5.7) to be non-adapted. We will see that if the non-adapted coefficients are dominated, in a sense to be precise later, by the adapted ones then we obtain potential matrices.

Definition 5.18 A nonnegative matrix U is said to be **weakly filtered** if there exists a filtration $\mathbf{F} = (\mathbb{E}_0, \cdots, \mathbb{E}_m)$ and nonnegative vectors $\mathfrak{a}_0, \cdots, \mathfrak{a}_m, \mathfrak{b}_0, \cdots, \mathfrak{b}_m$ such that

$$U = \sum_{\ell=0}^{m} \mathfrak{a}_\ell \cdot \mathbb{E}_\ell \cdot \mathfrak{b}_\ell, \tag{5.9}$$

and the following measurability condition holds

$$\mathfrak{a}_\ell, \mathfrak{b}_\ell \in \mathrm{Im}(\mathbb{E}_{\ell+1}),$$

where we take $\mathbb{E}_{m+1} = \mathbb{I}$.

We note that a weakly filtered matrix is not necessarily a symmetric matrix. We also point out that in the previous definition the filtration is not strict, so we can repeat some conditional expectations, giving in this way more freedom. Nevertheless, we will describe these matrices by means of a strict filtration in another way, see Proposition 5.20. On the other hand, in the definition of weakly filtered, we can assume without loss of generality that $\mathbb{E}_0 = \mathbb{M}_n$, where n is the size of U.

We can always normalize the representation using the next concept.

Definition 5.19 Given a conditional distribution $\mathbb{E}$ and $a \in \mathbb{R}^n$ we define $\hat{a}^{\mathbb{E}}$, the $\mathbb{E}$-**envelop** of a, as the element of $\mathrm{Im}(\mathbb{E})$

$$\hat{a}^{\mathbb{E}} = \inf\{h : h \in \mathrm{Im}(\mathbb{E}) \text{ and } h \geq a\}.$$

A formula for $\hat{a}^{\mathbb{E}}$ is obtained as follows. Let $\mathfrak{R} = \{A_1, \cdots, A_p\}$ be the partition associated with $\mathbb{E}$, and consider $j \in A_\ell$ and $\ell \in \{1, \cdots, p\}$, then

$$\hat{a}_j^{\mathbb{E}} = \max\{a_i : i \in A_\ell\}.$$

This operation is far from being linear, but it has some properties as if it were a projector

(1) $a = \hat{a}^{\mathbb{E}}$ if and only if $a \in \mathrm{Im}(\mathbb{E})$;
(2) $\widehat{a \vee c}^{\mathbb{E}} = \hat{a}^{\mathbb{E}} \vee \hat{c}^{\mathbb{E}}$;
(3) if a, c are nonnegative then $\widehat{ac}^{\mathbb{E}} \leq \hat{a}^{\mathbb{E}} \hat{c}^{\mathbb{E}}$;
(4) moreover if $0 \leq c \in \mathrm{Im}(\mathbb{E})$ then $\widehat{ac}^{\mathbb{E}} = \hat{a}^{\mathbb{E}} c$. This property is of particular utility when c is constant.

It is interesting to see the class of weakly filtered matrices as a perturbation of ultrametric matrices as it is shown by the following proposition.

Proposition 5.20 *Let U be a weakly filtered matrix, then there exists a filtration $\mathbf{F} = (\mathbb{E}_0, \cdots, \mathbb{E}_{2p})$ (non necessarily strict) and three sequences of nonnegative* *vectors* $\gamma_\ell, \eta_\ell, \mathfrak{z}_\ell$ *for* $\ell = 0, \cdots, 2p$ *such that*

$$
\begin{cases}
U = \sum_{\ell=0}^{2p} \gamma_\ell \eta_\ell \cdot \mathbb{E}_\ell \cdot \mathfrak{z}_\ell; \\
\mathbb{M}_n = \mathbb{E}_0 = \mathbb{E}_1 < \mathbb{E}_2 = \mathbb{E}_3 < \cdots < \mathbb{E}_{2p-2} = \mathbb{E}_{2p-1} < \mathbb{E}_{2p} = \mathbb{I}, \\
\textit{in particular } (\mathbb{E}_0, \mathbb{E}_2, \cdots, \mathbb{E}_{2p}) \textit{ is a strict filtration}; \\
(\gamma_\bullet) \textit{ is an adapted sequence of nonnegative vectors}; \\
\forall \ell = 0, \cdots, 2p \qquad \eta_\ell, \mathfrak{z}_\ell \in \mathrm{Im}(\mathbb{E}_{\ell+1}) \textit{ and } \quad 0 \le \eta_\ell \le 1, 0 \le \mathfrak{z}_\ell \le 1; \\
\forall q = 0, \cdots, p \qquad \eta_{2q} = \mathfrak{z}_{2q} = \mathbb{1}; \\
\forall q = 0, \cdots, p-1 \qquad \hat{\eta}_{2q+1}^{\mathbb{E}_{2q}} = \hat{\mathfrak{z}}_{2q+1}^{\mathbb{E}_{2q}} = \mathbb{1}; \\
\forall q = 0, \cdots, p-1 \qquad (\gamma_{2q+1} \eta_{2q+1} \mathfrak{z}_{2q+1})_i = 0, \textit{ if } \{i\} \textit{ is an atom of } \mathbb{E}_{2q}.
\end{cases}
$$
(5.10)

In particular $U = \sum_{\ell=0}^{p-1} \gamma_{2\ell} \cdot \mathbb{E}_{2\ell} + \gamma_{2\ell+1} \eta_{2\ell+1} \cdot \mathbb{E}_{2\ell} \cdot \mathfrak{z}_{2\ell+1} + \gamma_{2p} \cdot \mathbb{I}.$

Proof Assume that $U = \sum_{r=0}^{m} \mathfrak{a}_\ell \cdot \tilde{\mathbb{E}}_r \cdot \mathfrak{b}_r$ is a possible decomposition of U. The proof consists on grouping terms with equal conditional expectation

$$
\sum_{r=s}^{q} \mathfrak{a}_r \cdot \tilde{\mathbb{E}}_\ell \cdot \mathfrak{b}_r,
$$

with $\tilde{\mathbb{E}}_{s-1} < \tilde{\mathbb{E}} = \tilde{\mathbb{E}}_s = \tilde{\mathbb{E}}_q < \tilde{\mathbb{E}}_{q+1}$. Since $\mathfrak{a}_r, \mathfrak{b}_r \in \mathrm{Im}(\tilde{\mathbb{E}})$ for $r = s, \cdots, q-1$ and $\mathfrak{a}_q, \mathfrak{b}_q \in \mathrm{Im}(\tilde{\mathbb{E}}_{q+1})$, then

$$
\sum_{r=s}^{q} \mathfrak{a}_r \cdot \tilde{\mathbb{E}}_r \cdot \mathfrak{b}_r = \left(\sum_{r=s}^{q-1} \mathfrak{a}_r \mathfrak{b}_r \right) \cdot \tilde{\mathbb{E}} + \mathfrak{a}_q \cdot \tilde{\mathbb{E}} \cdot \mathfrak{b}_q.
$$

Collecting the different conditional expectations appearing in this procedure, we construct a strict filtration $\mathbf{F} = (\mathbb{E}_0 < \mathbb{E}_2 < \cdots < \mathbb{E}_{2p})$. By adding if necessary extra 0 terms we can assume that $\hat{\mathbb{E}}_0 = \mathbb{M}_n, \hat{\mathbb{E}}_{2p} = \mathbb{I}$. Each term of the form $\mathfrak{a} \cdot \mathbb{E} \cdot \mathfrak{b}$ can be transformed in

$$
\hat{\mathfrak{a}} \, \hat{\mathfrak{b}} \, \eta \cdot \mathbb{E} \cdot \mathfrak{z},
$$

with $\eta = \mathbb{1}$ if $\mathfrak{a} \in \mathrm{Im}(\mathbb{E})$. In case $\mathfrak{a} \notin \mathrm{Im}(\mathbb{E})$ the vector η is defined as, for $i \in I$

$$
(\eta)_i = \begin{cases} 1 & \text{if } (\hat{\mathfrak{a}})_i = 0 \\ (\mathfrak{a})_i / (\hat{\mathfrak{a}})_i & \text{otherwise} \end{cases}.
$$

where $\hat{\mathfrak{a}} = \hat{\mathfrak{a}}^{\mathbb{E}}$. We consider a similar definition for $\mathfrak{z}$. Notice that $\hat{\eta} = \hat{\mathfrak{z}} = \mathbb{1}$.

It is straightforward to show that conditions (5.10) are satisfied, except perhaps for the last one. Assume ℓ is even. Every term of the form $\gamma_\ell \mathbb{E}_\ell + \gamma_{\ell+1}\mathfrak{y}_{\ell+1}\cdot\mathbb{E}_{\ell}\cdot 3\ell+1$ can be modified to satisfy this condition. Indeed, it is enough to redefine $(\hat{\gamma}_\ell)_i = (\gamma_\ell)_i + (\gamma_{\ell+1}\mathfrak{y}_{\ell+1}3\ell+1)_i$ and $(\hat{\gamma}_{\ell+1})_i = 0$, to get a decomposition that satisfies all the requirements.

A decomposition of the form given in (5.10) will be called a **normal form** for a weakly filtered matrix.

In the next result, we detail decompositions for CBF, NBF and GUM matrices. We point out that some of the stated equivalences are not true if we do not impose the filtration to be dyadic (see Definition 5.9 and Remark 5.29).

Proposition 5.21 *The matrix V, of size n, is a permutation of a CBF matrix if and only if there exist a **dyadic** filtration $\mathbf{F} = (\mathbb{E}_0 = \mathbb{M}_n, \cdots, \mathbb{E}_m = \mathbb{I})$, two sequences of adapted vectors $\mathfrak{c}_0, \cdots, \mathfrak{c}_m$ $\mathfrak{g}_0, \cdots, \mathfrak{g}_m$ and a sequence of pairs of vectors $\{\mathfrak{t}_0, \mathfrak{u}_0\}, \cdots, \{\mathfrak{t}_m, \mathfrak{u}_m\}$ such that*

$$V = \sum_{\ell=0}^{m} [\mathfrak{c}_\ell \cdot \mathbb{E}_\ell + (\mathfrak{g}_\ell \mathfrak{t}_\ell) \cdot \mathbb{E}_\ell \cdot \mathfrak{u}_\ell], \tag{5.11}$$

together with the hypothesis

$$\begin{cases} \mathfrak{t}_m = \mathfrak{u}_m = \mathbb{1}, \mathfrak{g}_m = 0 \text{ and for } \ell = 0, \cdots, m-1; \\ \mathfrak{t}_\ell, \mathfrak{u}_\ell \in \mathbb{E}_{\ell+1}; \\ \mathfrak{t}_\ell, \mathfrak{u}_\ell \text{ take values on } \{0, 1\}; \\ (\mathbb{1} - \mathfrak{t}_\ell)(\mathbb{1} - \mathfrak{u}_\ell) = 0, \text{ and } \hat{\mathfrak{t}}_\ell^{\mathbb{E}_\ell} = \hat{\mathfrak{u}}_\ell^{\mathbb{E}_\ell} = \mathbb{1}; \\ \text{if } R \text{ is a nontrivial atom of } \mathbb{E}_\ell \text{ then } \mathfrak{t}_\ell \mathfrak{u}_\ell \mathbb{1}_R = 0; \\ \mathfrak{g}_\ell \mathfrak{t}_\ell \mathfrak{u}_\ell = 0. \end{cases} \tag{5.12}$$

For what follows, consider the incident matrix $\mathbb{F}_\ell$ and the counting vector $N_\ell = \mathbb{F}_\ell \mathbb{1} = (\text{diag}(\mathbb{E}_\ell))^{-1}$ associated with each $\mathbb{E}_\ell$ (see Definition 5.12). Also consider the vectors $C_\ell = \mathfrak{c}_\ell/N_\ell$ and $\Gamma_\ell = \mathfrak{g}_\ell/N_\ell$, for $\ell = 0, \cdots, m$. With these change of scale we have that (5.11) takes the form

$$V = \sum_{\ell=0}^{m} [C_\ell \cdot \mathbb{F}_\ell + (\Gamma_\ell \mathfrak{t}_\ell) \cdot \mathbb{F}_\ell \cdot \mathfrak{u}_\ell]. \tag{5.13}$$

Then, V is a GUM if and only if there is a decomposition like (5.11) (or equivalently (5.13)) satisfying (5.12) and

(i) for all $\ell = 0, \cdots, m$ it holds $0 \leq C_\ell$, $0 \leq \Gamma_\ell$;
(ii) for all $\ell = 0, \cdots, m-1$ the following perturbation condition holds

$$\Gamma_\ell \leq C_{\ell+1} + \Gamma_{\ell+1}. \tag{5.14}$$

Finally, V is an ultrametric matrix if and only if there is a decomposition with $g_\ell = 0$ for all ℓ (alternatively $\Gamma_\ell = 0$ for all ℓ).

Proof We will give only the main steps of the proof. We assume that V is a constant block form

$$V = \begin{pmatrix} A & \alpha \\ \beta & B \end{pmatrix}.$$

The proof is done inductively using decomposition (3.3), so we only give the first step. Assuming that the size of V is $n \geq 2$, consider the matrix

$$\Lambda = n\alpha \mathbb{M}_n + n(\beta - \alpha)\begin{pmatrix} 0_p \\ \mathbb{1}_q \end{pmatrix} \cdot \mathbb{M}_n \cdot \begin{pmatrix} \mathbb{1}_p \\ 0_q \end{pmatrix} = \begin{pmatrix} \alpha\mathbb{1}_p\mathbb{1}'_p & \alpha\mathbb{1}_p\mathbb{1}'_q \\ \beta\mathbb{1}_q\mathbb{1}'_p & \alpha\mathbb{1}_q\mathbb{1}'_q \end{pmatrix}.$$

If we define $\mathbb{E}_0 = \mathbb{M}_n$, $c_0 = n\alpha\mathbb{1}_n \in \mathrm{Im}(\mathbb{E}_0)$, $g_0 = n(\beta - \alpha)\mathbb{1}_n \in \mathrm{Im}(\mathbb{E}_0)$ and

$$t_0 = \begin{pmatrix} 0_p \\ \mathbb{1}_q \end{pmatrix} \in \mathrm{Im}(\mathbb{E}_1), \quad u_0 = \begin{pmatrix} \mathbb{1}_p \\ 0_q \end{pmatrix} \in \mathrm{Im}(\mathbb{E}_1),$$

where $\mathbb{E}_1$ is given by the partition $\{\{1, \cdots, p\}, \{p + 1, \cdots, n\}\}$. Then, with this notation, we have $\Lambda = c_0 \cdot \mathbb{E}_0 + (g_0 t_0) \cdot \mathbb{E}_0 \cdot u_0$. Notice that $0 = (\mathbb{1} - t_0)(\mathbb{1} - u_0)$ and $\hat{t}_0 = \hat{u}_0 = \mathbb{1}$. Moreover, we have $t_0 u_0 = 0$ and $g_0 t_0 u_0 = 0$.

On the other hand, $\Lambda = C_0 \mathbb{F}_0 + (\Gamma_0 t_0) \cdot \mathbb{F}_0 \cdot u_0$ where $C_0 = \alpha\mathbb{1}_n$, $\Gamma_0 = (\beta - \alpha)\mathbb{1}_n$ and $\mathbb{F}_0 = \mathbb{1}_n\mathbb{1}'_n$. The proof continues by noticing that

$$V - \Lambda = \begin{pmatrix} A - \alpha & 0 \\ 0 & B - \alpha \end{pmatrix},$$

together with the fact that $A - \alpha$, $B - \alpha$ are CBF matrices. When the size of V is $n = 1$, which will arrive eventually in the inductive argument, then there is not need for two terms and in this case we have

$$V = a\mathbb{E}_0 + 0\mathbb{1} \cdot \mathbb{I} \cdot \mathbb{1}.$$

This explains the two last conditions in (5.12).

In the case of a GUM, condition (5.14) for $\ell = 0$ follows at once and the proof continues inductively because $A - \alpha$, $B - \alpha$ are GUM.

Remark 5.22 The condition $t_\ell u_\ell = 0$ over the nontrivial atoms of $\mathbb{E}_\ell$ tell us that the perturbation on a CBF matrix is done in one of the sides, over or below the diagonal, at each level. On the other hand the condition $g_\ell t_\ell u_\ell = 0$ implies that $g_\ell = 0$ on the trivial atoms of $\mathbb{E}_\ell$. In this way the perturbation is done in a minimal

way and this has important consequences, for example in condition (5.14). Indeed, condition (5.14) should be verified only on the nontrivial atoms of $\mathbb{E}_\ell$, because the left hand side is 0 on the trivial atoms.

We point out the description of V in terms of incident matrices is simpler than the one in terms of conditional expectations. The main reason to use (normalized) conditional expectations is the algebraic and commutative properties they satisfy.

Example 5.23 Consider the CBF matrix

$$U = \begin{pmatrix} a & \alpha_2 & \alpha_1 & \alpha_1 \\ \beta_2 & b & \alpha_1 & \alpha_1 \\ \beta_1 & \beta_1 & c & \hat{\alpha}_2 \\ \beta_1 & \beta_1 & \hat{\beta}_2 & d \end{pmatrix}.$$

U is a NBF if the following constraints are satisfied:

$$\alpha_1 \le \beta_1, \ \alpha_1 \le \min\{\alpha_2, \hat{\alpha}_2\}, \ \beta_1 \le \min\{\beta_2, \hat{\beta}_2\},$$
$$\alpha_2 \le \beta_2, \ \hat{\alpha}_2 \le \hat{\beta}_2$$

and finally the diagonal dominates pointwise over each row and column, that is $\beta_2 \le \min\{a, b\}$, $\hat{\beta}_2 \le \min\{c, d\}$.

U is filtered with respect to the dyadic filtration whose chain of partitions is $\mathfrak{R}_0 = \{\{1, 2, 3, 4\}\} \prec \mathfrak{R}_1 = \{\{1, 2\}, \{3, 4\}\} \prec \mathfrak{R}_2 = \{\{1\}, \{2\}, \{3\}, \{4\}\}$ and can be written as

$$U = C_0 \cdot \mathbb{F}_0 + \Gamma_0 t_0 \cdot \mathbb{F}_0 \cdot u_0 + C_1 \cdot \mathbb{F}_1 + \Gamma_1 \cdot t_1 \cdot \mathbb{F}_1 \cdot u_1 + C_2 \cdot \mathbb{I},$$

where

$$C_0 = \begin{pmatrix} \alpha_1 \\ \alpha_1 \\ \alpha_1 \\ \alpha_1 \end{pmatrix}, \ \Gamma_0 = \begin{pmatrix} \beta_1 - \alpha_1 \\ \beta_1 - \alpha_1 \\ \beta_1 - \alpha_1 \\ \beta_1 - \alpha_1 \end{pmatrix}, \ t_0 = \begin{pmatrix} 0 \\ 0 \\ 1 \\ 1 \end{pmatrix}, \ u_0 = \begin{pmatrix} 1 \\ 1 \\ 0 \\ 0 \end{pmatrix}$$

$$C_1 = \begin{pmatrix} \alpha_2 - \alpha_1 \\ \alpha_2 - \alpha_1 \\ \hat{\alpha}_2 - \alpha_1 \\ \hat{\alpha}_2 - \alpha_1 \end{pmatrix}, \ \Gamma_1 = \begin{pmatrix} \beta_2 - \alpha_2 \\ \beta_2 - \alpha_2 \\ \hat{\beta}_2 - \hat{\alpha}_2 \\ \hat{\beta}_2 - \hat{\alpha}_2 \end{pmatrix}, \ t_1 = \begin{pmatrix} 0 \\ 1 \\ 0 \\ 1 \end{pmatrix}, \ u_1 = \begin{pmatrix} 1 \\ 0 \\ 1 \\ 0 \end{pmatrix},$$

and

$$C_2 = \begin{pmatrix} a - \alpha_2 \\ b - \alpha_2 \\ c - \hat{\alpha}_2 \\ d - \hat{\alpha}_2 \end{pmatrix}.$$

The constrains, described above, are translated into: Positivity of these vectors and the ones induced by (5.14), which in this case is just the following condition

$$\Gamma_0 = \begin{pmatrix} \beta_1 - \alpha_1 \\ \beta_1 - \alpha_1 \\ \beta_1 - \alpha_1 \\ \beta_1 - \alpha_1 \end{pmatrix} \le C_1 + \Gamma_1 = \begin{pmatrix} \beta_2 - \alpha_1 \\ \beta_2 - \alpha_1 \\ \hat{\beta}_2 - \alpha_1 \\ \hat{\beta}_2 - \alpha_1 \end{pmatrix}.$$

We point out that we can also choose, for example, $\Gamma_1 = (0, \beta_2 - \alpha_2, 0, \hat{\beta}_2 - \hat{\alpha}_2)'$, but in this case Γ_1 is not $\mathfrak{R}_1$-measurable. As we will see this measurability condition will play an important role.

Example 5.24 Consider the nonnegative CBF matrix

$$U = \begin{pmatrix} 2\ 2\ 2 \\ 2\ 2\ 1 \\ 2\ 1\ 2 \end{pmatrix}.$$

This matrix can be decomposed as in (5.10) except that no such decomposition can have all terms nonnegative. In particular no permutation of U is a NBF.

5.3.1   Algorithm for Weakly Filtered Matrices

We introduce here an algorithm to study the M-inverse problem for weakly filtered matrices, which was developed in [21] to study the inverse of $\mathbb{I} + U$.

Let U be a weakly filtered matrix with a decomposition as in (5.9)

$$U = \sum_{\ell=0}^{m} \mathfrak{a}_\ell \cdot \mathbb{E}_\ell \cdot \mathfrak{b}_\ell.$$

Consider the following algorithm defined by the backward recursion starting with the values $\lambda_m = \mu_m = \kappa_m = 1$, $\sigma_m = (1 + \mathbb{E}_m(\mathfrak{a}_m \mathfrak{b}_m))^{-1}$ and for $\ell = m-1, \cdots, 0$:

$$\begin{cases} \text{while } [\lambda_\ell \ge 0 \text{ and } \mu_\ell \ge 0] \text{ do} \\ \lambda_\ell = \lambda_{\ell+1}[1 - (\sigma_{\ell+1} \mathfrak{a}_{\ell+1}) \mathbb{E}_{\ell+1}(\kappa_{\ell+1} \mathfrak{b}_{\ell+1})] \\ \mu_\ell = \mu_{\ell+1}[1 - (\sigma_{\ell+1} \mathfrak{b}_{\ell+1}) \mathbb{E}_{\ell+1}(\kappa_{\ell+1} \mathfrak{a}_{\ell+1})] \\ \kappa_\ell = \mathbb{E}_{\ell+1}(\lambda_\ell) \\ \sigma_\ell = (1 + \mathbb{E}_\ell(\kappa_\ell \mathfrak{a}_\ell \mathfrak{b}_\ell))^{-1} \\ \text{end} \end{cases} \qquad (5.15)$$

The algorithm continues until some component of λ or μ is negative otherwise we arrive to $\ell = 0$.

While λ and μ are nonnegative it is easy to prove that

$$\kappa_{\ell-1} = \mathbb{E}_\ell(\kappa_\ell) - \frac{\mathbb{E}_\ell(\kappa_\ell \mathfrak{a}_\ell)\mathbb{E}_\ell(\kappa_\ell \mathfrak{b}_\ell)}{1 + \mathbb{E}_\ell(\kappa_\ell \mathfrak{a}_\ell \mathfrak{b}_\ell)}. \tag{5.16}$$

and also that $\kappa_\ell = \mathbb{E}_{\ell+1}(\lambda_\ell) = \mathbb{E}_{\ell+1}(\mu_\ell)$.

Proposition 5.25 *If in the algorithm all* λ_ℓ, μ_ℓ, $\ell = m, \cdots, 0$ *are nonnegative then* $\mathbb{I} + U$ *is nonsingular and its inverse is of the form* $\mathbb{I} - P$ *where*

$$P = \sum_{\ell=0}^{m} (\sigma_\ell \lambda_\ell \mathfrak{a}_\ell) \cdot \mathbb{E}_\ell \cdot (\mathfrak{b}_\ell \mu_\ell) \geq 0 .$$

Moreover, for $s = m, \cdots, 0$ *consider the weakly filtered matrices*

$$U_s = \sum_{\ell=s}^{m} \mathfrak{a}_\ell \cdot \mathbb{E}_\ell \cdot \mathfrak{b}_\ell .$$

Then the algorithm we propose is equivalent to have for each $s = m, \cdots, 0$

$$\begin{cases} \mathbb{I} + U_s \text{ is nonsingular, } (\mathbb{I} + U_s)^{-1}\mathbb{1} = \lambda_{s-1} \geq 0, \ (\mathbb{I} + U_s')^{-1}\mathbb{1} = \mu_{s-1} \geq 0, \\ \text{and } (\mathbb{I} + U_s)^{-1} = \mathbb{I} - P_s, \text{ where } P_s = \sum_{\ell=s}^{m} (\sigma_\ell \lambda_\ell \mathfrak{a}_\ell) \cdot \mathbb{E}_\ell \cdot (\mathfrak{b}_\ell \mu_\ell) . \end{cases} \tag{5.17}$$

In particular, we have $\lambda_{-1} = (\mathbb{I} - P)\mathbb{1}$, *and* $\mu_{-1} = (\mathbb{I} - P)'\mathbb{1}$, *where* λ_{-1}, μ_{-1} *are obtained from the first two formulae in (5.15) for* $\ell = -1$. *Therefore, if they are also nonnegative they are the right and left equilibrium potentials of* $\mathbb{I} + U$, *which it is then a bi-potential.*

The main algebraic tool we need to prove this proposition is the following result.

Lemma 5.26 *Let* $\mathbf{F} = (\mathbb{E}_0, \cdots, \mathbb{E}_m)$ *be a filtration and* K *a matrix that can be decomposed as*

$$K = \sum_{\ell=k}^{m} \mathfrak{c}_\ell \cdot \mathbb{E}_\ell \cdot \mathfrak{d}_\ell,$$

where the vectors $\mathfrak{c}_1, \cdots, \mathfrak{c}_m$ *and* $\mathfrak{d}_1, \cdots, \mathfrak{d}_m$ *are arbitrary. Consider a conditional expectation* $\mathbb{E}$ *smaller than* $\min\{\mathbb{E}_\ell : \mathfrak{c}_\ell \neq 0 \text{ and } \mathfrak{d}_\ell \neq 0\}$ *(where* $\min \emptyset = \mathbb{I}$*) and a vector* $\mathfrak{u}$. *Then*

$$K(\mathfrak{u} \cdot \mathbb{E}) = (K\mathfrak{u}) \cdot \mathbb{E},$$

and

$$(\mathbb{E}\cdot u)K = \mathbb{E}\cdot(K'u).$$

Proof The second equality is exactly the transpose of the first one, using K' instead of K. So it is enough to prove one of them. Let $x \in \mathbb{R}^n$ then

$$K\,(u\cdot\mathbb{E})x = KD_u\mathbb{E}x = K(u\mathbb{E}(x)) = \sum_{\ell=k}^{m} c_\ell\cdot\mathbb{E}_\ell\cdot\partial_\ell(u\mathbb{E}(x)) = \sum_{\ell=k}^{m} c_\ell\mathbb{E}_\ell(\partial_\ell u\mathbb{E}(x))$$

$$= \left(\sum_{\ell=k}^{m} c_\ell\mathbb{E}_\ell(\partial_\ell u) \right)\mathbb{E}(x),$$

where the last equality holds because $\mathbb{E}(x) \in \mathrm{Im}(\mathbb{E}_\ell)$ and Lemma 5.4. Thus we obtain

$$K\,(u\cdot\mathbb{E})x = (K(u)\cdot\mathbb{E})x$$

and the result holds.

Proof (Proposition 5.25) We will show (5.17) by a backward induction. For $s = m$ we look for a solution of the following problem

$$\mathbb{I} = (\mathbb{I} + a_m\cdot\mathbb{E}_m\cdot b_m)(\mathbb{I} - c\cdot\mathbb{E}_m\cdot\partial) = \mathbb{I} + a_m\cdot\mathbb{E}_m\cdot b_m - c\cdot\mathbb{E}_m\cdot\partial - (a_m\cdot\mathbb{E}_m\cdot b_m)(c\cdot\mathbb{E}_m\cdot\partial).$$

The last three terms simplify to

$$a_m\cdot\mathbb{E}_m\cdot b_m - c\cdot\mathbb{E}_m\cdot\partial - (a_m\mathbb{E}_m(b_m c))\cdot\mathbb{E}_m\cdot\partial.$$

We impose this quantity to be 0. So, we take $\partial = b_m$, which gives

$$a_m - c - a_m\mathbb{E}_m(b_m c) = 0.$$

As before, we multiply this equality by b_m and apply $\mathbb{E}_m$ to obtain

$$\mathbb{E}_m(b_m c) = \frac{\mathbb{E}_m(b_m a_m)}{1 + \mathbb{E}_m(b_m a_m)}.$$

This finally yields

$$c = \frac{a_m}{1 + \mathbb{E}_m(b_m a_m)} = a_m\sigma_m.$$

We obtain

$$(\mathbb{I} + U_m)^{-1} = \mathbb{I} - P_m = \mathbb{I} - \frac{a_m}{1 + \mathbb{E}_m(a_m b_m)}\cdot\mathbb{E}_m\cdot b_m\kappa_m = \mathbb{I} - \sigma_m a_m\cdot\mathbb{E}_m\cdot b_m\kappa_m$$

and

$$\lambda_{m-1} = \lambda_m(\mathbb{1} - \sigma_m \mathfrak{a}_m \mathbb{E}_m(\kappa_m \mathfrak{b}_m)) = (\mathbb{I} - P_m)\mathbb{1}.$$

Similarly, we have $\mu_{m-1} = (\mathbb{I} - P'_m)\mathbb{1}$. Therefore, the induction works for $s = m$.
We now show the step from $s + 1$ to s. For that purpose consider

$$\begin{aligned}
\mathbb{I} + U_s &= \mathbb{I} + U_{s+1} + \mathfrak{a}_s \cdot \mathbb{E}_s \cdot \mathfrak{b}_s = (\mathbb{I} + U_{s+1})(\mathbb{I} + (\mathbb{I} + U_{s+1})^{-1} \mathfrak{a}_s \cdot \mathbb{E}_s \cdot \mathfrak{b}_s) \\
&= (\mathbb{I} + U_{s+1})(\mathbb{I} + (\mathbb{I} + U_{s+1})^{-1}(\mathfrak{a}_s) \cdot \mathbb{E}_s \cdot \mathfrak{b}_s),
\end{aligned}$$

where we have used Lemma 5.26. If we denote by $\mathfrak{h} = (\mathbb{I} + U_{s+1})^{-1}(\mathfrak{a}_s)$, we have

$$\begin{aligned}
(\mathbb{I} + U_s)^{-1} &= (\mathbb{I} + \mathfrak{c} \cdot \mathbb{E}_s \cdot \mathfrak{b}_s)^{-1}(\mathbb{I} + U_{s+1})^{-1} = \left(\mathbb{I} - \tfrac{\mathfrak{h}}{1 + \mathbb{E}_s(\mathfrak{h}\mathfrak{b}_s)} \cdot \mathbb{E}_s \cdot \mathfrak{b}_s\right)(\mathbb{I} - P_{s+1}) \\
&= \mathbb{I} - \tfrac{\mathfrak{h}}{1 + \mathbb{E}_s(\mathfrak{h}\mathfrak{b}_s)} \cdot \mathbb{E}_s \cdot \mathfrak{b}_s - P_{s+1} - \tfrac{\mathfrak{h}}{1 + \mathbb{E}_s(\mathfrak{h}\mathfrak{b}_s)} \cdot \mathbb{E}_s \cdot (P'_{s+1}\mathfrak{b}_s).
\end{aligned}$$

Recall that

$$P_{s+1} = \sum_{\ell=s+1}^m (\sigma_\ell \lambda_\ell \mathfrak{a}_\ell) \cdot \mathbb{E}_\ell \cdot (\mathfrak{b}_\ell \mu_\ell)$$

and $\mathfrak{a}_s, \mathfrak{b}_s$ belong to $\text{Im}(\mathbb{E}_{s+1})$. Then we have $P_{s+1}(\mathfrak{a}_s) = \mathfrak{a}_s P_{s+1}(\mathbb{1}) = \mathfrak{a}_s \lambda_s$ and similarly $P'_{s+1}(\mathfrak{b}_s) = \mathfrak{b}_s P'_{s+1}(\mathbb{1}) = \mathfrak{b}_s \mu_s$. Also we get $\mathfrak{h} = \mathfrak{a}_s \lambda_s$, which implies that

$$\mathbb{E}_s(\mathfrak{h}\mathfrak{b}_s) = \mathbb{E}_s(\mathfrak{a}_s \lambda_s \mathfrak{b}_s) = \mathbb{E}_s(\mathfrak{a}_s \mathfrak{b}_s \mathbb{E}_{s+1}(\lambda_s)) = \mathbb{E}_s(\mathfrak{a}_s \mathfrak{b}_s \kappa_s).$$

Putting these elements together yields

$$(\mathbb{I} + U_s)^{-1} = \mathbb{I} - \sum_{\ell=s}^m (\sigma_\ell \lambda_\ell \mathfrak{a}_\ell) \cdot \mathbb{E}_\ell \cdot (\mathfrak{b}_\ell \mu_\ell).$$

Hence, we obtain that

$$\begin{aligned}
(\mathbb{I} - P_s)\mathbb{1} &= \lambda_s - \sigma_s \lambda_s \mathfrak{a}_s \mathbb{E}_s(\mathfrak{b}_s \mu_s) = \lambda_s[\mathbb{1} - \sigma_s \mathfrak{a}_s \mathbb{E}_s(\mathfrak{b}_s \mu_s)] \\
&= \lambda_s[\mathbb{1} - \sigma_s \mathfrak{a}_s \mathbb{E}_s(\mathfrak{b}_s \mathbb{E}_{s+1}(\mu_s))].
\end{aligned}$$

Here we have used that $\mathbb{E}_{s+1}(\mathfrak{b}_s \mu_s) = \mathfrak{b}_s \mathbb{E}_{s+1}(\mu_s)$. The only thing left to prove is $\mathbb{E}_{s+1}(\mu_s) = \mathbb{E}_{s+1}(\lambda_s)$. This follows from the identity, see Lemma 5.4 part (iv),

$$\mathbb{E}_{s+1}(\sigma_\ell \lambda_\ell \mathfrak{a}_\ell \mathbb{E}_\ell(\mathfrak{b}_\ell \mu_\ell)) = \mathbb{E}_{s+1}(\mathbb{E}_\ell(\sigma_\ell \lambda_\ell \mathfrak{a}_\ell)\mathfrak{b}_\ell \mu_\ell)$$

that holds for all $\ell \geq s + 1$. This implies $\mathbb{E}_{s+1}((\mathbb{I} - P_s)\mathbb{1}) = \mathbb{E}_{s+1}((\mathbb{I} - P'_s)\mathbb{1})$ and the definition of κ_s is consistent. The proof is now complete.

In this way we have a sufficient condition for $\mathbb{I} + U$ to be a bi-potential: The algorithm works for $\ell = m, \cdots, 0$ and all the λ, μ are nonnegative, including λ_{-1}, μ_{-1}. In this situation we have that λ (and μ) is a decreasing nonnegative sequence of vectors. Sufficient treatable conditions involve recurrence (5.16). Starting from $\kappa_m = 1$ we look for conditions that ensure this recurrence has a solution such that $\kappa_\ell \in [0, 1]$ for all $\ell = m, \cdots, -1$. We will do that in the next section.

5.3.2   Sufficient Conditions for a Weakly Filtered to be a Bi-potential

We now give a sufficient condition for a weakly filtered matrix U to be a bi-potential. Assume that U is written in normal form as in (5.10)

$$U = \sum_{\ell=0}^{2p} (\gamma_\ell \mathfrak{y}_\ell) \cdot \mathbb{E}_\ell \cdot \mathfrak{z}_\ell,$$

where we further assume the condition

$$(\mathbb{1} - \mathfrak{y}_\ell)(\mathbb{1} - \mathfrak{z}_\ell) = 0. \tag{5.18}$$

For even ℓ we have $\mathfrak{y}_\ell = \mathfrak{z}_\ell = \mathbb{1}$ and this condition is then satisfied. Observe that condition (5.18) is satisfied for example for any permutation of a CBF matrix, in particular for the GUM case (see (5.12)).

In what follows we consider the vectors $\tilde{\mathfrak{y}}_\ell = \mathfrak{y}_\ell(\mathbb{1} - \mathfrak{z}_\ell)$ and $\tilde{\mathfrak{z}}_\ell = \mathfrak{z}_\ell(\mathbb{1} - \mathfrak{y}_\ell)$. Due to condition (5.18) they reduce to

$$\tilde{\mathfrak{y}}_\ell = (\mathbb{1} - \mathfrak{z}_\ell), \tilde{\mathfrak{z}}_\ell = (\mathbb{1} - \mathfrak{y}_\ell).$$

Therefore $\tilde{\mathfrak{y}}_\ell \tilde{\mathfrak{z}}_\ell = 0$ and they are orthogonal.

Recall that $\mathbb{1}_A$ is the $\{0, 1\}$ valued vector that takes the value 1 at the coordinates in A. Thus $\mathbb{1}_{\mathfrak{y}_\ell > 0}$ is the indicator function of the sites i where $(\mathfrak{y}_\ell)_i > 0$. The important quantity for the next result is

$$\rho_\ell = \left(\mathbb{E}_\ell(\tilde{\mathfrak{y}}_\ell)\mathbb{1}_{\tilde{\mathfrak{y}}_\ell > 0}\right) \vee \left(\mathbb{E}_\ell(\tilde{\mathfrak{z}}_\ell)\mathbb{1}_{\tilde{\mathfrak{z}}_\ell > 0}\right).$$

Since $\mathbb{1}_{\tilde{\mathfrak{z}}_\ell > 0}\mathbb{1}_{\tilde{\mathfrak{y}}_\ell > 0} = 0$ we get that ρ_ℓ simplifies to

$$\rho_\ell = \mathbb{E}_\ell(\tilde{\mathfrak{y}}_\ell)\mathbb{1}_{\tilde{\mathfrak{y}}_\ell > 0} + \mathbb{E}_\ell(\tilde{\mathfrak{z}}_\ell)\mathbb{1}_{\tilde{\mathfrak{z}}_\ell > 0}. \tag{5.19}$$

As a final remark we note that $\rho_\ell = 0$ for even ℓ, which simply says that the perturbation occurs on the odd terms.

The next result gives sufficient conditions for a weakly filtered matrix to be a bi-potential.

Theorem 5.27 *Let U be a weakly filtered matrix with a normal form (see (5.10))*

$$U = \sum_{\ell=0}^{2p} (\gamma_\ell \eta_\ell) \cdot \mathbb{E}_\ell \cdot {}_{3\ell},$$

where in addition we assume $(\mathbb{1} - \eta_\ell)(\mathbb{1} - {}_{3\ell}) = 0$.
  The condition

$$\forall \ell \quad \rho_\ell \gamma_\ell \kappa_\ell \leq \mathbb{1} \tag{5.20}$$

ensures that the algorithm given in (5.15) yields $\lambda_\ell \geq 0, \mu_\ell \geq 0$ for $\ell = -1, \cdots, 2p$ and therefore according to Proposition 5.25 the matrix $\mathbb{I} + U$ is a bi-potential and its inverse is of the form $\mathbb{I} - P$, where

$$P = \sum_{\ell=0}^{2p} (\sigma_\ell \lambda_\ell \mathfrak{a}_\ell) \cdot \mathbb{E}_\ell \cdot (\mathfrak{b}_\ell \mu_\ell).$$

Each one of the following conditions is sufficient for (5.20) to hold

$$\forall \ell = 0, \cdots, 2p - 1 \quad \rho_\ell \gamma_\ell \leq \gamma_{\ell+1} + \gamma_{\ell+2}; \tag{5.21}$$

$$\forall \ell = 0, \cdots, 2p - 1 \quad \rho_\ell \gamma_\ell \leq \sum_{(\ell+1)/2 \leq r \leq p} \gamma_{2r}. \tag{5.22}$$

These two conditions have to be verified only for odd ℓ and since they are homogeneous in (γ_ℓ) we can replace them by (γ_ℓ/t) which proves the matrices $t\mathbb{I} + U$ are bi-potentials, for all $t > 0$. In particular, if U is nonsingular, then U is a bi-potential.

In the proof of this result we will need the following lemma, which is interesting on its own.

Lemma 5.28 *Assume x, y are nonnegative vectors and $\mathbb{E}$ is a conditional expectation. If $x\mathbb{E}(y) \leq \mathbb{1}$ then $\mathbb{E}(xy) \leq \mathbb{1}$. In particular $\mathbb{E}(y) \leq \mathbb{1}$ if and only if $\mathbb{E}(y)\mathbb{1}_{y>0} \leq \mathbb{1}$.*

Proof We first assume that y is positive. Since $x \leq 1/\mathbb{E}(y)$ and $\mathbb{E}$ is a nondecreasing operator, we have

$$\mathbb{E}(xy) \leq \mathbb{E}\left(\frac{1}{\mathbb{E}(y)}y\right) = \frac{\mathbb{E}(y)}{\mathbb{E}(y)} = \mathbb{1}.$$

For the general case consider $(y + \epsilon\mathbb{1})/(1 + \epsilon|x|_\infty)$ instead of y and pass to the limit $\epsilon \to 0$.

For the second part notice that obviously $\mathbb{E}(y) \leq \mathbb{1}$ implies $\mathbb{E}(y)\mathbb{1}_{y>0} \leq \mathbb{1}$. Reciprocally, consider $x = \mathbb{1}_{y>0}$ and apply the first part. The result follows by noticing that $y = y\mathbb{1}_{y>0}$.

Proof (Theorem 5.27) We first point out that conditions (5.18), (5.20), (5.21) and (5.22) are trivially satisfied for ℓ even because in this case $\eta_\ell = \mathfrak{z}_\ell = \mathbb{1}$ and $\rho_\ell = 0$.

In studying the algorithm we will take advantage of the special form of the decomposition of U where "even" terms are adapted and "odd" terms are measurable to the next conditional expectation.

We prove by induction that all λ's and μ's are nonnegative. So, assume we have proved the claim on $\ell, \cdots, 2n$. In particular $\kappa_\ell \geq 0$. The main hypothesis is $\rho_\ell\gamma_\ell\kappa_\ell \leq \mathbb{1}$, which is equivalent to $\mathbb{E}_\ell(\mathfrak{h}_\ell)\mathbb{1}_{\mathfrak{h}_\ell>0}\gamma_\ell\kappa_\ell \leq \mathbb{1}$. According to the previous lemma, this inequality gives $\mathbb{E}_\ell(\mathfrak{h}_\ell\mathbb{1}_{\mathfrak{h}_\ell>0}\gamma_\ell\kappa_\ell) \leq \mathbb{1}$. Since $\mathfrak{h}_\ell\mathbb{1}_{\mathfrak{h}_\ell>0} = \mathfrak{h}_\ell$ we conclude that

$$\mathbb{E}_\ell(\mathfrak{h}_\ell\gamma_\ell\kappa_\ell) \leq \mathbb{1} \tag{5.23}$$

and then $\mathbb{1} + \mathbb{E}_\ell(\gamma_\ell\kappa_\ell\eta_\ell\mathfrak{z}_\ell) = \mathbb{1} + \gamma_\ell\mathbb{E}_\ell(\kappa_\ell\eta_\ell\mathfrak{z}_\ell) \geq \gamma_\ell\mathbb{E}_\ell(\kappa_\ell\eta_\ell) \geq \gamma_\ell\mathfrak{z}_\ell\mathbb{E}_\ell(\kappa_\ell\eta_\ell)$. Equivalently

$$\sigma_\ell\gamma_\ell\mathfrak{z}_\ell\mathbb{E}_\ell(\kappa_\ell\eta_\ell) = \frac{\gamma_\ell\mathfrak{z}_\ell\mathbb{E}_\ell(\kappa_\ell\eta_\ell)}{1 + \gamma_\ell\mathbb{E}_\ell(\kappa_\ell\eta_\ell\mathfrak{z}_\ell)} \leq \mathbb{1},$$

which implies $\mu_{\ell-1} \geq 0$ and therefore $\kappa_{\ell-1} \geq 0$. Similarly we have

$$\sigma_\ell\gamma_\ell\eta_\ell\mathbb{E}_\ell(\kappa_\ell\mathfrak{z}_\ell) \leq \mathbb{1}$$

proving that $\lambda_{\ell-1} \geq 0$.

Therefore, the algorithm continues until $\ell = 0$, and moreover that $\lambda_{-1} \geq 0, \mu_{-1} \geq 0$. Thus $\mathbb{I} + U$ is a bi-potential.

(5.21)$\Rightarrow$(5.20). We need to prove the inequality $\rho_\ell\gamma_\ell\kappa_\ell \leq 1$ holds for all $\ell = 0, \cdots, 2p$. Since $\rho_\ell = 0$ for ℓ even, we need to prove this inequality only for ℓ odd. This is done again by backward induction and we prove simultaneously that for ℓ even we have

$$\rho_{\ell+1}\gamma_{\ell+1}\kappa_{\ell+1} \leq 1 \quad \text{and} \quad \gamma_{\ell+1}\kappa_{\ell+1} \leq \mathbb{1},$$

which is true for $\ell = 2p$ because $\gamma_{2p+1} = 0$. So, assume ℓ is even and the two properties hold on $\ell, \cdots, 2p$. In particular as before we have $\lambda_{\ell-1} \geq 0, \mu_{\ell-1} \geq 0$ and $\kappa_{\ell-1} \geq 0$. We need to show that $\rho_{\ell-1}\gamma_{\ell-1}\kappa_{\ell-1} \leq \mathbb{1}$ and $\gamma_{\ell-1}\kappa_{\ell-2} \leq \mathbb{1}$.

Let us concentrate first on proving $\rho_{\ell-1}\gamma_{\ell-1}\kappa_{\ell-1} \leq 1$. For that purpose we use that $\eta_\ell = \mathfrak{z}_\ell = 1$ and formula (5.16)

$$\kappa_{\ell-1} = \mathbb{E}_\ell(\kappa_\ell) - \gamma_\ell \frac{\mathbb{E}_\ell(\kappa_\ell \eta_\ell)\mathbb{E}_\ell(\kappa_\ell \mathfrak{z}_\ell)}{1 + \gamma_\ell \mathbb{E}_\ell(\kappa_\ell \eta_\ell \mathfrak{z}_\ell)} = \mathbb{E}_\ell(\kappa_\ell) - \gamma_\ell \frac{\mathbb{E}_\ell(\kappa_\ell)\mathbb{E}_\ell(\kappa_\ell)}{1 + \gamma_\ell \mathbb{E}_\ell(\kappa_\ell)} = \frac{\mathbb{E}_\ell(\kappa_\ell)}{1 + \gamma_\ell \mathbb{E}_\ell(\kappa_\ell)}.$$

At this moment let us recall hypothesis (5.21), that is $\gamma_{\ell+1} \geq \rho_{\ell-1}\gamma_{\ell-1} - \gamma_\ell$.

Assume for some coordinate i one has $(\gamma_{\ell+1})_i = 0$. Let R be the atom of i in $\mathbb{E}_\ell$. Then $(\gamma_{\ell+1})_j = 0$ for all $j \in R$ because $\gamma_{\ell+1}$ is $\mathbb{E}_{\ell+1} = \mathbb{E}_\ell$-measurable (here we use that ℓ is even). Thus,

$$\forall j \in R \ \ (\rho_{\ell-1}\gamma_{\ell-1})_j - (\gamma_\ell)_j \leq 0.$$

Using that $\frac{ax}{1+ax} \leq 1$ for nonnegative a, x, we obtain that

$$(\rho_{\ell-1}\gamma_{\ell-1}\kappa_{\ell-1})_j \leq (\kappa_{\ell-1}\gamma_\ell)_j = \left(\frac{\mathbb{E}_\ell(\kappa_\ell)}{1 + \gamma_\ell \mathbb{E}_\ell(\kappa_\ell)} \right)_j \leq 1,$$

which shows that for these coordinates the desired inequality holds.

The next step is to consider the case when $(\gamma_{\ell+1})_i > 0$ for all $i \in R$, which implies $(1/\gamma_{\ell+1})_i = (\mathbb{E}_\ell(1/\gamma_{\ell+1}))_i > 0$. Since $\kappa_\ell \leq 1/\gamma_{\ell+1}$ we get

$$\kappa_{\ell-1} \leq \frac{\mathbb{E}_\ell(1/\gamma_{\ell+1})}{1 + \gamma_\ell \mathbb{E}_\ell(1/\gamma_{\ell+1})}.$$

Using the measurability condition on the coefficients we conclude

$$(\rho_{\ell-1}\gamma_{\ell-1} - \gamma_\ell)\mathbb{E}_\ell\left(\frac{1}{\gamma_{\ell+1}}\right) = \mathbb{E}_\ell\left(\frac{\rho_{\ell-1}\gamma_{\ell-1} - \gamma_\ell}{\gamma_{\ell+1}}\right) \leq 1,$$

or equivalently

$$\rho_{\ell-1}\gamma_{\ell-1}\mathbb{E}_\ell(1/\gamma_{\ell+1}) \leq 1 + \gamma_\ell \mathbb{E}_\ell(1/\gamma_{\ell+1}).$$

This inequality shows the first part of what we needed to prove

$$\rho_{\ell-1}\gamma_{\ell-1}\kappa_{\ell-1} \leq 1.$$

We now prove $\gamma_{\ell-1}\kappa_{\ell-2} \leq 1$. As before we get the two inequalities (see (5.23))

$$\mathbb{E}_{\ell-1}(\mathfrak{h}_{\ell-1}\gamma_{\ell-1}\kappa_{\ell-1}) \leq 1, \ \ \mathbb{E}_{\ell-1}(\mathfrak{z}_{\ell-1}\gamma_{\ell-1}\kappa_{\ell-1}) \leq 1. \tag{5.24}$$

The desired inequality is proved from the equality

$$\gamma_{\ell-1}\kappa_{\ell-2} = 1 - \frac{[1 - \mathbb{E}_{\ell-1}(\gamma_{\ell-1}\kappa_{\ell-1}\tilde{\mathfrak{h}}_{\ell-1})][1 - \mathbb{E}_{\ell-1}(\gamma_{\ell-1}\kappa_{\ell-1}\tilde{\mathfrak{z}}_{\ell-1})]}{1 + \mathbb{E}_{\ell-1}(\gamma_{\ell-1}\kappa_{\ell-1}\mathfrak{h}_{\ell-1}\mathfrak{z}_{\ell-1})}, \tag{5.25}$$

because from (5.24) we obtain

$$\gamma_{\ell-1}\kappa_{\ell-2} \leq 1.$$

The equality (5.25) is straightforward to prove (from (5.16)) in the particular case where $\gamma_{\ell-1}\mathfrak{h}_{\ell-1}\mathfrak{z}_{\ell-1} = 0$, which is the case for a GUM. In order to prove (5.25) in general, and for notational simplicity on the next computations, we omit the index $\ell - 1$ and consider the operator $e(\bullet) = \mathbb{E}(\gamma\kappa\bullet)$. In this computation we only use the linearity and positivity of e. We start from $\tilde{\mathfrak{h}} = \mathfrak{h}(1 - \mathfrak{z}) = 1 - \mathfrak{z}$ to obtain

$$e(\mathfrak{h})e(\mathfrak{z}) = e(\tilde{\mathfrak{h}} + \mathfrak{h}\mathfrak{z})e(\tilde{\mathfrak{z}} + \mathfrak{h}\mathfrak{z}) = e(\tilde{\mathfrak{h}})e(\tilde{\mathfrak{z}}) + e(\mathfrak{h}\mathfrak{z})e(\tilde{\mathfrak{h}} + \tilde{\mathfrak{z}} + \mathfrak{h}\mathfrak{z}).$$

Since $\tilde{\mathfrak{h}} + \tilde{\mathfrak{z}} + \mathfrak{h}\mathfrak{z} = 1 - (1 - \mathfrak{h})(1 - \mathfrak{z})$ and the hypothesis $(1 - \mathfrak{h})(1 - \mathfrak{z}) = 0$, we get the equality

$$e(\mathfrak{h})e(\mathfrak{z}) = e(\tilde{\mathfrak{h}})e(\tilde{\mathfrak{z}}) + e(\mathfrak{h}\mathfrak{z})e(1).$$

Furthermore we obtain

$$e(1) - e(\tilde{\mathfrak{h}})e(\tilde{\mathfrak{z}}) = (1 + e(\mathfrak{h}\mathfrak{z}))e(1) - e(\mathfrak{h})e(\mathfrak{z}),$$

and using again $e(1) = e(\tilde{\mathfrak{h}}) + e(\tilde{\mathfrak{z}}) + e(\mathfrak{h}\mathfrak{z})$ we have

$$1 + e(\mathfrak{h}\mathfrak{z}) - (1 - e(\tilde{\mathfrak{h}}))(1 - e(\tilde{\mathfrak{z}})) = (1 + e(\mathfrak{h}\mathfrak{z}))e(1) - e(\mathfrak{h})e(\mathfrak{z}).$$

Dividing both sides by $1 + e(\mathfrak{h}\mathfrak{z})$ (note this quantity is at least one) we arrive to

$$\begin{aligned}
1 - \frac{(1-e(\tilde{\mathfrak{h}}))(1-e(\tilde{\mathfrak{z}}))}{1+e(\mathfrak{h}\mathfrak{z})} &= e(1) - \frac{e(\mathfrak{h})e(\mathfrak{z})}{1+e(\mathfrak{h}\mathfrak{z})} \\
&= \mathbb{E}_{\ell-1}(\gamma_{\ell-1}\kappa_{\ell-1}) - \frac{\mathbb{E}_{\ell-1}(\gamma_{\ell-1}\kappa_{\ell-1}\mathfrak{h}_{\ell-1})\mathbb{E}_{\ell-1}(\gamma_{\ell-1}\kappa_{\ell-1}\mathfrak{z}_{\ell-1})}{1+\mathbb{E}_{\ell-1}(\gamma_{\ell-1}\kappa_{\ell-1}\mathfrak{h}_{\ell-1}\mathfrak{z}_{\ell-1})} \\
&= \gamma_{\ell-1}\left[\mathbb{E}_{\ell-1}(\kappa_{\ell-1}) - \gamma_{\ell-1}\frac{\mathbb{E}_{\ell-1}(\kappa_{\ell-1}\mathfrak{h}_{\ell-1})\mathbb{E}_{\ell-1}(\gamma_{\ell-1}\kappa_{\ell-1}\mathfrak{z}_{\ell-1})}{1+\mathbb{E}_{\ell-1}(\gamma_{\ell-1}\kappa_{\ell-1}\mathfrak{h}_{\ell-1}\mathfrak{z}_{\ell-1})}\right] \\
&= \gamma_{\ell-1}\kappa_{\ell-2},
\end{aligned}$$

where the last equality follows from the facts that $\gamma_{\ell-1}$ is $\mathbb{E}_{\ell-1}$-measurable and the general iteration (5.16). Thus we have proved (5.25).

(5.22)$\Rightarrow$(5.20). Consider any collection of vectors (ξ_r : r even number in $0, \cdots$, $2p$) solution of the inequalities

$$\rho_{r-1}\gamma_{r-1} \vee \xi_{r-2} \leq \gamma_r + \xi_r,$$

for r an even number on $2, \cdots, 2p$. The smallest nonnegative solution of this iteration is given by

$$\xi_r = \sup\left\{0, \ \rho_1\gamma_1 - \sum_{s=1}^{r/2}\gamma_{2s}, \cdots, \rho_{2l+1}\gamma_{2l+1} - \sum_{s=l+1}^{r/2}\gamma_{2s}, \cdots, \rho_{r-1}\gamma_{r-1} - \gamma_r\right\}.$$

Then, (5.22) can be expressed as $\xi_{2p} = 0$. On the other hand notice that $\xi_r \in \text{Im}(\mathbb{E}_r)$.

We denote by $d_s = 1/\mathbb{E}_s(\kappa_s)$. We shall prove by induction that $\rho_s\gamma_s\kappa_s \leq \mathbb{1}$ and $\xi_r \leq d_r$ for even r. As before we consider only ℓ even. Note first that all λ's, μ's and κ's are nonnegative on $\ell, \cdots, 2p$. Recall that

$$\kappa_{\ell-1} = \frac{\mathbb{E}_\ell(\kappa_\ell)}{\mathbb{1} + \gamma_\ell\mathbb{E}_\ell(\kappa_\ell)} = \frac{1}{d_\ell + \gamma_\ell},$$

which is nonnegative. Since $\rho_{\ell-1}\gamma_{\ell-1} \leq \gamma_\ell + \xi_\ell \leq \gamma_\ell + d_\ell$, we get

$$\rho_{\ell-1}\gamma_{\ell-1}\kappa_{\ell-1} \leq \mathbb{1},$$

proving the first inequality.

Now, using (5.16) and the induction hypothesis we have

$$\kappa_{\ell-2} \leq \mathbb{E}_{\ell-1}(\kappa_{\ell-1}) = \mathbb{E}_{\ell-1}((d_\ell + \gamma_\ell)^{-1}).$$

On the other hand $\xi_{\ell-2} \in \text{Im}(\mathbb{E}_{\ell-2})$ and by construction $\xi_{\ell-2}^{-1} \geq (\gamma_\ell + \xi_\ell)^{-1}$ which implies that $\xi_{\ell-2}^{-1} \geq \mathbb{E}_{\ell-1}(\gamma_\ell + \xi_\ell)^{-1}$. Finally, from

$$d_{\ell-2}^{-1} \leq \mathbb{E}_{\ell-1}((\gamma_\ell + d_\ell)^{-1})$$

we conclude that $\xi_{\ell-2} \leq d_{\ell-2}$. The result is proven.

Remark 5.29 Consider the matrix

$$U_\beta = \begin{pmatrix} 1 & 0 & 0 & 0 \\ 0 & 1 & 0 & 0 \\ \beta & \beta & 1 & 0 \\ \beta & \beta & 0 & 1 \end{pmatrix} = (\Gamma_0 t_0)\cdot\mathbb{F}_0\cdot u_0 + \mathbb{I},$$

where $t_0 = \mathbb{1}_{\{3,4\}}$, $u_0 = \mathbb{1}_{\{1,2\}}$, $\Gamma_0 = \beta\mathbb{1}' \leq C_1 = \mathbb{1}$ and $\mathbb{F}_0 = \mathbb{1}\mathbb{1}'$. Let us decompose U in normal form as in (5.10)

$$U = \gamma_0\mathbb{E}_0 + \gamma_1\mathfrak{y}_1\cdot\mathbb{E}_0\cdot\mathfrak{z}_1 + \gamma_2\mathbb{I}$$

with $\gamma_0 = 0$, $\gamma_1 = 4\beta$, $\mathfrak{y}_1 = t_0$, $\mathfrak{z}_1 = u_0$, $\gamma_2 = \mathbb{1}$ and $\mathbb{E}_0 = \mathbb{M}_4$.

It is straightforward to check that $U_\beta^{-1} = U_{-\beta}$. Then for all $\beta \geq 0$ the matrix U_β is an inverse M-matrix. Moreover, U_β is a bi-potential if and only if $0 \leq \beta \leq 1/2$. We notice that for $\beta > 0$ the matrix U_β is not a GUM. When $\beta \geq 0$ the condition (5.14), that is

$$\Gamma_0 \leq C_1 + \Gamma_1$$

is equivalent to $\beta \leq 1$. This condition does not ensure that U is a bi-potential (this happens because the filtration is not dyadic). Nevertheless, the analogue sufficient condition in terms of the normalized factors (5.21)

$$\rho_0 \gamma_0 \leq c_1 + \gamma_1$$

is equivalent to $\beta \leq 1/2$, which is the right one.

Corollary 5.30 *Let U be a GUM. Then for all $t > 0$ the matrix $U + t\mathbb{I}$ is a bi-potential and*

$$(U + t\mathbb{I})^{-1} = \frac{1}{t}(\mathbb{I} - P_t)$$

for some doubly substochastic matrix P_t.

Moreover, U has a left and a right equilibrium potentials. If U is non singular then it is the inverse of a row and column diagonally dominant M-matrix.

Proof The main idea is to use Theorem 5.27 and to show that condition (5.21) holds for a GUM. Recall that U is weakly filtered with a decomposition in terms of a dyadic filtration $\{\mathbb{E}_0 = \mathbb{M}_n < \mathbb{E}_1 < \cdots < \mathbb{E}_{m-1} < \mathbb{E}_m = \mathbb{I}\}$ (see Proposition 5.21)

$$U = \sum_{s=0}^{m} c_s \cdot \mathbb{E}_s + (g_s t_s) \cdot \mathbb{E}_s \cdot u_s.$$

The coefficients t_s, u_s take values on $\{0, 1\}$, they satisfy $(\mathbb{1} - t_s)(\mathbb{1} - u_s) = 0$, and the smallest $\mathbb{E}_s$ variable that dominates them is $\hat{t}_s = \hat{u}_s = \mathbb{1}$. Also, $g_s = 0$ on the trivial atoms of $\mathbb{E}_s$ and on the nontrivial atoms R we have $t_s u_s \mathbb{1}_R = 0$ (see (5.12))

The main restriction is that the coefficients of this decomposition are nonnegative and satisfy (5.14) which we recall is

$$\forall s = 0, \cdots, m - 1 \quad \Gamma_s \leq C_{s+1} + \Gamma_{s+1},$$

where $\Gamma_s = \frac{g_s}{N_s}$ and $C_s = \frac{c_s}{N_s}$, being N_s the counting vector associated with $\mathbb{E}_s$. Hence, this condition is simplified to

$$\forall s = 0, \cdots, m - 1 \quad \frac{N_{s+1}}{N_s} g_s \leq c_{s+1} + g_{s+1}. \tag{5.26}$$

Recall that on the trivial atoms of $\mathbb{E}_s$ this condition holds because $g_s = 0$.

Now, let us put the decomposition of U in the form given in Theorem 5.27

$$U = \sum_{\ell=0}^{2m} (\gamma_\ell \eta_\ell) \cdot \tilde{\mathbb{E}}_{\ell \cdot 3\ell}$$

with

$$\forall s = 0, \cdots, m-1 \quad \begin{cases} \gamma_{2s} = \mathfrak{c}_s; \\ \gamma_{2s+1} = \mathfrak{g}_s; \\ \mathfrak{y}_{2s} = \mathfrak{z}_{2s} = \mathbb{1}; \\ \mathfrak{y}_{2s+1} = \mathfrak{t}_s; \\ \mathfrak{z}_{2s+1} = \mathfrak{u}_s, \end{cases}$$

and $\gamma_{2m} = \mathfrak{c}_m + \mathfrak{g}_m \mathfrak{t}_m \mathfrak{u}_m = \mathfrak{c}_m$. Also $\tilde{\mathbb{E}}_{2s} = \tilde{\mathbb{E}}_{2s+1} = \mathbb{E}_s$. So condition (5.21), which has to be verified for odd indexes, is implied by condition (5.26) after proving that

$$\forall s = 0, \cdots, m-1 \quad \rho_{2s+1} \leq \frac{N_{s+1}}{N_s},$$

holds for any dyadic filtration $\mathbb{E}_0, \cdots, \mathbb{E}_m$. Indeed, recall the definition of ρ_{2s+1} (see (5.19))

$$\rho_{2s+1} = ([\mathbb{E}_s(\mathbb{1} - \mathfrak{u}_s)](\mathbb{1} - \mathfrak{u}_s)) + ([\mathbb{E}_s(\mathbb{1} - \mathfrak{t}_s)](\mathbb{1} - \mathfrak{t}_s)),$$

where we have used the fact that $\mathfrak{t}_s, \mathfrak{u}_s$ are $\{0,1\}$ valued and then $\mathbb{1}_{\mathbb{1} - \mathfrak{u}_s > 0} = \mathbb{1} - \mathfrak{u}_s, \mathbb{1}_{\mathbb{1} - \mathfrak{t}_s > 0} = \mathbb{1} - \mathfrak{t}_s$.

If $R = \{i\}$ is a trivial atom of $\mathbb{E}_s$, then it is also a trivial atom in $\mathbb{E}_{s+1}$ and $(\mathfrak{u}_s)_i = (\mathfrak{t}_s)_i = 1$. It is straightforward to show that $(\rho_{2s+1})_i = 0$ and the desired inequality holds for those sites.

Now, every nontrivial atom R in $\mathbb{E}_s$ splits into two atoms P, Q in $\mathbb{E}_{s+1}$. Since $\mathfrak{t}_s, \mathfrak{u}_s$ take only values on $\{0, 1\}$, we get for $i \in R$

$$(\mathbb{E}_s(\mathbb{1} - \mathfrak{u}_s))_i = \frac{\#\{j \in R : (\mathfrak{u}_s)_j = 0\}}{(N_s)_i}, \quad (\mathbb{E}_s(\mathbb{1} - \mathfrak{t}_s))_i = \frac{\#\{j \in R : (\mathfrak{t}_s)_j = 0\}}{(N_s)_i}.$$

On the other hand $\mathfrak{t}_s \in \mathrm{Im}(\mathbb{E}_{s+1})$ implies that $\mathfrak{t}_s$ is constant on the atoms of $\mathbb{E}_{s+1}$. Since $\mathfrak{t}_s$ takes values on $\{0, 1\}$, we conclude $\mathfrak{t}_s \mathbb{1}_R = a \mathbb{1}_P + b \mathbb{1}_Q$ with $a, b \in \{0, 1\}$.

Now, recall that $\hat{\mathfrak{t}}_s$ is the smallest variable $\mathbb{E}_s$-measurable that dominates $\mathfrak{t}_s$, which we assume is $\hat{\mathfrak{t}}_s = \mathbb{1}$. On the other hand $\hat{\mathfrak{t}}_s \mathbb{1}_R = a \vee b \, \mathbb{1}_R$, from where we get $a \vee b = 1$. Similarly, $\mathfrak{u}_s \mathbb{1}_R = c \mathbb{1}_P + d \mathbb{1}_Q$, with $c, d \in \{0, 1\}$ and $c \vee d = 1$. Using that $\mathfrak{t}_s \mathfrak{u}_s \mathbb{1}_R = 0$ (see condition (5.12)), we deduce that one of the two mutually exclusive cases hold $\mathfrak{t}_s \mathbb{1}_R = \mathbb{1}_P, \mathfrak{u}_s \mathbb{1}_R = \mathbb{1}_Q$ or $\mathfrak{t}_s \mathbb{1}_R = \mathbb{1}_Q, \mathfrak{u}_s \mathbb{1}_R = \mathbb{1}_P$. Hence, in the first case we obtain

$$(\rho_{2s+1})_i = \frac{\#P}{(N_s)_i}(\mathbb{1}_P)_i + \frac{\#Q}{(N_s)_i}(\mathbb{1}_Q)_i = \frac{(N_{s+1})_i}{(N_s)_i}.$$

The other case is completely analogous and we conclude the proof.

Remark 5.31 Notice that in the last part of the proof we have actually showed that ρ_{2s+1} and $\frac{N_{s+1}}{N_s}$ agree on the nontrivial atoms of $\mathbb{E}_s$. This shows that (5.21) and (5.26) are equivalent and therefore in the context of positive CBF the condition (5.21) is equivalent to have a GUM.

Example 5.32 Let us study condition (5.21) for weakly filtered matrices of size 4 associated with the strict filtration $\mathbf{F} = (\mathbb{E}_0, \mathbb{E}_1, \mathbb{E}_2)$ where

$$\mathbb{E}_0 = \mathbb{M}_4 = \frac{1}{4}\begin{pmatrix} 1 & 1 & 1 & 1 \\ 1 & 1 & 1 & 1 \\ 1 & 1 & 1 & 1 \\ 1 & 1 & 1 & 1 \end{pmatrix}, \ \mathbb{E}_1 = \begin{pmatrix} \mathbb{M}_1 & 0 & 0 \\ 0 & \mathbb{M}_1 & 0 \\ 0 & 0 & \mathbb{M}_2 \end{pmatrix} = \begin{pmatrix} 1 & 0 & 0 & 0 \\ 0 & 1 & 0 & 0 \\ 0 & 0 & 1/2 & 1/2 \\ 0 & 0 & 1/2 & 1/2 \end{pmatrix}, \text{ and } \mathbb{E}_2 = \mathbb{I}.$$

We take as example

$$\gamma_0 = 4a\mathbb{1}, \gamma_1 = 4b\mathbb{1}, \gamma_2 = (c, d, 2e, 2e)', \gamma_3 = (0, 0, 2f, 2f)', \gamma_4 = (g, h, i, j)',$$
$$t_1 = (1, 1, 1, 1)', u_1 = (1/3, 0, 1, 1)', t_3 = (1, 1, 0, 1)', u_3 = (1, 1, 1, 0)'.$$

Notice that u_1 does not take values on $\{0, 1\}$. By construction, for $\ell = 0, 1$ we have

$$\begin{cases} \gamma_{2\ell}, \gamma_{2\ell+1} \text{ are } \mathbb{E}_\ell\text{-measurable}, t_{2\ell+1}, u_{2\ell+1} \text{ are } \mathbb{E}_{\ell+1}\text{-measurable}; \\[2mm] \hat{t}_{2\ell+1}^{\mathbb{E}_\ell} = \hat{u}_{2\ell+1}^{\mathbb{E}_\ell} = \mathbb{1}, (\mathbb{1} - t_{2\ell+1})(\mathbb{1} - u_{2\ell+1}) = 0; \\[2mm] \gamma_3 t_3 u_3 = 0 \text{ on the trivial atoms of } \mathbb{E}_1 \text{ (see (5.10)).} \end{cases}$$

We assume that all the parameters are strictly positive. The matrix we consider is

$$U = \gamma_0 \mathbb{E}_0 + \gamma_1 t_1 \cdot \mathbb{E}_0 \cdot u_1 + \gamma_2 \cdot \mathbb{E}_1 + \gamma_3 t_3 \cdot \mathbb{E}_1 \cdot u_3 + \gamma_4 \cdot \mathbb{I},$$

which turns out to be

$$U = \begin{pmatrix} g + c + a + b/3 & a & a + b & a + b \\ a + b/3 & h + d + a & a + b & a + b \\ a + b/3 & a & i + a + b + e & a + b + e \\ a + b/3 & a & f + a + b + e & j + a + b + e \end{pmatrix}.$$

Notice that g, c and h, d appear only on the diagonal and moreover they appear as $g + c$ and $h + d$. This indicates that we can rewrite the decomposition with fewer parameters. In fact it is enough to consider $\gamma_4 = (0, 0, i, j)'$, that is $g = h = 0$, and c will play the role of $g + c$ (similarly d with respect to $h + d$). This simplification gives

$$
U = \begin{pmatrix}
c + a + b/3 & a & a + b & a + b \\
a + b/3 & d + a & a + b & a + b \\
a + b/3 & a & i + a + b + e & a + b + e \\
a + b/3 & a & f + a + b + e & j + a + b + e
\end{pmatrix}.
$$

Notice that there is no constant block of size 1×3 and there is only one constant block of size 2×2. Thus, U is not a positive CBF. In particular, U is not a GUM.

Let us see what restrictions are imposed by condition (5.21). For that we have to compute ρ_1, ρ_3

$$
\rho_1 = \frac{1}{4}(5/3, 5/3, 0, 0)', \quad \rho_3 = \frac{1}{2}(0, 0, 1, 1)'.
$$

The restrictions on the parameters are

$$
b \leq \frac{3}{5}\min\{c, d\} \quad \text{and} \quad f \leq \min\{i, j\}. \tag{5.27}
$$

The last constrain is natural because is exactly the condition that U is row and column pointwise diagonally dominant at sites 3 and 4 (a property that every bi-potential satisfies). In addition, the first condition implies that U is pointwise diagonally dominant at sites 1 and 2. Nevertheless, the exact conditions for U being a pointwise diagonally dominant at sites 1 and 2 are respectively $b \leq 3c/2$ and $b \leq d$.

The exact conditions that ensure U is a bi-potential are of two types. As before the necessary conditions that come from the diagonally dominance of U

$$
b \leq \min\{3c/2, d\} \quad \text{and} \quad f \leq \min\{i, j\}.
$$

Given the structure of U it is possible to show that under these conditions U is nonsingular and its inverse is a row diagonally dominant M-matrix (use Schur's decomposition or MAPLE). The condition for U^{-1} to be column diagonally dominant is

$$
b \leq \frac{3cd}{3c + 2d}.
$$

Notice that this condition is strictly weaker than $b \leq \frac{3}{5}\min\{c, d\}$, except when $c = d$, in which case both conditions agree. So, in this example (5.21) is close to be necessary for U to be a bi-potential.

5.4 Spectral Functions, M-Matrices and Ultrametricity

This section is inspired by the work of N. Bouleau (see [7]). We shall establish some stability properties for the class of M-matrices under spectral functions. Given a regular function f and a matrix A we consider $f(A)$ in the functional sense. That is, if $f(x) = \sum_{n \geq 0} a_n x^n$ and the spectral radius of A is smaller than the radius of convergence of the series, then $f(A)$ is defined to be $\sum_{n \geq 0} a_n A^n$. One interesting case, which is not cover with this definition is the function defined on $\mathbb{R}^+$ by $f(x) = x^\alpha$ for $\alpha > 0$. In this situation one can use the following fact: There exists a measure μ_α such that for all $x > 0$

$$x^\alpha = \int_0^\infty e^{-rx} d\mu_\alpha(r).$$

Then $A^\alpha = \int_0^\infty e^{-rA} d\mu_\alpha(r)$, whenever this integral makes sense. If A is symmetric, we can use the spectral theorem to define $f(A)$ for more general functions. This is done as follows. Assume that $A = PDP^{-1}$, where D is a diagonal matrix. Then $f(A)$ is defined as $f(A) = Pf(D)P^{-1}$, where $f(D)$ is the diagonal matrix whose i^{th} diagonal element is $f(D_{ii})$.

In this section, we shall study the following problem: Characterize the symmetric semidefinite matrices A with a spectral decomposition given by

$$A = \sum_{s=0}^{m} \theta_s (\mathbb{E}_s - \mathbb{E}_{s-1}), \tag{5.28}$$

where $\mathbf{F} = (\mathbb{E}_0, \mathbb{E}_1, \cdots, \mathbb{E}_m)$ is a filtration. Here we assume also that $\mathbb{E}_{-1} = 0, \mathbb{E}_m = \mathbb{I}$ and $0 \leq \theta_0 \leq \cdots \leq \theta_m$ are real numbers. Without loss of generality we can take $\theta_m = 1$. The assumption that θ is increasing will play an important role in what follows. Indeed, integrating by parts yields that $A = \mathbb{I} - \sum_{s=0}^{m-1} a_s \mathbb{E}_s$, where $a_s = \theta_{s+1} - \theta_s$ are nonnegative. Thus, $\mathbb{I} - A$ is a strongly filtered operator in the sense of Definition (5.11) and A is a Z-matrix. Since $A\mathbb{1} = \theta_0 \geq 0$ the diagonal elements of A are nonnegative and A is row diagonally dominant. So A is an M-matrix as soon as it is nonsingular, which is equivalent to $\theta_0 > 0$. Indeed, we have

$$A^{-1} = \sum_{s=0}^{m} \frac{1}{\theta_s} (\mathbb{E}_s - \mathbb{E}_{s-1}),$$

and therefore we obtain

$$A^{-1} = \mathbb{I} - \sum_{s=0}^{m-1} \left(\frac{1}{\theta_{s+1}} - \frac{1}{\theta_s} \right) \mathbb{E}_s = \mathbb{I} + \sum_{s=0}^{m-1} \left(\frac{1}{\theta_s} - \frac{1}{\theta_{s+1}} \right) \mathbb{E}_s$$

is a nonnegative matrix.

Before introducing the main result of this section, we need a definition. Recall that $\langle \, , \, \rangle$ is the standard inner product on $\mathbb{R}^n$ and $(y)_+$ is the positive part of the vector y. As usual if the spectral decomposition of A is $A = \sum_{s=0}^{m} \theta_s (F_s - F_{s-1})$, then $g(A)$ is defined as $\sum_{s=0}^{m} g(\theta_s)(F_s - F_{s-1})$. Here $(F_s : s = -1, \cdots, m)$ is an increasing sequence of projections such that $F_m = \mathbb{I}$, $F_{-1} = 0$.

Definition 5.33 An $n \times n$ symmetric matrix B is said to be a **Dirichlet** matrix if for all $x \in \mathbb{R}^n$ we have

$$\langle Bx, (x - \mathbb{1})_+ \rangle \leq 0.$$

Moreover B is said **Dirichlet-Markov** if it satisfies in addition $B\mathbb{1} = 0$.

The next result is an improvement of Theorem 2.27 given in Chap. 2, on the relation between M-matrices and generators of semigroups.

Theorem 5.34 *Let A be a matrix. The following relations hold:*

(i) A is a Z-matrix if and only if $P_t = e^{-tA}$ is a nonnegative semigroup;
(ii) A is a row diagonally dominant Z-matrix if and only if $P_t = e^{-tA}$ is a nonnegative semigroup and for all $t \geq 0$ we have $P_t \mathbb{1} \leq \mathbb{1}$.

Hence, when A is nonsingular we obtain: A is a row diagonally dominant M-matrix if and only if (P_t) is a substochastic semigroup.
If A is symmetric the following are equivalent:

(iii) $-A$ is Dirichlet;
(iv) A is a row diagonally dominant Z-matrix, that is the entries of A satisfy

> *(iv.1) for all i, $A_{ii} \geq 0$;*
> *(iv.2) for all $i \neq j$, $A_{ij} \leq 0$;*
> *(iv.3) $A\mathbb{1} \geq 0$.*

(v) $S_t = e^{-tA}$ is a nonnegative semigroup and for all $t \geq 0$ we have $S_t \mathbb{1} \leq \mathbb{1}$.

Moreover, $-A$ is Dirichlet-Markov if and only if (S_t) is a Markov semigroup, which is also equivalent to (iv) where (iv.3) is strengthen to $A\mathbb{1} = 0$.

Recall that a semigroup is said substochastic if it is nonnegative and $P_t \mathbb{1} \leq \mathbb{1}$, with strict inequality at some coordinate. When $P_t \mathbb{1} = \mathbb{1}$ the semigroup is called Markov (see Definition 2.26). We also point out that condition (iv.1) is deduced from $(iv.2)$ and $(iv.3)$, but it is included in the statement for the sake of completeness.

Proof Let us show the equivalence in (i). From the formula

$$P_t = \lim_{n \to \infty} \left(\mathbb{I} - \frac{t}{n} A \right)^n,$$

we deduce that P_t is nonnegative if A is a Z-matrix. Conversely, the following formula

$$\frac{P_t - \mathbb{I}}{t} = -A + t \sum_{r=2}^{\infty} \frac{(-t)^{r-2}}{r!} A^r,$$

and the positivity of (P_t) show that A is a Z-matrix by taking the limit as $t \to 0$. This shows (i).

(ii). From the relation

$$\frac{dP_t \mathbb{1}}{dt} = -P_t A \mathbb{1},$$

we deduce that if $A \mathbb{1} \geq 0$ then $P_t \mathbb{1}$ is decreasing. As $P_0 \mathbb{1} = \mathbb{I} \mathbb{1} = \mathbb{1}$ we deduce that $P_t \mathbb{1} \leq \mathbb{1}$. Conversely, assume that (P_t) is a nonnegative semigroup and for all $t \geq 0$ it holds $P_t \mathbb{1} \leq \mathbb{1}$, then A is a Z-matrix and

$$0 \geq \frac{dP_t \mathbb{1}}{dt}\Big|_{t=0} = -P_0 A \mathbb{1} = -A \mathbb{1},$$

showing that A is row diagonally dominant.

For the rest of the proof we assume that A is symmetric. From (ii), we deduce (iv) and (v) are equivalent. We now show that under (iv) the matrix A is positive semidefinite. In fact, for $\epsilon > 0$ the matrix $A_\epsilon = A + \epsilon \mathbb{I}$ is a symmetric Z-matrix for which $A_\epsilon \mathbb{1} > 0$. This implies again that A_ϵ is an M-matrix and therefore is positive definite. We conclude that A is positive semidefinite.

Let us prove that if A satisfies (iv) then $-A$ is a Dirichlet matrix. That is, given $x \in \mathbb{R}^n$, we need to prove that $\langle Ax, (x - \mathbb{1})_+ \rangle \geq 0$. For that purpose consider $J = \{ j \in I : x_j > 1 \}$. Since

$$\langle Ax, (x - \mathbb{1})_+ \rangle = \langle A(x - \mathbb{1}), (x - \mathbb{1})_+ \rangle + \langle A\mathbb{1}, (x - \mathbb{1})_+ \rangle$$

and $A\mathbb{1} \geq 0$, it is enough to show that $\langle A(x - \mathbb{1}), (x - \mathbb{1})_+ \rangle \geq 0$. This quantity is

$$\langle A(x - \mathbb{1}), (x - \mathbb{1})_+ \rangle = \sum_{i,j \in J} A_{ij}(x_j - 1)(x_i - 1) + \sum_{i \in J, j \in J^c} A_{ij}(x_j - 1)(x_i - 1).$$

The first term in the last equality is nonnegative, because the matrix A_{JJ} is positive semidefinite. The second term is also nonnegative because for $i \in J, j \in J^c$ we have $x_i - 1 \geq 0, x_j - 1 \leq 0, A_{ij} \leq 0$. This proves that $-A$ is a Dirichlet matrix.

Assume now that $-A$ is a Dirichlet matrix. We shall prove that A satisfies (iv). For that purpose take $x = ae(i)$, with $a > 1$ and $e(i)$ the i-th vector of the canonical basis in $\mathbb{R}^n$. Then

$$a(a - 1)A_{ii} = \langle Ax, (x - 1)_+ \rangle \geq 0,$$

proving $(iv.1)$.

Consider now $x = ae(i) + be(j)$, with $i \neq j$ and $a > 1$, $b < 0$. Since

$$0 \leq \langle Ax, (x - 1)_+ \rangle = (a - 1)(aA_{ii} + bA_{ij}),$$

we conclude that $A_{ij} \leq 0$, by taking b sufficiently large. This shows $(iv.2)$ holds.

Now, let us take $x = (a - 1)e(i) + \mathbb{1}$ for $a > 1$. From $\langle Ax, (x - 1)_+ \rangle \geq 0$, we deduce that

$$(a - 1)A_{ii} + \sum_j A_{ij} \geq 0.$$

This inequality shows that $(iv.3)$ holds, by passing to the limit $a \downarrow 1$.

The next result is taken from [7]. We shall use the notion of complete maximum principle (CMP) introduced in Definition 2.5, Chap. 2. Recall that a nonnegative matrix V satisfies the CMP if for all $x \in \mathbb{R}^n$, $x_i \geq 0$ implies that $(Vx)_i \leq 1$, then also $Vx \leq \mathbb{1}$.

Theorem 5.35 *Consider a symmetric positive semidefinite matrix A that satisfies $A\mathbb{1} = 0$. We define $P_t = e^{-At}$, for $t \geq 0$, the semigroup associated with $-A$. Then the following are equivalent*

(i) *A has a spectral decomposition as in (5.28), that is*

$$A = \sum_{s=0}^{m} \theta_s(\mathbb{E}_s - \mathbb{E}_{s-1}),$$

where $\mathbf{F} = (\mathbb{E}_0, \mathbb{E}_1, \cdots, \mathbb{E}_m)$ is a filtration and $0 \leq \theta_0 \leq \cdots \leq \theta_m$;
(ii) *for all $t \geq 0$ the matrix $A_t = -\mathbb{1}_{(t,\infty)}(A)$ is Dirichlet-Markov;*
(iii) *for any increasing, nonnegative left continuous function $\phi : \mathbb{R}^+ \to \mathbb{R}^+$, the matrix $\phi(0)\mathbb{I} - \phi(A)$ is Dirichlet-Markov;*
(iv) *for any $p \geq 1$ the matrix $-(A^p)$ is Dirichlet-Markov;*
(v) *for all t the matrix $-P_t^{-1}$ is Dirichlet;*
(vi) *the semigroup (P_t) is Markovian and for all t the matrix P_t satisfies the CMP;*
(vii) *$-P_1^{-1}$ is a Dirichlet matrix;*
$(viii)$ *P_1 is nonnegative and satisfies the CMP.*

Let us state the following result for projection matrices. Recall that a projection E is a symmetric idempotent matrix.

Lemma 5.36 *Assume E is a projection matrix. Then E is a conditional expectation if and only if for all $t \geq 0$ the semigroup $S_t = e^{-t}\mathbb{I} + (1 - e^{-t})E$ is Markov.*

Proof If (S_t) is Markovian then passing to the limit $t \to \infty$ we conclude that E is nonnegative and $E\mathbb{1} = \mathbb{1}$. Since E is a projection we conclude it is a conditional expectation (see Definition 5.1).

Conversely, if E is a conditional expectation then S_t is positive-preserving and $S_t\mathbb{1} = \mathbb{1}$, proving the result.

Proof (Theorem 5.35) Consider $(F_t : t \in \mathbb{R})$ the resolution of the identity associated with the semidefinite matrix A. That is (F_t) is an increasing right continuous with left limits family of projections, such that $F_{0-} = 0$ (because A is positive semidefinite), $F_t = \mathbb{I}$ for all $t \geq \|A\|$ and

$$A = \int_0^\infty u \, dF_u.$$

(F_t) is a piecewise constant family with a jump at t if and only if t is an eigenvalue of A and the size of the jump $F_t - F_{t-}$ is the projection onto their corresponding eigenspace. Since $A\mathbb{1} = 0$ we obtain that 0 is an eigenvalue and the corresponding eigenspace contains at least the constant vectors (hence $F_0 > 0$).

Let us prove the equivalence between (i) and (ii). The matrix $A_t = -\mathbb{1}_{(t,\infty)}(A) = F_t - \mathbb{I}$ is negative semidefinite and the semigroup that induces is

$$S_s = e^{sA_t} = e^0 F_t + e^{-s}(\mathbb{I} - F_t) = e^{-s}\mathbb{I} + (1 - e^{-s})F_t.$$

Then, (S_s) is a Markov semigroup if and only if A_t is Dirichlet-Markov according to Theorem 5.34 and this semigroup is Markov if and only if F_t is a conditional expectation according to Lemma 5.36. proving the desired equivalence.

It is clear that (iii) implies (ii) and we next prove the converse. Integration by parts yields

$$\phi(0)\mathbb{I} - \phi(A) = \int_0^\infty \phi(0) - \phi(u) \, dF_u = \int_0^\infty (F_u - \mathbb{I}) \, d\phi(u) = \int_0^\infty A_u \, d\phi(u).$$

Then we obtain a representation for the quadratic form

$$\langle (\phi(0)\mathbb{I} - \phi(A))x, (x - \mathbb{1})_+ \rangle = \int_0^\infty \langle A_u x, (x - \mathbb{1})_+ \rangle \, d\phi(u) \leq 0,$$

and $\phi(0)\mathbb{I} - \phi(A)$ is Dirichlet. Since $(\phi(0)\mathbb{I} - \phi(A))\mathbb{1} = 0$ we deduce (iii) holds.

Clearly (*iii*) implies (*iv*) and we prove they are equivalent by showing that (*iv*) implies (*i*). In fact, consider the following family of matrices

$$G(t, r, p) = \exp\left(-t\left(\frac{A}{r}\right)^p\right)$$

for $t > 0, 0 \leq r \leq \|A\|$. According to Theorem 5.34 $G = G(t, r, p)$ is a positive matrix and $G\mathbb{1} = \mathbb{1}$. Using the Dominated Convergence Theorem and the fact that

$$\exp\left(-t\left(\frac{x}{r}\right)^p\right) \underset{p \to \infty}{\to} \mathbb{1}_{[0,r)}(x) + e^{-t}\mathbb{1}_{\{r\}}(x),$$

we obtain that

$$G(t, r) = \int_0^{\|A\|} \mathbb{1}_{[0,r)}(u) + e^{-t}\mathbb{1}_{\{r\}}(u) dF_u = F_{r-} + e^{-t}(F_r - F_{r-}),$$

is a nonnegative matrix and $G(t, r)\mathbb{1} = \mathbb{1}$. By passing to the limit as $t \downarrow 0$, we obtain that for each r the projection F_r is nonnegative and $F_r\mathbb{1} = \mathbb{1}$, proving that it is a conditional expectation and therefore (*i*) holds.

We prove now that (*i*) implies (*v*). For any $t \geq 0$ the function $\phi(x) = e^{tx}$ is increasing on x, which yields that $\phi(0)\mathbb{I} - \phi(A) = \mathbb{I} - P_t^{-1}$ is a Dirichlet-Markov matrix (by (*iii*)) and a fortiori $-P_t^{-1}$ is a Dirichlet matrix.

(*v*) implies (*vi*). From Theorem 5.34 we obtain that P_t^{-1} is row diagonally dominant M-matrix. Hence, P_t is a potential, in particular it is a nonnegative matrix. Since $A\mathbb{1} = 0$ we conclude that $P_t\mathbb{1} = \mathbb{1}$ and therefore (P_t) is a Markov semigroup. On the other hand from Theorem 2.9 P_t satisfies the CMP and (*vi*) is proved. The converse, that is (*vi*) implies (*v*), is a consequence of the same cited theorem. We mention here that this argument also shows the equivalence between (*vii*) and (*viii*).

Clearly (*vi*) is stronger than (*viii*) and then to finish the proof we need, for example, to show that (*i*) is deduced from (*viii*). We have that $e^A = P_1^{-1}$ is a row diagonally dominant symmetric M-matrix. Then for any $\alpha \in [0, 1]$ the fractional power $e^{\alpha A}$ is also a row diagonally dominant M-matrix (see Corollary 5.39) and therefore

$$\left(e^{-te^{\alpha A}} : t \geq 0\right)$$

is a nonnegative substochastic semigroup. Recall that $A\mathbb{1} = 0$, which implies $e^{\alpha A}\mathbb{1} = \mathbb{1}$. We deduce that

$$\left(e^{-t(e^{\alpha A} - \mathbb{I})} : t \geq 0\right)$$

is a Markovian semigroup. On the other hand since $e^{\alpha A}$ is a row diagonally dominant M-matrix we conclude its inverse $e^{-\alpha A}$ is nonnegative and $e^{-\alpha A}\mathbb{1} = \mathbb{1}$. This implies that the semigroup (e^{-tA}) is a Markovian semigroup. Indeed, if $[t]$ is the integer part of t and $\alpha = t - [t]$ we get

$$e^{-tA} = e^{-[t]A}e^{-(t-[t])A} = (e^{-A})^{[t]}e^{-\alpha A},$$

proving that e^{-tA} is nonnegative and $e^{-tA}\mathbb{1} = \mathbb{1}$. Hence, according to Theorem 5.34, A is a row diagonally dominant M-matrix. With these properties at hand we shall prove the following claim by induction

$$\begin{cases} Q_t(k,\alpha) = \exp\left(\frac{-t}{\alpha^{k+1}}\left(e^{\alpha A} - \sum_{r=0}^{k} \frac{\alpha^r}{r!} A^r \right) \right) & \text{is a Markov semigroup} \\[4mm] A^{k+1} & \text{is a row diagonally dominant } M\text{-matrix} \end{cases}$$

(5.29)

Clearly, this property holds for $k = 0$. Assume it also holds for $k - 1$, that is $(Q_t(k-1,\alpha) : t \geq 0)$ is a Markov semigroup and A^k is a row diagonally dominant M-matrix. In particular $(e^{-sA^k} : t \geq 0)$ is a Markov semigroup (recall that $A\mathbb{1} = 0$ by hypothesis). Since

$$\begin{aligned} Q_t(k,\alpha) &= \exp\left(\frac{-t}{\alpha^{k+1}}\left(e^{\alpha A} - \sum_{r=0}^{k} \frac{\alpha^r}{r!} A^r \right) \right) \\ &= \left[\exp\left(\frac{-t/\alpha}{\alpha^k}\left(e^{\alpha A} - \sum_{r=0}^{k-1} \frac{\alpha^r}{r!} A^r \right) \right) \right]\left[\exp\left(\frac{-t}{\alpha^{k+1}} \frac{\alpha^k}{k!} A^k \right) \right] \\ &= Q_{t/\alpha}(k-1,\alpha) \exp\left(\frac{-t}{\alpha^{k+1}} \frac{\alpha^k}{k!} A^k \right) \end{aligned}$$

is the product of two nonnegative matrices. Then $Q_t(k,\alpha)$ is also a nonnegative matrix. Since $Q_t(k,\alpha)\mathbb{1} = \mathbb{1}$ we deduce that $(Q_t(k,\alpha) : t \geq 0)$ is a Markov semigroup. Now, if we take the limit as $\alpha \downarrow 0$, we obtain that

$$Q_t(k,0^+) = e^{-\frac{t}{(k+1)!}A^{k+1}}$$

is a Markov semigroup, which implies that A^{k+1} is also a row diagonally dominant M-matrix. Proving the claim (5.29).

Since $A\mathbb{1} = 0$, we conclude that for all $k \geq 1$ the matrix $-A^k$ is Dirichlet-Markov, that is (iv) holds, which is equivalent to (i), and the theorem is proved.

5.4.1   Stability of M-Matrices Under Completely Monotone Functions

The results we have obtained in the previous discussion have interesting consequences in the stability of M-matrices under Completely monotone functions.

Definition 5.37 A function $\psi : (0, \infty) \to \mathbb{R}$ is said to be a **Completely monotone function** if there exists a positive finite measure μ carried by $(0, \infty)$ such that for all $x > 0$

$$\psi(x) = \int_0^\infty e^{-rx} \, \mu(dr).$$

The measure μ is finite if and only if $\psi(0^+)$ is finite, and in this case we say that ψ is a Completely monotone function on $\mathbb{R}^+$.

Notice that it is assumed μ has support on $(0, \infty)$, which rules out in particular the Dirac measure at 0. This is important because functions associated with a measure proportional to a Dirac measure at 0 are constant. In summary, in our discussion a Completely monotone function is not constant.

Also we point out that a Completely monotone function is a C^∞ function that satisfies the alternating sign pattern

$$\forall n \geq 0 \; \forall x > 0 \quad (-1)^n \psi^{(n)}(x) \geq 0.$$

Theorem 5.38 *Let A be a row diagonally dominant M-matrix and $\psi : \mathbb{R}^+ \to \mathbb{R}$ a function such that for all $t > 0$ fixed, the function $G(x) = e^{t\psi(x)}$ is a Completely monotone function. Then, $\psi(0)\mathbb{I} - \psi(A)$ is a row diagonally dominant M-matrix. This is the case if ψ itself is a Completely monotone function.*

Proof Consider the finite measure μ_t such that

$$e^{t(\psi(x)-\psi(0))} = \int_0^\infty e^{-rx - t\psi(0)} \mu_t(dr).$$

From Theorem 5.34 we know that $P_r = e^{-rA}$ is a substochastic semigroup. Then

$$S_t = e^{t(\psi(A)-\psi(0)\mathbb{I})} = e^{-t\psi(0)} \int P_r \, \mu_t(dr)$$

is clearly a nonnegative semigroup and

$$S_t \mathbb{1} = e^{-t\psi(0)} \int P_r \mathbb{1} \, \mu_t(dr) \leq e^{-t\psi(0)} \int \mu_t(dr) \mathbb{1} = e^{-t\psi(0)} e^{t\psi(0)} \mathbb{1} = \mathbb{1}.$$

Since, for $t > 0$, the function $e^{t(\psi(\bullet))}$ is not constant then μ_t is not proportional to the Dirac measure at 0. This implies that there exists a coordinate where $(P_r \mathbb{1})_i$ is strictly smaller than 1, on a set of positive measure $\mu_t(dr)$, showing that (S_t) is a substochastic semigroup. From the previous lemma we conclude that $\psi(0)\mathbb{I} - \psi(A)$ is a row diagonally dominant M-matrix.

The last part of the result is shown as soon as we prove that if ψ is a Completely monotone function then for all $t > 0$ fixed $e^{t\psi(x)}$ is a Completely monotone function. This is straightforward from the expansion

$$e^{t\psi(x)} = \sum_{n=0}^{\infty} \frac{t^n}{n!} \psi(x)^n.$$

Indeed, since $\psi(x) = \int e^{-rx} \mu(dr)$ we obtain

$$\psi(x)^n = \int \cdots \int e^{-r_1 x} \cdots e^{-r_n x} \, \mu(dr_1) \cdots \mu(dr_n) = \int e^{-rx} \mu_n(dr),$$

where $r = r_1 + \cdots + r_n$ and

$$\mu_n(dr) = \int \mathbb{1}_{0 \le r_2, \cdots, 0 \le r_n} \mathbb{1}_{r_2 + \cdots + r_n \le r} \, \mu(d(r - r_2 - \cdots - r_n)) \mu(dr_2) \cdots \mu(dr_n).$$

The result follows from

$$e^{t\psi(x)} = \int e^{-rx} \sum_{n=0}^{\infty} \frac{t^n}{n!} \mu_n(dr).$$

Corollary 5.39 *Assume that A is a row diagonally dominant M-matrix and* $\alpha \in [0, 1]$. *Then* A^{α} *is also a row diagonally dominant M-matrix.*

Proof The function $G(x) = e^{-x^{\alpha}}$ is a Completely monotone function.

Chapter 6
Hadamard Functions of Inverse M-Matrices

There are remarkable properties relating inverse M-matrices and Hadamard functions. In the first part of this chapter we study stability for the class of inverse M-matrices under Hadamard functions. We prove that the class of GUM matrices is the largest class of bi-potential matrices stable under Hadamard increasing functions.

We show that for any real $r \geq 1$ and any inverse M-matrix W, the r-power $W^{(r)}$, in the sense of Hadamard, is also an inverse M-matrix. This was conjectured for $r = 2$ by Neumann in [54], and solved for integer $r \geq 1$ by Chen in [11] and for general real numbers greater than 1 in [12]. We present the proof of this fact based on our paper [23]. As a complement to this result, we mention the characterization of matrices whose p-th Hadamard power is an inverse M-matrix, for large p, given in Johnson and Smith [39].

We study the invariance of the class of inverse M-matrices under strictly increasing convex functions and introduce the concept of class $\mathscr{T}$ that, along with the algorithm developed in Chap. 5 gives sufficient conditions for this invariance.

The last part of the chapter is devoted to the class of inverse tridiagonal M-matrices, which in probability theory correspond to potentials of random walks.

In general the Hadamard product of two potentials is not an inverse M-matrix. For example take the following ultrametric matrices (actually B is a permutation of A)

$$A = \begin{pmatrix} 2 & 2 & 2 & 2 \\ 2 & 4 & 4 & 4 \\ 2 & 4 & 6 & 6 \\ 2 & 4 & 6 & 8 \end{pmatrix} \text{ and } B = \begin{pmatrix} 2 & 2 & 2 & 2 \\ 2 & 8 & 6 & 4 \\ 2 & 6 & 6 & 4 \\ 2 & 4 & 4 & 4 \end{pmatrix}.$$

Verify that $A \odot B$ is not an inverse M-matrix. A difficult problem is to give conditions under which the Hadamard product of two inverse M-matrices is again an inverse M-matrix. There are a few results in this direction (see for example

© Springer International Publishing Switzerland 2014
C. Dellacherie et al., *Inverse M-Matrices and Ultrametric Matrices*, Lecture Notes in Mathematics 2118, DOI 10.1007/978-3-319-10298-6_6

[58]). In this chapter, we show that if W, X are inverses of irreducible tridiagonal M-matrices, then their Hadamard product $W \odot X$ is the inverse of an irreducible and tridiagonal M-matrix (see Theorem 6.51).

The decompositions given in this Chapter, as well as the ones in [48, 50], allow us to study the Hadamard powers of inverses of irreducible and tridiagonal M-matrices. We show that for any $r > 0$, the r-th power of any of these matrices is again an inverse of an irreducible tridiagonal M-matrix (see Theorem 6.48). Thus, for $r \geq 1$ the r-th power of a potential associated with a transient irreducible random walk is again a potential of another transient irreducible random walk. This provides a means of generating new random walks by using the Hadamard powers of random walk potentials. Furthermore, formula (6.20) below gives some probabilistic information on the random walk associated with the powers of such matrices.

A transient Markov chain is completely determined by its potential. It is a difficult problem to translate structural restrictions on the Markov chain into its potential. This is done for simple random walks in Sect. 6.4, where we show that any of these potentials is the Hadamard product of two ultrametric matrices, each one associated further with one dimensional random walks. For example, if W is a potential of a symmetric random walk, then W must satisfy the restrictions

$$W_{ij} = \frac{W_{in} W_{1j}}{W_{1n}}, \ 1 \leq i \leq j \leq n, \text{ and } W_{\bullet n} \uparrow, W_{1 \bullet} \downarrow .$$

6.1 Definitions

A main concept for this chapter is the notion of Hadamard functions of matrices.

Definition 6.1 Given a function f and a matrix A, the matrix $f(A)$ is defined as $f(A)_{ij} = f(A_{ij})$. We shall say that $f(A)$ is a Hadamard function of A.

By abuse of language such a function f will be called a Hadamard function.

Of particular interest in this chapter is $A^{(\alpha)}$ which is the Hadamard transformation of A under $f(x) = x^{\alpha}$.

We note that our purpose is to study Hadamard functions of matrices and not spectral functions of matrices, which are quite different. For spectral functions of matrices, there are profound and beautiful results for the same classes of matrices we consider here. See for example the work of Bouleau [7] for operators (part of which we have included in Sect. 5.4). For M-matrices see the works of Varga [56], Micchelli and Willoughby [49], Ando [2], Fiedler and Schneider [31], and the work of Bapat, Catral and Neumann [3] for M-matrices and inverse M-matrices.

The class of CBF matrices (and their permutations) is stable under Hadamard functions. On the other hand the class of NBF, and therefore also the class of GUM matrices, is stable under nonnegative nondecreasing Hadamard functions. We leave it to the reader to prove that if U is weakly filtered with a decomposition like (5.10) where η and $\mathfrak{z}$ are 0, 1 valued then $f(U)$ is also weakly filtered for every nonnegative function f.

Nevertheless, there are examples of weakly filtered matrices U for which $f(U)$ is not weakly filtered, for nonnegative nondecreasing functions f.

Example 6.2 Consider the matrix

$$U = \alpha \mathbb{F}_0 + \mathfrak{a} \cdot \mathbb{F}_0 \cdot \mathfrak{b} + \beta \cdot \mathbb{I},$$

where $\mathbb{F}_0 = \mathbb{1}\mathbb{1}'$. We have α is a constant and the vectors $\mathfrak{a}$, $\mathfrak{b}$, β are all $\mathfrak{F}$-measurable. Then U is weakly filtered and moreover

$$U = \alpha + \mathfrak{a}\mathfrak{b}' + \beta \cdot \mathbb{I}. \tag{6.1}$$

Take $\alpha = \beta = 0$ and $\mathfrak{a} = (2, 3, 5, 7)'$ and $\mathfrak{b} = (11, 13, 17, 19)'$. In this case, all the entries of U are different. As f runs all possible nonnegative nondecreasing functions $f(U)$ runs over a full dimension subset of the 4×4 nonnegative matrices. This implies that some of them can not be written as in (6.1), because in this representation we have at most 13 free variables. However, it is still possible that each $f(U)$ is decomposable as in (5.9) using maybe a different filtration. A more detailed analysis shows that this is not the case. For every filtration the class of weakly filtered matrices is a subset of the class of nonnegative matrices, of dimension at most 15. For example, if we choose the filtration $\mathfrak{N} \prec \{\{1, 2\}, \{3, 4\}\} \prec \mathfrak{F}$ then every matrix V, weakly filtered with respect to this filtration, satisfy $V_{13} - V_{23} = V_{14} - V_{24}$. As the number of possible filtrations is finite we conclude the existence of the desired example.

6.2 Hadamard Convex Functions and Powers

Theorem 6.3 *Let us assume W is a potential and $f : \mathbb{R}_+ \to \mathbb{R}_+$ is a nonnegative strictly increasing convex function. Then, $f(W)$ is nonsingular, $\det(f(W)) > 0$ and it has a right equilibrium potential. If W is a bi-potential then $f(W)$ also has a left equilibrium potential. Furthermore, if $f(0) = 0$ we have $M = W^{-1} f(W)$ is an M-matrix.*

Note that $H = f(W)^{-1}$ is not necessarily a Z-matrix, that is for some $i \neq j$ it can happen that $H_{ij} > 0$, as the following example will show. Therefore the existence of a right equilibrium potential, which is

$$\forall i \quad H_{ii} + \sum_{j \neq i} H_{ij} \geq 0$$

does not imply the inverse is row diagonally dominant, that is

$$\forall i \quad H_{ii} \geq \sum_{j \neq i} |H_{ij}|.$$

Example 6.4 Consider the matrix

$$P = \begin{pmatrix} 0 & \frac{1}{2} & 0 \\ \frac{1}{2} & 0 & \frac{1}{2} \\ 0 & \frac{1}{2} & 0 \end{pmatrix}$$

Then $W = (\mathbb{I} - P)^{-1}$ is a bi-potential. Consider the strictly increasing convex function $f(x) = x^2 - \cos(x) + 1$. A numerical computation gives

$$(f(W))^{-1} \approx \begin{pmatrix} 0.3590 & -0.0975 & 0.0027 \\ -0.0975 & 0.2372 & -0.0975 \\ 0.0027 & -0.0975 & 0.3590 \end{pmatrix},$$

which is not a Z-matrix. In Theorem 6.23 below, we supply a sufficient condition under which a bi-potential is stable under increasing convex functions.

Proof (Theorem 6.3) We first assume that $f(0) = 0$. Since W is a potential then $W^{-1} = k(\mathbb{I} - P)$ for some constant $k > 0$ and a substochastic matrix P. Without loss of generality we can assume $k = 1$, because it is enough to consider kW instead of W and $\tilde{f}(x) = f(x/k)$ instead of f.

Consider $M = W^{-1} f(W)$. Take $i \neq j$ and compute

$$M_{ij} = (W^{-1} f(W))_{ij} = (1 - p_{ii}) f(W_{ij}) - \sum_{k \neq i} p_{ik} f(W_{kj}) = f(W_{ij}) - \sum_k p_{ik} f(W_{kj}).$$

Since $1 - p_{ii} - \sum_{k \neq i} p_{ik} \geq 0$, which is equivalent to $\sum_k p_{ik} \leq 1$, and f is convex we obtain

$$\sum_k p_{ik} f(W_{kj}) = \left(1 - \sum_k p_{ik}\right) f(0) + \sum_k p_{ik} f(W_{kj}) \geq f\left(\sum_k p_{ik} W_{kj}\right) = f(W_{ij}).$$

The last equality follows from the fact that $W^{-1} = \mathbb{I} - P$. This shows that $M_{ij} \leq 0$. Consider now a positive vector x such that $y' = x' W^{-1} > 0$ (see [38], Theorem 2.5.3). Then

$$x' M = x' W^{-1} f(W) = y' f(W) > 0,$$

which implies, by the same cited theorem, that M is an M-matrix. In particular M is nonsingular and $\det(M) > 0$. So, $f(W)$ is nonsingular and $\det(f(W)) > 0$. Let ρ be the unique solution of $f(W)\rho = \mathbb{1}$. We shall prove ρ is the right equilibrium potential of $f(W)$, so we just need to prove $\rho \geq 0$. We have

$$M\rho = W^{-1} f(W)\rho = W^{-1}\mathbb{1} = \lambda^W \geq 0,$$

then $\rho = M^{-1}\lambda^W \geq 0$, because M^{-1} is a nonnegative matrix. This means that $f(W)$ possesses a right equilibrium potential. Since $f(W)$ is non singular, we also have a solution μ to the equation $\mu' f(W) = \mathbb{1}$, which we do not know if it is nonnegative. Then, the first part is proven under the extra hypothesis $f(0) = 0$.

Assume now $a = f(0) > 0$, and consider $g(x) = f(x) - a$, which is a strictly increasing convex function. Obviously $f(W) = g(W) + a\mathbb{1}\mathbb{1}'$, so the candidates for equilibrium potentials (except maybe for their sign) of $f(W)$ are

$$\lambda^{f(W)} = \frac{1}{1 + a\bar{\lambda}^{g(W)}}\lambda^{g(W)}, \quad \mu^{f(W)} = \frac{1}{1 + a\bar{\mu}^{g(W)}}\mu^{g(W)},$$

where $\lambda^{g(W)} \geq 0$ and $\bar{\mu}^{g(W))} = \mathbb{1}'\mu^{g(W))} > 0$. Since $\lambda^{g(W)} \geq 0$ we have that $f(W)$ has a right equilibrium potential. We need to prove that $f(W)$ is nonsingular, and $\det(f(W)) > 0$. This follows immediately from the equality

$$f(W) = g(W)(\mathbb{I} + a\lambda^{g(W)}\mathbb{1}').$$

Indeed, we have

$$f(W)^{-1} = g(W)^{-1} - \frac{a}{1 + a\bar{\lambda}^{g(W)}}\lambda^{g(W)}(\mu^{g(W)})' \text{ and}$$

$$\det(f(W)) = \det(g(W))(1 + a\bar{\lambda}^{g(W)}),$$

from which the result is proven.

In the bi-potential case use W' instead of W to obtain the existence of a nonnegative left equilibrium potential for $f(W)$.

We recall that $W^{(\alpha)}$ is the Hadamard transformation of W under $f(x) = x^\alpha$. In particular $W^{(2)} = W \odot W$. It was conjectured by Neumann in [54] that $W^{(2)}$ is an inverse M-matrix if W is so. This was solved by Chen in his beautiful article [11], for any positive integer power of W. His proof depends on the following interesting result: W is an inverse M-matrix if and only if its adjugate is a Z-matrix, and each proper principal submatrix is an inverse M-matrix. Chen generalizes his result to real numbers $\alpha \geq 1$ in [12]. This proposition will be proven here with a different technique based on the idea of equilibrium potential and homogeneity of the power function (some result for fractional powers can be found in our article [25]).

This result has the following probabilistic interpretation. If W is the potential of a transient continuous time Markov process then $W^{(\alpha)}$ is also the potential of a transient continuous time Markov process. This means that if a potential is realized by an electrical network then it is possible to construct another electrical network whose potential is a power of the initial one. In Theorem 6.7 we show the same is true for a potential of a Markov chain, which represents an improvement with respect to the case of a potential. An interesting open question is what is the relation between the Markov chain associated with W and the one associated with $W^{(\alpha)}$.

Theorem 6.5 *Assume W is an inverse M-matrix and $\alpha \geq 1$. Then $W^{(\alpha)}$ is also an inverse M-matrix. If W is a potential then $W^{(\alpha)}$ is a potential. If W is a bi-potential then $W^{(\alpha)}$ is a bi-potential.*

The proof of this theorem is based on the following result.

Lemma 6.6 *Assume W is a bi-potential with block decomposition*

$$W = \begin{pmatrix} A & b \\ c' & d \end{pmatrix}.$$

Then, for all $\alpha \geq 1$ there exists a nonnegative vector η such that

$$A^{(\alpha)}\eta = b^{(\alpha)}.$$

Proof We first perturb the matrix W to have a positive matrix. Consider $\epsilon > 0$ and the positive matrix $W_\epsilon = W + \epsilon \mathbb{1}\mathbb{1}'$. It is straightforward to prove that

$$W_\epsilon^{-1} = W^{-1} - \frac{\epsilon}{1 + \epsilon \bar{\lambda}^W} \lambda^W (\mu^W)',$$

where $\bar{\lambda}^W = \mathbb{1}'\lambda^W$ is the total mass of λ^W, which coincides with the total mass of μ^W the left equilibrium potential. Then, W_ϵ is a bi-potential and its equilibrium potentials are given by

$$\lambda^{W_\epsilon} = \frac{1}{1 + \epsilon \bar{\lambda}^W} \lambda^W, \quad \mu^{W_\epsilon} = \frac{1}{1 + \epsilon \bar{\mu}_W} \mu_W.$$

We decompose the inverse of W_ϵ as

$$W_\epsilon^{-1} = \begin{pmatrix} \Lambda_\epsilon & \zeta_\epsilon \\ \varrho_\epsilon' & \theta_\epsilon \end{pmatrix},$$

and notice that $A_\epsilon \zeta_\epsilon + \theta_\epsilon b_\epsilon = 0$ which implies that

$$b_\epsilon = A_\epsilon \lambda_\epsilon,$$

with $\lambda_\epsilon = -\frac{1}{\theta_\epsilon}\zeta_\epsilon \geq 0$. We also mention here that λ_ϵ has total mass $\mathbb{1}'\lambda_\epsilon \leq 1$. This follows from the fact that W_ϵ^{-1} is column diagonally dominant.

Take now the matrix $V_\epsilon = D_{b_\epsilon}^{-1} A_\epsilon$. It is straightforward to check that V is an inverse M-matrix and its equilibrium potentials are

$$\lambda^{V_\epsilon} = \lambda_\epsilon, \quad \mu^{V_\epsilon} = D_{b_\epsilon}\mu^{A_\epsilon}.$$

Thus, V_ϵ is a bi-potential and we can apply Theorem 6.3 to get $V_\epsilon^{(\alpha)}$ possesses a right equilibrium potential $\eta_\epsilon \geq 0$, that is, for all i

$$\sum_j (V_\epsilon^{(\alpha)})_{ij}(\eta_\epsilon)_j = 1,$$

which is equivalent to

$$\sum_j \frac{(A_\epsilon)_{ij}^\alpha}{(b_\epsilon)_i^\alpha}(\eta_\epsilon)_j = 1.$$

Hence

$$A_\epsilon^{(\alpha)} \eta_\epsilon = b_\epsilon^{(\alpha)}.$$

Recall that the matrix $A^{(\alpha)}$ is nonsingular. Since obviously $A_\epsilon^{(\alpha)} \to A^{(\alpha)}$ as $\epsilon \to 0$, we get

$$\eta_\epsilon \to \eta = (A^{(\alpha)})^{-1} b^{(\alpha)},$$

and the lemma is proved.

Proof (Theorem 6.5) Consider first the case where W is a bi-potential. We already know that $W^{(\alpha)}$ is nonsingular and that it has left and right equilibrium potentials. Therefore, in order to prove that $W^{(\alpha)}$ is a bi-potential, it is enough to prove that $(W^{(\alpha)})^{-1}$ is a Z-matrix: For $i \neq j$ we have to show $((W^{(\alpha)})^{-1})_{ij} \leq 0$. An argument based on permutations shows that it is enough to prove the claim for $i = 1, j = n$, where n is the size of W.

Decompose $W^{(\alpha)}$ and its inverse as follows

$$W^{(\alpha)} = \begin{pmatrix} A^{(\alpha)} & b^{(\alpha)} \\ (c^{(\alpha)})' & d^\alpha \end{pmatrix} \text{ and } (W^{(\alpha)})^{-1} = \begin{pmatrix} \Omega & -\beta \\ -\gamma' & \delta \end{pmatrix}.$$

We need to show that $\beta \geq 0$. We notice that $\delta = \frac{\det(A^{(\alpha)})}{\det(W^{(\alpha)})} > 0$ and that

$$-A^{(\alpha)}\beta + \delta b^{(\alpha)} = 0,$$

which implies that

$$b^{(\alpha)} = A^{(\alpha)} \left(\frac{\beta}{\delta}\right).$$

Therefore, $\frac{\beta}{\delta} = \eta \geq 0$, where η is the vector given in Lemma 6.6. Thus $\beta \geq 0$ and the result is proven for the case W is a bi-potential.

Now consider $W = M^{-1}$ is the inverse of an M-matrix. Using Theorem 2.5.3 in [38], we get the existence of two positive diagonal matrices D, E such that DME is a strictly row and column diagonally dominant M-matrix. Thus $V = E^{-1}WD^{-1}$ is a bi-potential, from where it follows that $V^{(\alpha)}$ is a bi-potential. Hence, $W^{(\alpha)} = E^{(\alpha)}V^{(\alpha)}D^{(\alpha)}$ is the inverse of an M-matrix. Here we have used strongly that for a power function $f(xy) = f(x)f(y)$. The rest of the result is proven in a similar way.

The next result gives some extra information on the inverse of $W^{(\alpha)}$ in the context of Markov chains.

Theorem 6.7 *Assume that $W^{-1} = \mathbb{I} - P$ where P is substochastic. Then, for all $\alpha \geq 1$ there exists a substochastic matrix $Q(\alpha)$ such that $(W^{(\alpha)})^{-1} = \mathbb{I} - Q(\alpha)$. If in addition P' is substochastic then $Q(\alpha)'$ is also substochastic.*

Proof By hypothesis we have $W^{-1} = \mathbb{I} - P$, where $P \geq 0$ and $P\mathbb{1} \leq \mathbb{1}$. We notice that the diagonal of W dominates pointwise every row, that is for all i, j

$$W_{ii} \geq W_{ji}.$$

Also we notice that $W = \mathbb{I} + PW$ and therefore $W_{ii} \geq 1$ for all i.

According to Theorem 6.5 we know that $H = (W^{(\alpha)})^{-1}$ is a row diagonally dominant M-matrix. The only thing left to prove is that the diagonal elements of H are dominated by one, that is $H_{ii} \leq 1$ for all i. It is enough to prove this for $i = n$, where n is the size of W.

Consider the following decompositions

$$W = \begin{pmatrix} A & b \\ c' & d \end{pmatrix} \quad W^{-1} = \begin{pmatrix} \Lambda & -\omega \\ -\eta' & \theta \end{pmatrix} \quad (W^{(\alpha)})^{-1} = \begin{pmatrix} \Omega & -\beta \\ -\gamma' & \delta \end{pmatrix}$$

$$W^{-1}W^{(\alpha)} = \begin{pmatrix} \Xi & -\zeta \\ -\chi' & \rho \end{pmatrix}$$

A straightforward computation gives

$$\theta = \rho\delta + \chi'\beta \geq \rho\delta.$$

Since by hypothesis $\theta \leq 1$, to conclude that $\delta \leq 1$ is enough to prove that $\rho \geq 1$. We have

$$\rho = (1 - p_{nn})W_{nn}^{\alpha} - \sum_{j \neq n} p_{nj}W_{jn}^{\alpha} = W_{nn}^{\alpha} - \sum_{j} p_{nj}W_{jn}^{\alpha} = W_{nn}^{\alpha} - \sum_{j} p_{nj}W_{jn}W_{jn}^{\alpha-1}.$$

On the other hand we have $W_{jn}^{\alpha-1} \leq W_{nn}^{\alpha-1}$ and $\sum_j p_{nj} W_{jn} = W_{nn} - 1$ from where we deduce

$$\rho \geq W_{nn}^{\alpha-1} \geq 1.$$

The rest of the result is proven by using W' instead of W.

6.2.1 Some Examples and a Conjecture

The main purpose of this section is to prove that the exponential of a bi-potential is a bi-potential. We shall see that this invariance is not true for the class of potentials. In fact, we shall give an example of a potential matrix W for which $exp(W)$ is not an inverse M-matrix. Thus, also the class of inverse M-matrices does not satisfies this invariance. We end this section with a discussion of a conjecture about a class of functions that should leave invariant the class of bi-potentials.

Let us start studying the exponential of a matrix in the sense of Hadamard. As any Hadamard function of a matrix, the exponential is defined entrywise as, for all i, j

$$(exp(W))_{ij} = e^{W_{ij}}.$$

As for numbers, the Hadamard exponential of W, can be obtained as the following limit

$$exp(W) = \lim_{n\to\infty} \left(\mathbb{1}\mathbb{1}' + \frac{1}{n}W \right)^{(n)}.$$

Note the similarity and differences with the exponential of W in the spectral sense, noted e^W, which can be defined as

$$e^W = \lim_{n\to\infty} \left(\mathbb{I} + \frac{1}{n}W \right)^n.$$

Here, the matrix $\mathbb{I}$ replace the matrix of ones $\mathbb{1}\mathbb{1}'$ and the standard power replace the power in the sense of Hadamard.

So, given a bi-potential W we know that $X = exp(W)$ is nonsingular and it possesses a left and a right equilibrium potentials μ^X and λ^X, respectively (see Theorem 6.3). Since powers of bi-potentials are bi-potentials, we shall deduce X is a bi-potential once the following lemma is established.

Lemma 6.8 *Assume that W is a bi-potential, then for all $a \geq 0$ the matrix $V = W + a\mathbb{1}\mathbb{1}'$ is again a bi-potential.*

Proof Consider $\mu = \mu^W$ and $\lambda = \lambda^W$ the left and right equilibrium potentials of W. From the equality

$$(W + a\mathbb{1}\mathbb{1}')(W^{-1} - \alpha\lambda\mu') = \mathbb{I} + (a - \alpha(1 + a\bar{\lambda}))\mathbb{1}\mu',$$

we deduce that $(W + a\mathbb{1}\mathbb{1}')^{-1} = W^{-1} - \frac{a}{1+a\bar{\lambda}}\lambda\,\mu'$ (recall that $\bar{\lambda} = \mathbb{1}'\lambda > 0$).

Hence, $N = (W + a\mathbb{1}\mathbb{1}')^{-1}$ is a Z-matrix. Since $W + a\mathbb{1}\mathbb{1}'$ is nonnegative, we deduce that N is an M-matrix. The only thing left to be proved is that N is a row and column diagonally dominant matrix, or equivalently, that the signed left and right equilibrium potentials of V are indeed nonnegative. This property holds because these equilibrium potentials are

$$\lambda^V = \frac{1}{1 + a\bar{\lambda}^W}\,\lambda^W \geq 0, \; \mu^V = \frac{1}{1 + a\bar{\mu}^W}\,\mu^W \geq 0.$$

Corollary 6.9 *If W is a bi-potential then $\exp(W)$ is also a bi-potential.*

With respect to the invariance of potentials or inverse M-matrices under the exponential function we use the following example taken from [25].

Example 6.10 Let us show the Hadamard exponential does not preserve the class of inverse M-matrices and the class of potential matrices. For that purpose consider the row diagonally dominant M-matrix

$$M = \begin{pmatrix} 10 & -5 & -4 \\ -1 & 4 & -2 \\ -4 & -1 & 10 \end{pmatrix}.$$

Its inverse $W = M^{-1}$ is a potential matrix, but the entry $(3, 2)$ of the inverse of $\exp(W)$ is approximately 0.3467 which shows that $\exp(W)$ is not an inverse M-matrix.

The exponential of a bi-potential is a special bi-potential. Indeed, if $U = \exp(W)$ then we deduce that for all $r > 0$ the matrix $U^{(r)}$ is also a bi-potential. This follows directly from the fact that $U^{(r)} = \exp(rW)$ and the fact that rW is again a bi-potential.

Corollary 6.11 *Let $W > 0$ be a bi-potential. A sufficient condition for $W^{(r)}$ to be a bi-potential for all $r > 0$, is that $\log(W) + a$ is a bi-potential for some $a \geq 0$.*

The class of bi-potentials are invariant under any power greater or equal to one. This class is also invariant under $e^x - 1$ and $x^2 + x$, a fact that can be found in the cited paper, as well as the following result.

Proposition 6.12 *If $\psi : \mathbb{R}_+ \to \mathbb{R}_+$ is a smooth function that preserves the class of bi-potentials, then necessarily the function*

$$x \rightarrow \log(\psi(e^x)) \quad \textit{is convex on the entire real line.} \tag{6.2}$$

If ψ is assumed to be also strictly increasing and convex then this condition is sufficient for ψ to leave invariant the class of bi-potentials of size 3.

We point out that condition (6.2) is not sufficient to leave invariant the class of bi-potentials of size 4 (see an example in the cited paper). It seems that as the size of the matrices grows, it is required to impose more and more conditions on a function in order that it leaves invariant the class of bi-potentials of that size.

Conjecture 6.13 A natural conjecture is the class of bi-potentials should be invariant under polynomials with nonnegative coefficients and then by a limit procedure this class should be invariant under absolutely monotone functions. Recall that a function $\psi : \mathbb{R}_+ \rightarrow \mathbb{R}$ is absolutely monotone if for all $n \geq 0$ an all $x > 0$

$$\frac{d^n}{d^n x} \psi(x) \geq 0.$$

Under these conditions ψ has an analytic extension to $\mathbb{C}$ and all the coefficients in its Taylor expansion are nonnegative, that is $\psi(x) = \sum_{n=0}^{\infty} a_n x^n$, where $a_n \geq 0$ for all n. In favour of this conjecture, we notice that $e^x, e^x - 1, x^2 + x$ and x^k, for $k \geq 1, k \in \mathbb{N}$ are all absolutely monotone and they leave invariant the class of bi-potentials. Also notice that every absolutely monotone function satisfies (6.2).

6.3 A Sufficient Condition for Hadamard Invariance: Class $\mathscr{T}$

In this section, we shall prove the class of GUM matrices is the largest class of potentials stable under nondecreasing nonnegative Hadamard functions. For that purpose, let us introduce the following concept where $bi\,\mathscr{P}$ denotes the class of bi-potentials.

Definition 6.14 For any nonnegative matrix W we define

$$\tau(W) = \inf\{t \geq 0 : \mathbb{I} + tW \notin bi\,\mathscr{P}\}. \tag{6.3}$$

This quantity is invariant under permutations. We point out that if W is a positive matrix then $\tau(W) > 0$ due to Taylor's theorem. Notice that $(\tau(W))^{-1}$ measures the minimal amount of diagonal increase on W to have a bi-potential.

We shall study some properties of this function τ. In particular we are interested in matrices for which $\tau(W) = \infty$.

Proposition 6.15 *Assume W is a nonnegative matrix, which is nonsingular and $\tau(W) = \infty$, then W is a bi-potential.*

Proof It is straightforward from the observation that

$$t(\mathbb{I}+tW)^{-1} \underset{t\to\infty}{\to} W^{-1}.$$

Remark 6.16 We shall prove later on that the converse is also true: if W is a bi-potential, then $\tau(W) = \infty$.

Proposition 6.17 *Assume U is a GUM and $f : \mathbb{R}_+ \to \mathbb{R}_+$ is a nondecreasing function. Then $f(U)$ is a GUM. In particular $\tau(f(U)) = \infty$, and if $f(U)$ is nonsingular then $f(U)$ is a bi-potential. A sufficient condition for $f(U)$ to be nonsingular is that U is nonsingular and f is increasing.*

Proof It is clear that $f(U)$ is a GUM matrix and therefore $\tau(f(U)) = \infty$. Then, from Proposition 6.15 we have that $f(U)$ is a bi-potential as long as it is nonsingular. If U is nonsingular then it does not contain a row (or column) of zeros and there are not two equal rows (or columns). This condition is stable under increasing nonnegative functions, so the result follows (see Theorem 3.9).

One of the main results of this section is a sort of converse of the previous one. We shall prove that if $\tau(f(W)) = \infty$ for all nondecreasing nonnegative functions f, then W must be a GUM (see Theorem 6.20). Also we shall prove that every nonnegative matrix W which is a permutation of a CBF satisfies this property.

We shall use Lemma 2.32 and Lemma 2.34 in Chap. 2, which has important consequences in our discussion. In particular, any principal submatrix of a potential (bi-potential, inverse M-matrix) is again a potential (respectively bi-potential, inverse M-matrix). Also, if W is a potential then $tW + \mathbb{I}$ is a potential for any $t > 0$ (respectively bi-potential, inverse M-matrix). The next result shows that the class of matrices verifying $\tau(W) = \infty$ is a generalization of the class of bi-potentials.

Corollary 6.18 *Let W be a bi-potential then $\tau(W) = \infty$.*

Corollary 6.19 *Let W be a nonnegative matrix, then*

$$\tau(W) = \sup\{t \geq 0 : \mathbb{I} + tW \in bi\,\mathscr{P}\}$$

Proof It is clear that $\tau(W) \leq \sup\{t \geq 0 : \mathbb{I} + tW \in bi\,\mathscr{P}\}$. On the other hand if $\mathbb{I} + tW$ is a bi-potential then we get $\mathbb{I} + sW$ is a bi-potential for all $0 \leq s \leq t$. This fact and the definition of $\tau(W)$, implies the result.

Theorem 6.20 *Let W be a nonnegative matrix such that $\tau(f(W)) = \infty$ for all nondecreasing nonnegative functions f. Then, W must be a GUM.*

Example 6.21 Given $a, b, c, d \in \mathbb{R}_+$ consider the non-singular matrix

$$W = \begin{pmatrix} 1 & 0 & 0 & 0 \\ 0 & 1 & 0 & 0 \\ a & b & 1 & 0 \\ c & d & 0 & 1 \end{pmatrix}.$$

For all nondecreasing nonnegative functions f and all $t > 0$ the matrix $(\mathbb{I} + tf(W))^{-1}$ is an M-matrix, while W is not a GUM. Moreover, W is not a permutation of an CBF. This shows that the last theorem does not hold if we replace in the definition of τ (see (6.3)), the class of bi-potentials by the class of inverse M-matrices, .

Proof Let n be the dimension of W. Notice that if W is a nonnegative matrix whose diagonal dominates pointwise each row (and column), that is

$$\forall i, j \quad W_{ii} \geq W_{ij},$$

then it is a GUM if and only if $n \leq 2$ or every principal submatrix of size 3 is a GUM (see Theorem 3.8).

Since by hypothesis the matrix $\mathbb{I} + tW$ is a bi-potential, it is pointwise diagonally dominant, that is

$$1 + tW_{ii} \geq tW_{ij},$$

and we deduce that $W_{ii} \geq W_{ij}$ (by taking $t \to \infty$). This proves the result when $n \leq 2$. So, in the sequel we assume $n \geq 3$.

Consider any principal submatrix A of W, of size 3×3. Since $\mathbb{I} + tf(A)$ is a principal submatrix of $\mathbb{I} + tf(W)$, we deduce that $\mathbb{I} + tf(A)$ is a bi-potential (as long as $\mathbb{I} + tf(W)$ is a bi-potential). If the result holds for 3×3 matrices, we deduce that A is GUM implying that W is also a GUM.

Thus, in what follows we consider that W is a 3×3 matrix that satisfies the hypothesis of the Theorem. After a suitable permutation we can further assume that

$$W = \begin{pmatrix} a & b_1 & b_2 \\ c_1 & d & \alpha \\ c_2 & \beta & e \end{pmatrix},$$

where $\alpha = \min\{W_{ij} : i \neq j\} = \min\{W\}$ and $\beta = \min\{W_{ji} : W_{ij} = \alpha, i \neq j\}$.

Since W is a pointwise row and column diagonally dominant matrix, we have $\min\{a, d, e\} \geq \alpha$. Take f nondecreasing such that $f(\alpha) = 0$ and $f(x) > 0$ for $x > \alpha$. Then,

$$\mathbb{I} + f(W) = \begin{pmatrix} 1 + f(a) & f(b_1) & f(b_2) \\ f(c_1) & 1 + f(d) & 0 \\ f(c_2) & f(\beta) & 1 + f(e), \end{pmatrix},$$

is by hypothesis a bi-potential whose inverse we denote by

$$\begin{pmatrix} \delta & -\rho_1 & -\rho_2 \\ -\theta_1 & \gamma_1 & -\gamma_2 \\ -\theta_2 & -\gamma_3 & \gamma_4 \end{pmatrix}.$$

In particular we obtain

$$\begin{pmatrix} 1 + f(d) & 0 \\ f(\beta) & 1 + f(e) \end{pmatrix}^{-1} = \begin{pmatrix} \gamma_1 & -\gamma_2 \\ -\gamma_3 & \gamma_4 \end{pmatrix} - \frac{1}{\delta} \begin{pmatrix} \theta_1 \\ \theta_2 \end{pmatrix} \begin{pmatrix} \rho_1 \\ \rho_2 \end{pmatrix}',$$

and deduce that

$$0 = \gamma_2 = \theta_1 \rho_2. \tag{6.4}$$

- **Case** $\rho_2 = 0$. We deduce that $f(b_2) = 0$ (think in terms of blocks), and then

$$b_2 = \alpha, \text{ and } c_2 \geq \beta, \tag{6.5}$$

where the last conclusion follows from the definition of β. Therefore we have that

$$W = \begin{pmatrix} a & b_1 & \alpha \\ c_1 & d & \alpha \\ c_2 & \beta & e \end{pmatrix}, \tag{6.6}$$

and we should prove that W is GUM.

Now consider another nondecreasing function g such that $g(\beta) = 0$ and $g(x) > 0$ for $x > \beta$. Then,

$$\mathbb{I} + g(W) = \begin{pmatrix} 1 + g(a) & g(b_1) & 0 \\ g(c_1) & 1 + g(d) & 0 \\ g(c_2) & 0 & 1 + g(e) \end{pmatrix}.$$

Its inverse is of the form

$$\begin{pmatrix} \tilde{\delta} & -\tilde{\rho}_1 & 0 \\ -\tilde{\theta}_1 & \tilde{\gamma}_1 & 0 \\ -\tilde{\theta}_2 & -\tilde{\gamma}_3 & \tilde{\gamma}_4 \end{pmatrix}.$$

As before we deduce that $0 = \tilde{\gamma}_3 = \tilde{\theta}_2 \tilde{\rho}_1$.

- **Subcase** $\tilde{\theta}_2 = 0$. In this situation, we have $g(c_2) = 0$, which implies $c_2 = \beta$. Thus, we conclude that

$$W = \begin{pmatrix} a & b_1 & \alpha \\ c_1 & d & \alpha \\ \beta & \beta & e \end{pmatrix}.$$

By permuting rows and columns $1, 2$, if necessary, we can assume that $b_1 \leq c_1$. Consider the situation where $c_1 < \beta$, of course implicitly we should have $\alpha < \beta$. Under a suitable nondecreasing transformation h we have

$$\mathbb{I} + h(W) = \begin{pmatrix} 1 + h(a) & 0 & 0 \\ 0 & 1 + h(d) & 0 \\ h(\beta) & h(\beta) & 1 + h(e) \end{pmatrix},$$

and its inverse is

$$\begin{pmatrix} \frac{1}{1+h(a)} & 0 & 0 \\ 0 & \frac{1}{1+h(d)} & 0 \\ -\frac{h(\beta)}{(1+h(a))(1+h(e))} & -\frac{h(\beta)}{(1+h(d))(1+h(e))} & \frac{1}{1+h(e)} \end{pmatrix}.$$

The sum of the third row is then

$$\frac{1}{1 + h(e)} \left(1 - h(\beta) \left(\frac{1}{1 + h(a)} + \frac{1}{1 + h(d)} \right) \right),$$

and this quantity can be made negative by choosing an appropriate function h. The idea is to make $h(\beta) \to \infty$ and

$$\frac{h(\beta)}{\max\{h(a), h(d)\}} \to 1.$$

Therefore, $c_1 \geq \beta$ and W is a GUM.

- **Subcase $\tilde{\rho}_1 = 0$.** We have $g(b_1) = 0$ and then $b_1 \leq \beta$. Take again a nondecreasing function, denoted by ℓ, such that

$$\mathbb{I} + \ell(W) = \begin{pmatrix} 1 + \ell(a) & 0 & 0 \\ \ell(c_1) & 1 + \ell(d) & 0 \\ \ell(c_2) & 0 & 1 + \ell(e) \end{pmatrix}.$$

The inverse of this matrix is

$$\begin{pmatrix} \frac{1}{1+\ell(a)} & 0 & 0 \\ -\frac{\ell(c_1)}{(1+\ell(a))(1+\ell(d))} & \frac{1}{1+\ell(d)} & 0 \\ -\frac{\ell(c_2)}{(1+\ell(a))(1+\ell(e))} & 0 & \frac{1}{1+\ell(e)} \end{pmatrix}.$$

The sum of the first column is

$$\frac{1}{1 + \ell(a)} \left(1 - \frac{\ell(c_1)}{(1 + \ell(d))} - \frac{\ell(c_2)}{(1 + \ell(e))} \right),$$

which can be made negative by repeating a similar argument as before, if both $c_1 > \beta$ and $c_2 > \beta$.

So, if $c_1 > \beta$ we conclude that $c_2 \le \beta$, but we know that $c_2 \ge \beta$ (see (6.5)) and we deduce that $c_2 = \beta$. Thus, $\alpha \le b_1 \le \beta < c_1$ and

$$
W = \begin{pmatrix} a & b_1 & \alpha \\ c_1 & d & \alpha \\ \beta & \beta & e \end{pmatrix},
$$

which is a GUM.

Therefore we can continue under the hypothesis $c_1 \le \beta \le c_2$.

· **Subsubcase $b_1 < \beta$.** Again we must have $\alpha < \beta$. Under this condition we have that $c_2 > \alpha$. Using a nondecreasing function k we get

$$
\mathbb{I} + k(W) = \begin{pmatrix} 1+k(a) & 0 & 0 \\ k(c_1) & 1+k(d) & 0 \\ k(c_2) & k(\beta) & 1+k(e) \end{pmatrix},
$$

and its inverse is

$$
\begin{pmatrix} \frac{1}{1+k(a)} & 0 & 0 \\[2mm] -\frac{k(c_1)}{(1+k(a))(1+k(d))} & \frac{1}{1+k(d)} & 0 \\[2mm] -\frac{k(c_2)(1+k(d))-k(\beta)k(c_1)}{(1+k(a))(1+k(d))(1+k(e))} & -\frac{k(\beta)}{(1+k(d))(1+k(e))} & \frac{1}{1+k(e)} \end{pmatrix}.
$$

The sum of the third row is

$$
\frac{1}{(1+k(e))} \left(1 - \frac{k(c_2)}{1+k(a)} + \frac{k(\beta)k(c_1)}{(1+k(a))(1+k(d))} - \frac{k(\beta)}{1+k(d)} \right).
$$

If $c_1 < \beta$ we can assume that $k(c_1) = 0$ and this sum can be made negative by choosing large k. Thus we must have $c_1 = \beta$, in which case the sum under study is proportional to

$$
1 - \frac{k(c_2)}{1+k(a)} + \frac{k(\beta)^2}{(1+k(a))(1+k(d))} - \frac{k(\beta)}{1+k(d)}. \tag{6.7}
$$

If $c_2 = \beta$ then

$$
W = \begin{pmatrix} a & b_1 & \alpha \\ \beta & d & \alpha \\ \beta & \beta & e \end{pmatrix}
$$

is a GUM. So we must analyze the case where $c_2 > \beta$ in (6.7). We will arrive to a contradiction by taking an asymptotic as before. Consider a fixed number $\lambda \in (0, 1)$. Choose a family of functions $(k_r : r \in \mathbb{N})$ such that as r grows toward ∞

$$k_r(\beta) \to \infty, \ \frac{k_r(\beta)}{k_r(c_2)} \to \lambda, \ \frac{k_r(c_2)}{k_r(a)} \to 1, \ \frac{k_r(d)}{k_r(a)} \to \phi,$$

where $\phi = 1$ if $d > \beta$, and $\phi = \lambda$ if $d = \beta$. The asymptotic of (6.7) is then

$$1 - 1 + \frac{\lambda^2}{\phi} - \frac{\lambda}{\phi}.$$

This quantity is strictly negative for the two possible values of ϕ, which is a contradiction, and therefore $c_2 = \beta$.

To finish with the **Subcase** $\tilde{\rho}_1 = 0$ we consider the following condition. **Subsubcase** $b_1 = \beta$. We recall that we are under the restrictions $c_1 \leq \beta \leq c_2$ and

$$W = \begin{pmatrix} u & \beta & u \\ c_1 & d & \alpha \\ c_2 & \beta & e \end{pmatrix}.$$

Notice that if $c_2 = \beta$ then W is GUM. So for the rest of this subcase we assume $c_2 > \beta$. Also if $c_1 = \alpha$ we can permute 1 and 2 to get

$$\Pi W \Pi' = \begin{pmatrix} d & \alpha & \alpha \\ \beta & a & \alpha \\ \beta & c_2 & e \end{pmatrix},$$

which is also in NBF, and W is a GUM. Thus we can assume that $c_1 > \alpha$, and again of course we have $\alpha < \beta$.

Take a nondecreasing function m such that

$$\mathbb{I} + m(W) = \begin{pmatrix} 1 + m(a) & m(\beta) & 0 \\ m(c_1) & 1 + m(d) & 0 \\ m(c_2) & m(\beta) & 1 + m(e) \end{pmatrix},$$

We take the asymptotic under the following restrictions:

$$\frac{m(\beta)}{m(a)} \to \lambda \in (0, 1), \ \frac{m(c_1)}{m(a)} \to \lambda, \ \frac{m(e)}{m(a)} \to 1, \ \frac{m(c_2)}{m(a)} \to 1, \ \frac{m(d)}{m(a)} \to \phi,$$

where $\phi = 1$ if $d > \beta$, and it is λ if $d = \beta$. The limiting matrix for $\frac{1}{m(a)}(\mathbb{I} + m(W))$ is

$$V = \begin{pmatrix} 1 & \lambda & 0 \\ \lambda & \phi & 0 \\ 1 & \lambda & 1 \end{pmatrix}.$$

The determinant of V is $\Delta = \phi - \lambda^2 > 0$. Thus, V is a bi-potential since it is nonsingular and a limit of bi-potential matrices. The inverse of V is given by

$$V^{-1} = \frac{1}{\Delta}\begin{pmatrix} \phi & -\lambda & 0 \\ -\lambda & 1 & 0 \\ -(\phi - \lambda^2) & 0 & \phi - \lambda^2 \end{pmatrix},$$

and the sum of the first column is

$$\frac{\lambda^2 - \lambda}{\Delta} < 0,$$

which is a contradiction.

This finishes with the subcase $\rho_2 = 0$ and we return to (6.4) to consider now the following case

- **Case $\theta_1 = 0$.** Under this condition we get $c_1 = \alpha$ and

$$W = \begin{pmatrix} a & b_1 & b_2 \\ \alpha & d & \alpha \\ c_2 & \beta & e \end{pmatrix}.$$

Consider the transpose of W and permute on it 2 and 3, to obtain the matrix

$$\tilde{W} = \begin{pmatrix} a & c_2 & \alpha \\ b_2 & e & \alpha \\ b_1 & \beta & d \end{pmatrix},$$

where now $b_1 \geq \beta$. Clearly the matrix $\tilde{W}$ satisfies the hypothesis of the Theorem and has the shape of (6.6), that is we are in the "case $\rho_2 = 0$" which we already know implies that $\tilde{W}$ is GUM. Therefore W itself is GUM.

Recall that $\tau(W) = \inf\{t \geq 0 : \mathbb{I} + tW \notin bi\mathscr{P}\}$ is the infimum of t for which either $\mathbb{I} + tW$ is singular, or its inverse is not a Z-matrix, or it is not row or column diagonally dominant. Below we distinguish a special class of matrices where the last property gives the active restriction.

Definition 6.22 A nonnegative matrix W is said to be in **class** $\mathscr{T}$ if $\tau(W) = \infty$ or

$$\tau(W) < \infty, \ \mathbb{I} + \tau(W)\,W \text{ is nonsingular and}$$

$$\tau(W) = \inf\{t > 0 : \ (\mathbb{I} + tW)^{-1}\mathbb{1} \not\geq 0 \text{ or } \mathbb{1}'(\mathbb{I} + tW)^{-1} \not\geq 0\}.$$

Theorem 6.23 *Let W is a bi-potential and $f : \mathbb{R}_+ \to \mathbb{R}_+$ be a strictly increasing convex function. The matrix $f(W)$ is a bi-potential if and only if $f(W)$ belongs to class $\mathscr{T}$.*

Proof We can assume that $f(0) = 0$. We have $M_t = W^{-1}(\mathbb{I} + tf(W)) = W^{-1} + tW^{-1}f(W)$ is an M-matrix, for all $t \geq 0$. Indeed, as in the proof of Theorem 6.3 we have that M_t is a Z-matrix and we take a vector x such that $y' = x'W^{-1} > 0$ (see [38], Theorem 2.5.3). Thus $x'M_t = y'(\mathbb{I} + tf(W)) \geq y' > 0$ and using again the cited theorem in [38] we get M_t is an M-matrix. Therefore $\mathbb{I} + tf(W)$ is nonsingular for all t. Let λ_t and μ_t be the signed equilibrium potentials of $\mathbb{I} + tf(W)$.

Assume first that $f(W)$ is in class $\mathscr{T}$ which means that

$$\tau(f(W)) = \inf\{t > 0 : \ \lambda_t \not\geq 0 \text{ or } \mu_t \not\geq 0\}.$$

We prove that for all $t \geq 0$, λ_t, μ_t are nonnegative. Since

$$M_t\lambda_t = W^{-1}\mathbb{1} = \lambda^W \geq 0,$$

we obtain $\lambda_t = M_t^{-1}\lambda^W \geq 0$, because M_t^{-1} is a nonnegative matrix. Similarly, we show that $\mu_t \geq 0$. Since $f(W)$ is in class $\mathscr{T}$ we deduce $\tau(f(W)) = \infty$ and given that $f(W)$ is nonsingular we get from Proposition 6.15 that $f(W)$ is a bi-potential.

Conversely if $f(W)$ is a bi-potential then $\tau(f(W)) = \infty$ and the result follows.

6.3.1 Class $\mathscr{T}$ and Weakly Filtered Matrices

The aim of this section is to relate the concepts of class $\mathscr{T}$ and weakly filtered matrices.

Theorem 6.24 *Assume that W is a weakly filtered matrix. Then W belongs to the class $\mathscr{T}$ and moreover*

$$\tau(W) = \inf\{t > 0 : \ (\mathbb{I} + tW)^{-1}\mathbb{1} \not> 0 \text{ or } \mathbb{1}'(\mathbb{I} + tW)^{-1} \not> 0\}$$

In particular if $\tau(W) < \infty$ then $\mathbb{I} + \tau(W)W$ is a bi-potential.

Remark 6.25 If W is in class $\mathscr{T}$ and $\tau(W) < \infty$, then $\mathbb{I} + \tau(W)\,W$ is nonsingular. Since the set of nonsingular matrices is open then $\mathbb{I} + t\,W$ is nonsingular for $t > \tau(W)$, sufficiently close to $\tau(W)$.

Proof Consider a decomposition of W as

$$W = \sum_{s=0}^{\ell} a_s \cdot \mathbb{E}_s \cdot b_s,$$

where a_s, b_s are nonnegative $\mathbb{E}_{s+1}$-measurable.

First, for $p = 0, \ldots, \ell$ consider the matrices

$$W(p) = \sum_{s=p}^{\ell} a_s \cdot \mathbb{E}_s \cdot b_s.$$

We notice that $W(0) = W$. We shall prove that $\tau_p = \tau(W(p))$ is nondecreasing in p and $\tau_\ell = \infty$.

We rewrite the algorithm developed in Sect. 2.5 for $\mathbb{I} + t W$. This takes the form $\lambda_\ell(t) = \mu_\ell(t) = \kappa_\ell(t) = 1$, $\sigma_\ell(t) = (1 + t\, a_\ell b_\ell)^{-1}$ and for $p = \ell - 1, \cdots, 0$:

$$\begin{aligned}
\lambda_p(t) &= \lambda_{p+1}(t)[1 - \sigma_{p+1}(t)\, t\, a_{p+1}\mathbb{E}_{p+1}(\kappa_{p+1}(t)b_{p+1})]; \\
\mu_p(t) &= \mu_{p+1}(t)[1 - \sigma_{p+1}(t)\, t\, b_{p+1}\mathbb{E}_{p+1}(\kappa_{p+1}(t)a_{p+1})]; \\
\kappa_p(t) &= \mathbb{E}_{p+1}(\lambda_p(t)) = \mathbb{E}_{p+1}(\mu_p(t)); \\
\sigma_p(t) &= (1 + \mathbb{E}_p(\kappa_p(t)\, t\, a_p b_p))^{-1}.
\end{aligned} \tag{6.8}$$

Also $\lambda_{-1}(t)$, $\mu_{-1}(t)$ are defined similarly. If $\lambda_s(t)$, $\mu_s(t)$, $\sigma_s(t) : \ s = \ell, \ldots, p$ are well defined then

$$(\mathbb{I} + t W(p))^{-1} = \mathbb{I} - N(p, t),$$

where

$$N(p, t) = \sum_{s=p}^{\ell} \sigma_s(t)\lambda_s(t)\, t\, a_s \cdot \mathbb{E}_s \cdot b_s \mu_s(t). \tag{6.9}$$

If $\lambda_s(t)$, $\mu_s(t)$, $\sigma_s(t) : \ s = \ell, \ldots, p$ are nonnegative then $N(p, t) \geq 0$, and $(\mathbb{I} + t W(p))$ is an inverse M-matrix. Moreover, $\lambda_{p-1}(t)$ and $\mu_{p-1}(t)$ are the signed right and left equilibrium potentials of $(\mathbb{I} + t W(p))$

$$(\mathbb{I} + t W(p))\lambda_{p-1}(t) = \mathbb{1}, \text{ and } \mu'_{p-1}(t)(\mathbb{I} + t W(p)) = \mathbb{1}'.$$

So, if they are nonnegative, we have $\mathbb{I} + t W(p)$ is a bi-potential. In particular we have that

$$(\mathbb{I} + t a_\ell\, \mathbb{E}_\ell\, b_\ell)^{-1} = (\mathbb{I} + t W(\ell))^{-1} = \mathbb{I} - t(1 + t\, a_\ell b_\ell)^{-1} a_\ell\, \mathbb{E}_\ell\, b_\ell.$$

Since $\mathbb{E}_\ell = \mathbb{I}$ we obtain that $\lambda_{\ell-1} = \mu_{\ell-1} = (1 + t\, a_\ell b_\ell)^{-1}$. This means that $\mathbb{I} + t\,W(\ell)$ is a bi-potential for all $t \geq 0$. Therefore $\tau_\ell = \infty$ and the result is true for $W(\ell)$. This implies in particular that $\tau_{\ell-1} \leq \tau_\ell$. Assume the following inductive hypothesis holds:

- $\tau_{p+1} \leq \cdots \leq \tau_\ell$;
  and for $q = p + 1, \ldots, \ell$
- $\tau_q = \inf\{t > 0 : \lambda_{q-1}(t) \not\geq 0 \text{ or } \mu_{q-1}(t) \not\geq 0\} = \inf\{t > 0 : \lambda_{q-1}(t) \not> 0 \text{ or } \mu_{q-1}(t) \not> 0\}$;
- $\lambda_s(t), \mu_s(t)$, for $s = \ell, \ldots, q - 1$, are strictly positive for $t \in [0, \tau_q)$;
- If $\tau_q < \infty$ we have $\mathbb{I} + \tau_q W(q)$ is a bi-potential.

We first consider the case $\tau_{p+1} = \infty$ and fix $t \geq 0$. From Lemma 2.34, $\mathbb{I} + t\,W(p + 1)$ is a bi-potential and its equilibrium potential are strictly positive, that is $\lambda_p(t) > 0, \mu_p(t) > 0$. We conclude that $\lambda_s(t), \mu_s(t), \sigma_s(t) : s = \ell, \ldots, p$ are nonnegative and therefore $\mathbb{I} + t\,W(p)$ is nonsingular. Its inverse is $\mathbb{I} - N(p, t)$, where $N(p, t) \geq 0$ is given by (6.9). Hence, $\mathbb{I} + t\,W(p)$ is an inverse M-matrix. We conclude that

$$\tau_p = \inf\{t > 0 : \mathbb{I} + t\,W(p) \notin bi\,\mathcal{P}\} = \inf\{t > 0 : \lambda_{p-1}(t) \not\geq 0 \text{ or } \mu_{p-1}(t) \not\geq 0\}.$$

So, if $\tau_p = \infty$ we have, from Lemma 2.34 that

$$\lambda_{p-1}(t) > 0, \ \mu_{p-1}(t) > 0,$$

and the induction step holds in this case.

Now if $\tau_p < \infty$, we already have $\mathbb{I} + \tau_p W(p)$ is nonsingular and by continuity we have $\mathbb{I} + \tau_p W(p)$ is a bi-potential. We shall prove later on that $\lambda_{p-1}(t), \mu_{p-1}(t)$ are strictly positive in $[0, \tau_p)$.

We analyze now the case $\tau_{p+1} < \infty$. We first notice that in the algorithm the only possible problem is with the definition of $\sigma_p(t)$. Since $\sigma_p(\tau_{p+1}) > 0$ the algorithm is well defined, by continuity, for steps $\ell, \ldots, p$ on an interval $[0, \tau_{p+1} + \epsilon]$, for small enough $\epsilon > 0$. This proves that the matrix $\mathbb{I} + t\,W(p)$ is nonsingular in that interval, and that λ_{p-1}, μ_{p-1} exist in the same interval.

Now, for a sequence $t_n \downarrow \tau_{p+1}$ either $\lambda_p(t_n)$ or $\mu_p(t_n)$ has a negative component. Since there are a finite number of components we can assume without loss of generality that for a fixed component i we have $(\lambda_p(t_n))_i < 0$. Then, by continuity we get that $(\lambda_p(\tau_{p+1}))_i = 0$, which implies (by the algorithm) that $(\lambda_{p-1}(\tau_{p+1}))_i = 0$.

Now, assume that for some $t > \tau_{p+1}$ the matrix $\mathbb{I} + t\,W(p)$ is a bi-potential. Again by Lemma 2.34 we will have that $\mathbb{I} + \tau_{p+1} W(p)$ is a bi-potential and its equilibrium potential is positive $\lambda_{p-1}(\tau_{p+1}) > 0$, which is a contradiction. Therefore we conclude that $\tau_p \leq \tau_{p+1}$.

The conclusion of this discussion is that the matrix $\mathbb{I} + t W(p)$, for $t \in [0, \tau_{p+1}]$, is nonsingular and its inverse is $\mathbb{I} - N(p, t)$, with $N(p, t) \geq 0$. That is $\mathbb{I} + t W(p)$ is an inverse M-matrix and therefore

$$\tau_p = \inf\{t > 0 : \mathbb{I} + t W(p) \notin bi\mathscr{P}\} = \inf\{t > 0 : \lambda_{p-1}(t) \not\geq 0 \text{ or } \mu_{p-1}(t) \not\geq 0\},$$

and by continuity $\mathbb{I} + \tau_p W(p)$ is a bi-potential.

To finish the proof we need to show that τ_p coincides with

$$S = \inf\{t > 0 : \lambda_{p-1}(t) \not> 0 \text{ or } \mu_{p-1}(t) \not> 0\}.$$

It is clear that $S \leq \tau_p$. If $S < \tau_p$ then, due to Lemma 2.34, we have that both $\lambda_{p-1}(S) > 0$ and $\mu_{p-1}(S) > 0$, which is a contradiction and then $S = \tau_p$. This shows that $\lambda_{p-1}(t), \mu_{p-1}(t)$ are strictly positive for $t \in [0, \tau_p)$, and the induction is proven.

Remark 6.26 It is possible to prove that $\kappa_p(\tau_p) > 0$ when $\tau_p < \infty$.

We finish this section with two interesting results. The first one is an immediate consequence of Theorem 6.24. The second result follows from the first one, the Theorem 6.23 and the fact that CBF is stable under Hadamard functions.

Theorem 6.27 *If W is a nonnegative CBF matrix then W is in class $\mathscr{T}$.*

Theorem 6.28 *Assume W is a CBF bi-potential matrix, and $f : \mathbb{R}_+ \to \mathbb{R}_+$ is a nonnegative strictly increasing convex function. Then $f(W)$ is a bi-potential.*

Remark 6.29 In the last theorem we could replace CBF by weakly filtered, but we have to make the extra hypothesis that $f(W)$ is also weakly filtered (see Example 6.2).

6.4 Potentials of Random Walks and Hadamard Products

In this section we shall study potentials of nearest neighbor random walks on $I = \{1, \cdots, n\}$, that is, general birth and death chains on I. These potentials will be shown to be Hadamard product of two ultrametric potentials on the same set. This representation is intimately related to a result of Gantmacher and Krein [34], which shows that for a nonsingular symmetric matrix A its inverse A^{-1} is an irreducible and tridiagonal matrix if and only if A is the Hadamard product of, what they called, a weak type D-matrix and a flipped weak type D-matrix. In McDonald et al. [48], the authors extend this result when A is non-symmetric and A^{-1} is a Z-matrix (see for related results Nabben [50]). In this situation, an extra diagonal matrix is needed.

We shall prove that the inverse of a tridiagonal irreducible M-matrix can be written as the Hadamard product of two (or just one in some extreme cases) special nonsingular ultrametric matrices, after a suitable change by two diagonal matrices, which can be taken to be the identity if the M-matrix is symmetric and row diagonally dominant (see Corollary 6.47). Conversely, if $W = U \odot V$ is the Hadamard product of two nonsingular ultrametric matrices U, V associated with random walks then W^{-1} is a tridiagonal M-matrix (see Theorem 6.38). We also give necessary and sufficient conditions in terms of U, V in order that W^{-1} is row diagonally dominant. We discuss the uniqueness of this decomposition and study the special case where the random walk associated with W loses mass at one or two ends. We note that the above decomposition can be simplified if one allows one or both of them to be singular (see Theorems 6.34 and 6.38).

If W is a potential of a random walk then W is determined by two monotone sequences of positive numbers $0 < x_1 \leq x_2 \cdots \leq x_n, \ 0 < y_n \leq y_{n-1} \cdots \leq y_1$ such that

$$W_{ij} = x_{i \wedge j} \, y_{i \vee j} \text{ and } x_i = \frac{W_{in}}{W_{1n}} x_1, \ y_j = \frac{W_{1j}}{W_{11}} y_1.$$

So, W must satisfy the structural equation and monotonicity

$$W_{ij} = \frac{W_{in} W_{1j}}{W_{1n}}, \ 1 \leq i \leq j \leq n, \text{ and } W_{\bullet n} \uparrow, W_{1 \bullet} \downarrow . \tag{6.10}$$

This condition is close to be sufficient for having a potential of a random walk (see Corollary 6.40 and formula (6.24)). Therefore, in applications if one wants to model a random walk by specifying its potential, a restriction like (6.10) must be imposed. Notice also that Theorem 6.44 discriminates between the ultrametric case and the non ultrametric one by the disposition of the sites where the chain will lose mass (roots). So, as a consequence of this result, if the model is not ultrametric we expect to have at least two non consecutive roots.

Notably, restriction (6.10) is stable under Hadamard positive powers, which is in accordance with the fact that potentials are stable under Hadamard powers. On the other hand (6.10) is also stable under Hadamard products, an indication that the product of two inverse tridiagonal M-matrices is again an inverse tridiagonal M-matrix. The probabilistic consequences of this fact and how the restriction that both graphs are linear intervenes in this property, remain as open questions.

To end this introduction, let us state a general remark on potentials whose inverses are tridiagonal matrices. Recall that a potential W is a pointwise column diagonally dominant matrix, that is $W_{ij} \leq W_{jj}$. Moreover, we have (see Chap. 2 Lemma 2.36)

$$W_{ij} = f_{ij}^W W_{jj},$$

where f_{ij}^W is the probability that the underline Markov chain ever visits j when starting from i (by convention $f_{ii}^W = 1$). We notice that if there exists a path (of positive probability) connecting k with a root i, which does not contain j, then

$f_{kj}^W < 1$ and so $W_{kj} < W_{jj}$. On the other hand, if every path (of positive probability) that connects k with any root also contains j, then $f_{kj}^W = 1$ and $W_{kj} = W_{jj}$.

So, if W^{-1} is irreducible and tridiagonal, then for $i \leq j$, we have

$$\mathscr{R}(W) \cap [1, j) = \emptyset \Rightarrow f_{ij}^W = 1 \text{ and}$$
$$\mathscr{R}(W) \cap [1, j) \neq \emptyset \Rightarrow f_{ij}^W < 1.$$

Finally, if W is the inverse of a tridiagonal irreducible M-matrix then $\mathscr{G}^W$, the incidence graph of W^{-1} is linear: $(i, j) \in \mathscr{G}^W$, for $i \neq j$, if and only if $|i - j| = 1$.

6.4.1   Linear Ultrametric Matrices

The following special class of ultrametric matrices will play an important role.

Definition 6.30 A symmetric matrix U, of size n, is said to be a **linear ultrametric matrix** (LUM) if there exist $k \in \{1, \cdots, n\}$ and positive numbers $x_1, \cdots, x_n$ such that

$$x_1 \geq x_2 \geq \cdots x_{k-1} \geq x_k \leq x_{k+1} \leq \cdots \leq x_n \qquad (6.11)$$

and U_{ij} is given by

$$U_{ij} = \min\{x_s : i \wedge j \leq s \leq i \vee j\} = \begin{cases} x_{i \vee j} & \text{if } i, j \leq k \\ x_{i \wedge j} & \text{if } i, j \geq k \ . \\ x_k & \text{otherwise} \end{cases}$$

We shall say that U is in class LUM(k) to emphasize the dependence on k. We call $x_1, \cdots, x_n$ the characteristics of U.

Matrices in LUM(1) are a special case of D-matrices introduced by Markham [43]. In the same vein, LUM(n) are a special case of flipped D-matrices. Also, matrices in LUM(k) are special case of cyclops with eye $k+$ in the notation of [48]. In particular, U can be described by blocks as

$$U = \begin{pmatrix} A & x_k E \\ x_k E' & B \end{pmatrix},$$

where A is a LUM(k) matrix of size k determined by $x_1 \geq \cdots \geq x_k > 0$, B is a LUM(1) matrix of size $n - k$ determined by $0 < x_{k+1} \leq \cdots \leq x_n$, E is a matrix of ones of the appropriate size and $x_k \leq x_{k+1}$. Every linear ultrametric matrix is an ultrametric matrix.

Remark 6.31 If $J \subseteq \{1, \cdots k\}$ has cardinal $p \geq 1$ and $U \in \mathrm{LUM}(k)$ then $U_J \in \mathrm{LUM}(p)$ Similarly, if $J \subseteq \{k, \cdots, n\}$, then $U_J \in \mathrm{LUM}(1)$.

The following theorem collects known results about these matrices and shows the connection between them and symmetric random walks.

Theorem 6.32   *(i) Assume that $U \in LUM(k)$. Then, U is nonsingular if and only if all inequalities in (6.11) are strict.*

(ii) Assume that $U \in LUM(k)$ is nonsingular. Then, $U^{-1} = \kappa(\mathbb{I} - P)$ for some constant κ and a symmetric irreducible substochastic and tridiagonal matrix P that loses mass only at k.

(iii) Assume that P is a symmetric irreducible substochastic and tridiagonal matrix. Then $U = (\mathbb{I} - P)^{-1}$ is an ultrametric matrix if and only if one of the two cases occurs: for some k

    (iii.1) $\mathscr{R}(U) = \{k\}$, in which case $U \in LUM(k)$;

    (iii.2) $\mathscr{R}(U) = \{k, k+1\}$, that is the roots are adjacent. In this case $U = V_1 \odot V_2$ where $V_1 \in LUM(k)$, $V_2 \in LUM(k+1)$ are nonsingular.

Part *(iii.2)* will be generalized in Theorem 6.44 to the case where the potential has more than 2 roots or has two roots in general position. A formula for the tridiagonal matrix U^{-1} when $U \in \mathrm{LUM}(k)$, $2 \leq k \leq n - 1$, is given by

$$
\begin{cases}
U_{11}^{-1} = -U_{12}^{-1} = -U_{21}^{-1} = \frac{1}{U_{11}-U_{22}}; \\
\forall\, i \leq n - 1 \quad U_{i,i+1}^{-1} = U_{i+1,i}^{-1} = \frac{-1}{|U_{i+1,i+1}-U_{ii}|}; \\
\forall\, 2 \leq i \leq n, i \neq k \quad U_{ii}^{-1} = -U_{i,i-1}^{-1} - U_{i,i+1}^{-1}; \\
U_{kk}^{-1} = \frac{U_{k+1,k+1}}{U_{kk}U_{k+1,k+1}-U_{kk}^2} + \frac{1}{U_{k-1,k-1}-U_{kk}}.
\end{cases}
\tag{6.12}
$$

When $U \in \mathrm{LUM}(1)$, we have

$$
\begin{cases}
U_{11}^{-1} = \frac{1}{U_{11}} + \frac{1}{U_{22}-U_{11}}; \\
\forall i \leq n - 1 \quad U_{i,i+1}^{-1} = U_{i+1,i}^{-1} = \frac{-1}{U_{i+1,i+1}-U_{ii}}; \\
\forall\, 2 \leq i \leq n \quad U_{ii}^{-1} = -U_{i,i-1}^{-1} - U_{i,i+1}^{-1}.
\end{cases}
\tag{6.13}
$$

In both cases we assume implicitly that $U_{n,n+1}^{-1} = 0$. A similar formula holds when $U \in \mathrm{LUM}(n)$.

For the proof of Theorem 6.32 we need the following useful result about principal submatrices of inverse tridiagonal irreducible M-matrices. Some parts of this lemma are already proven for general inverse M-matrices (see Lemma 2.32 in Chap. 2).

Lemma 6.33 *Assume $W = M^{-1}$ is the inverse of an irreducible tridiagonal M-matrix indexed by $I = \{1, \cdots, n\}$. Let $J = \{\ell_1 < \cdots < \ell_p\} \subset I$ and $X = W_J$ be a principal submatrix of W. Then X^{-1} is an irreducible tridiagonal M-matrix. If W*

is a potential (Markov potential), then X is also a potential (respectively a Markov potential). Moreover, if $\lambda = \lambda^W$ is the right equilibrium potential of W then,

$$\lambda^X \geq \lambda_J - M_{JJ^c} M_{J^c J^c}^{-1} \lambda_{J^c} \geq \lambda_J. \tag{6.14}$$

Then, the set $\ell(\mathcal{R}(X)) = \{\ell_i : i \in \mathcal{R}(X)\}$ contains $\mathcal{R}(W) \cap J$.

Consider $i = \ell_s$ for some $s = 1, \cdots, p$. If $\{i, i+1\} \subseteq J$ then $X_{s,s+1}^{-1} = M_{i,i+1}$. Furthermore, if $\{i-1, i, i+1\} \subseteq J$ then $X_{s,s+t}^{-1} = M_{i,i+t}$ for $t \in \{-1, 0, 1\}$. Finally, if $\{1, 2\} \subseteq J$ we have $X_{1k}^{-1} = M_{1k}$ for $k \in \{1, 2\}$ (a similar relation holds when $\{n-1, n\} \subset J$).

Proof According to Lemma 2.32, in Chap. 2, X^{-1} is an M-matrix. If W is a potential (Markov potential) then X is also a potential (respectively Markov potential) and (6.14) also holds. Since $X > 0$ then X^{-1} is irreducible (see (2.25)).

The only thing left to be proven in the first part is the fact that X^{-1} is tridiagonal. Using the inverse by block formula we obtain

$$X^{-1} = M_{JJ} - M_{JJ^c} M_{J^c J^c}^{-1} M_{J^c J}. \tag{6.15}$$

For the rest of the proof, we denote by $Y = M_{J^c J^c}^{-1}$.

The set J induces a partition on $J^c = \{l_1 < \cdots < l_{n-p}\}$, in at least 1 atom, given by the sets $[\ell_s, \ell_{s+1}] \cap J^c$, where $\ell_s, \ell_{s+1} \in J$ are consecutive in this subset, together with the sets $[1, \ell_1] \cap J^c$ and $[\ell_p, n] \cap J^c$. Denote the nonempty atoms by $\mathscr{A}_1, \cdots, \mathscr{A}_r$. The fact that M is tridiagonal implies that $M_{\mathscr{A}_a \mathscr{A}_b} = 0$ for $a \neq b$. This block structure of $M_{J^c J^c}$ is also present in Y. Therefore we have the formula, for $\ell_s = i < j = \ell_t$

$$X_{st}^{-1} = M_{ij} - C_{i+1, j-1} M_{i, i+1} Y_{qr} M_{j-1, j},$$

where $C_{i+1, j-1} = 1$ if $l_q = i+1, l_r = j-1$ belong to the same atom in J^c. When $i+1$ or $j-1$ do not belong to J^c or they belong to different atoms we take $C_{i+1, j-1} = 0$.

So, if there exists $k \in J$ such that $i < k < j$ we conclude that $X_{st}^{-1} = 0$ and therefore X^{-1} is a tridiagonal M-matrix.

Similarly, we obtain that

$$X_{ss}^{-1} = M_{ii} - C_{i+1, i+1} M_{i, i+1} Y_{qq} M_{i+1, i} - C_{i-1, i-1} M_{i, i-1} Y_{q'q'} M_{i-1, i}, \tag{6.16}$$

from where the last part of the lemma follows (here $l_{q'} = i - 1$ when this element belongs to J^c).

Proof (Theorem 6.32)

(i) The property follows from the fact that a positive ultrametric matrix is nonsingular if and only if all rows are different (Theorem 3.5 in Chap. 3 or [20, 47]).

(ii) Every nonsingular ultrametric matrix is a potential. When $U \in LUM(k)$ the k-th column is constant and therefore $U^{-1}\mathbb{1} = \frac{1}{x_k}e(k)$, where $e(k)$ is the k-th vector of the canonical basis in $\mathbb{R}^n$. Thus, the unique root of U is k. The fact that U^{-1} is tridiagonal follows from Theorem 4.7, because the tree matrix extension of U is supported by a path or linear tree (see also Theorem 4.10 in [50]).

(iii) Assume that U^{-1} is tridiagonal, that is, its incidence graph is a path. For a nonsingular ultrametric matrix, all the roots are connected (see Theorem 4.10). and so U can has one root or two adjacent roots. Conversely, assume that U has only one root at k. For $i \leq j \leq k$ we have $U_{ij} = f_{ij}^U U_{jj} = U_{jj}$. Similarly, if $k \leq i \leq j$ we have $U_{ij} = U_{ii}$. Finally, for the case $i \leq k \leq j$ one has $U_{ij} = U_{kj} = U_{jk} = U_{kk}$. Thus, $U \in LUM(k)$ with characteristics $x_i = U_{ii}$.

Finally, assume that U has two consecutive roots at $k, k+1$, then according to Theorem 4.10 in [50] U is ultrametric. The fact that $U = V_1 \odot V_2$ with $V_1 \in LUM(k), V_2 \in LUM(k+1)$ is a particular case of what we will prove in Theorem 6.44.

6.4.2 Hadamard Products of Linear Ultrametric Matrices

Theorem 6.34 *Let M be a tridiagonal irreducible M-matrix, of size n, with inverse $W = M^{-1}$. Then, there exist two positive diagonal matrices D, E such that $(DWE)^{-1}$ is an irreducible, symmetric, tridiagonal, row diagonally dominant M-matrix and*

$$DWE = U \odot V \tag{6.17}$$

for some $U \in LUM(1)$ and $V \in LUM(n)$. If M is row diagonally dominant we can take $D = \mathbb{I}$ and if M is symmetric we can take $D = E$.

All diagonal matrices D, E for which $X = DWE$ is a symmetric potential are constructed in the following way. Take $\rho \in \mathbb{R}^n$ any nonnegative nonzero vector and define $D = D(\rho)$ as

$$\forall i \quad D_{ii} = \frac{1}{(W\rho)_i}. \tag{6.18}$$

Next, define $E = E(a, \rho)$ as the solution of the iteration: $E_{11} = a > 0$ arbitrary and

$$\forall i \geq 2 \quad E_{ii} = \frac{D_{ii}W_{i,i-1}}{D_{i-1,i-1}W_{i-1,i}}E_{i-1,i-1}. \tag{6.19}$$

The right equilibrium potential of X is $\lambda^X = E^{-1}\rho$.

In particular, if we choose $\rho = e(1) = (1,0,\cdots,0)'$ *the first vector in the canonical basis of* $\mathbb{R}^n$, *and consider* $\overline{D} = D(e(1))$, $\overline{E} = E(1,e(1))$ *then* $\overline{D}W\overline{E} = \overline{U}$ *is a symmetric potential whose inverse is an irreducible tridiagonal row diagonally dominant M-matrix and its right equilibrium potential is* $\lambda^{\overline{U}} = \overline{E}^{-1}e(1) = e(1)$. *Hence,* $\overline{U} \in LUM(1)$.

The last part of the previous theorem states that for some $\overline{D}$ and $\overline{E}$ the matrix $\overline{D}W\overline{E} \in$ LUM(1). This is a special case of the general decomposition given in (6.17) where $V = \mathbb{1}\mathbb{1}'$ the matrix full of ones. The importance of this special decomposition is that every potential of a random walk can be changed to a linear ultrametric matrix through two diagonal matrices.

Using this transformation and formula (6.13) we obtain a formula for W^{-1} in terms of W, when W^{-1} is an irreducible tridiagonal M-matrix. This formula is

$$
\begin{cases}
W_{ij}^{-1} = 0 \text{ if } |i-j| > 1; \\
\forall i \le n-1 \quad
\begin{cases}
W_{i,i+1}^{-1} = \dfrac{-W_{i,i+1}W_{i1}W_{i+1,1}}{W_{i+1,i+1}W_{i+1,i}W_{i1}^2 - W_{ii}W_{i,i+1}W_{i+1,1}^2} \\
W_{i+1,i}^{-1} = W_{i,i+1}^{-1}\dfrac{W_{i+1,i}}{W_{i,i+1}}
\end{cases} ; \\
\forall 2 \le i \le n-1 \quad W_{ii}^{-1} = -W_{i,i-1}^{-1}\dfrac{W_{i-1,1}}{W_{i1}} - W_{i,i+1}^{-1}\dfrac{W_{i+1,1}}{W_{i,1}}; \\
W_{11}^{-1} = -W_{12}^{-1}\dfrac{W_{2n}}{W_{1n}}, \quad W_{nn}^{-1} = -W_{n,n-1}^{-1}\dfrac{W_{n-1,1}}{W_{n,1}}.
\end{cases}
\tag{6.20}
$$

Remark 6.35 Recall that each diagonal entry of an ultrametric matrix dominates its corresponding column (and row) and this property is stable under Hadamard products. Thus, if $W = U \odot V$ is the Hadamard product of two ultrametric matrices, then its diagonal entries dominate the corresponding columns.

Proof (Theorem 6.34) Assume first M is a symmetric tridiagonal irreducible row diagonally dominant M-matrix. Consider $W = M^{-1}$. The proof is done by induction on n the size of M. For $n = 1$ the result is obvious. So assume the result holds when the size is at most $n - 1$. So, assume M has size n. Without loss of generality we can assume that $M = \mathbb{I} - P$ where P is a substochastic tridiagonal matrix. We decompose M and W by blocks as following

$$
M = \begin{pmatrix} 1 - P_{11} & -\zeta' \\ -\zeta & N \end{pmatrix} \text{ and } W = \begin{pmatrix} W_{11} & a' \\ a & T \end{pmatrix}.
$$

Here $\zeta' = (P_{12}, 0, \cdots, 0) \ge 0$, $\zeta'\mathbb{1} \le 1 - P_{11}$, $a' = (W_{12}, \cdots, W_{1n})$ and N is a tridiagonal, symmetric row diagonally dominant M-matrix and moreover $N\mathbb{1} - \zeta \ge 0$. From Lemma 6.33 T is also a Markov potential and T^{-1} is tridiagonal. Then, by the induction hypothesis there exist two ultrametric matrices R, S, where $R \in$ LUM(1) and $S \in$ LUM$(n - 1)$ such that $T = R \odot S$. We denote by $0 < x_2 \le x_3 \le \cdots \le x_n$ and $y_2 \ge y_3 \ge \cdots \ge y_n > 0$ the numbers defining R and S respectively.

The fact that P is tridiagonal and symmetric implies that for $i \geq 2$

$$W_{i1} = f_{i2}^W W_{21} = f_{i2}^W W_{12} = f_{i2}^W f_{12}^W W_{22} = f_{12}^W W_{i2}.$$

We take $x_1 = x_2 f_{12}^W$ so $0 < x_1 \leq x_2$. Then $W_{21} = x_1 y_2$, $W_{31} = x_1 y_3, \cdots, W_{n1} = x_1 y_n$. Finally, we take $y_1 = W_{11}/x_1 \geq W_{21}/x_1 = y_2$. We define $U \in \mathrm{LUM}(1)$ associated with $0 < x_1 \leq x_2 \leq x_3 \leq \cdots \leq x_n$, and $V \in \mathrm{LUM}(n)$ associated with $y_1 \geq y_2 \geq y_3 \geq \cdots \geq y_n > 0$ to get $W = U \odot V$ and the result is proven in this case.

Now, consider M a general irreducible tridiagonal M-matrix. Then, there exists a positive diagonal matrix F (see [38], Theorem 2.5.3) such that $L = MF$ is a row diagonally dominant M-matrix. If M is a row diagonally dominant M-matrix we can take $F = \mathbb{I}$. Clearly, L is also tridiagonal and irreducible. Now, we look for a diagonal matrix G such that GL is also symmetric. The condition is that $G_{ii}L_{i,i-1} = G_{i-1,i-1}L_{i-1,i}$ for $i = 2, \cdots, n$. We take $G_{11} = F_{11}$ and define inductively, for $i \geq 2$

$$G_{ii} = G_{i-1,i-1} \frac{L_{i-1,i}}{L_{i,i-1}}.$$

The diagonal of G is positive, by construction. Thus, $H = GMF$ is a symmetric tridiagonal irreducible M-matrix, and

$$H\mathbb{1} = G(MF\mathbb{1}) \geq 0.$$

Hence, H is also row diagonally dominant and there exist two ultrametric matrices $U \in \mathrm{LUM}(1)$, $V \in \mathrm{LUM}(n)$ such that $F^{-1}M^{-1}G^{-1} = H^{-1} = U \odot V$. Thus, we take $D = F^{-1}$ and $E = G^{-1}$.

If M is symmetric we obtain for $i \geq 2$

$$G_{ii} = G_{i-1,i-1} \frac{L_{i-1,i}}{L_{i,i-1}} = G_{i-1,i-1} \frac{M_{i-1,i}F_{ii}}{M_{i,i-1}F_{i-1,i-1}} = G_{i-1,i-1} \frac{F_{ii}}{F_{i-1,i-1}}.$$

The solution is $G = F$, which implies $H = FMF$ is a symmetric tridiagonal row diagonally dominant M-matrix and we obtain $D = E = F^{-1}$.

Take D, E diagonal matrices with positive diagonal elements. Assume that $X = DWE$ is a symmetric potential and consider λ its right equilibrium potential. Then, $DWE\lambda = \mathbb{1}$ and $\rho = E\lambda$ is a nonzero, nonnegative vector. Obviously, we get that $D = D(\rho)$ as defined in (6.18). Since X is symmetric, we get that E must satisfy (6.19) and finally, we obtain $\lambda = E^{-1}\rho$.

Conversely, assume that D, E are constructed as in (6.18) and (6.19). The matrix $X = DWE$ is nonsingular and its inverse is $X^{-1} = E^{-1}MD^{-1}$. Thus, X^{-1} is an irreducible tridiagonal M-matrix. On the other hand, $XE^{-1}\rho = DW\rho = \mathbb{1}$, which means that X^{-1} is a row diagonally dominant matrix.

Now, let us prove X is symmetric. This is equivalent to prove that $E^{-1}MD^{-1}$ is symmetric, which follows from the fact that for an irreducible tridiagonal M-matrix M with inverse W it holds (see formula (6.20))

$$\frac{M_{i,i+1}}{M_{i+1,i}} = \frac{W_{i,i+1}}{W_{i+1,i}}.$$

Finally, if we take $\rho = e(1)$ then $\overline{U} = \overline{D}W\overline{E}$ is a symmetric potential whose inverse is an irreducible symmetric tridiagonal row diagonally dominant M-matrix. Hence, $(\overline{U})^{-1} = \kappa(\mathbb{I} - P)$, for some constant κ and an irreducible symmetric tridiagonal substochastic matrix P. Since the right equilibrium potential of $\overline{U}$ is $\lambda^{\overline{U}} = e(1)$ we get that P loses mass only at the first row. Then, from Theorem 6.32 we conclude that $\overline{U}$ is a linear ultrametric matrix that belongs to LUM(1).

In the next result we show uniqueness of the decomposition $W = U \odot V$ up to a multiplicative constant.

Proposition 6.36 *Assume that* $W = U \odot V = \tilde{U} \odot \tilde{V}$, *for some* $U, \tilde{U} \in LUM(1)$ *and* $V, \tilde{V} \in LUM(n)$. *Then, there exists a constant* $a > 0$ *such that* $\tilde{U} = aU$ *and* $\tilde{V} = \frac{1}{a}V$.

Proof Let

$$0 < x_1 \leq \cdots \leq x_n, \, 0 < \tilde{x}_1 \leq \cdots \leq \tilde{x}_n,$$
$$y_1 \geq \cdots \geq y_n > 0, \, \tilde{y}_1 \geq \cdots \geq \tilde{y}_n > 0,$$

be the collection of numbers defining $U, \tilde{U}, V$ and $\tilde{V}$ respectively. We define $a = \tilde{x}_1/x_1$. We notice that for all i we have $W_{i1} = x_1 y_i = \tilde{x}_1 \tilde{y}_i$ and therefore for all i we obtain $y_i = a\tilde{y}_i$ which implies that $V = a\tilde{V}$. Similarly, we obtain that $\tilde{U} = aU$ and the result is shown.

Example 6.37 When M is not row diagonally dominant, a decomposition like $W = U \odot V$ may not exist, and the use of the diagonal matrices is necessary. Take for example

$$W = \begin{pmatrix} 35 & 20 & 25 \\ 20 & 16 & 20 \\ 25 & 20 & 35 \end{pmatrix},$$

whose inverse is the irreducible tridiagonal M-matrix

$$M = \begin{pmatrix} 0.1 & -0.125 & 0 \\ -0.125 & 0.375 & -0.125 \\ 0 & -0.125 & 0.1 \end{pmatrix}.$$

We point out that M is row diagonally dominant only for the second row. A simple inspection shows that W is not the Hadamard product of linear ultrametric matrices (see Remark 6.35 and Theorem 6.38 (ii) below).

Theorem 6.38 *Let $U \in LUM(1)$, $V \in LUM(n)$ of size n and consider $W = U \odot V$.*

(i) W is nonsingular if and only if for all $i = 1, \cdots, n-1$

$$U_{i+1,i+1} V_{ii} > U_{ii} V_{i+1,i+1}. \tag{6.21}$$

This condition is equivalent to for all $i = 1, \cdots, n-1$ $W_{1,i} W_{i+1,n} > W_{1n} W_{i,i+1}$. In particular, W is nonsingular if U and V are nonsingular.

In what follows we assume that W is nonsingular with inverse $M = W^{-1}$.

(ii) M is an irreducible tridiagonal M-matrix, which is row diagonally dominant at rows $1, n$.

(iii) M is strictly diagonally dominant at row 1 if and only if

$$(U_{22} - U_{11})(U_{33} V_{22} - U_{22} V_{33}) > 0. \tag{6.22}$$

Since $V_{22} \geq V_{33} > 0$ then a sufficient condition for this to happen is that $U_{33} > U_{22} > U_{11}$, which is the case when U is nonsingular. That is, if U is nonsingular M is strictly row diagonally dominant at row 1.

Similarly, M is strictly diagonally dominant at row n if and only if

$$(V_{n-1,n-1} - V_{nn})(U_{n-1,n-1} V_{n-2,n-2} - U_{n-2,n-2} V_{n-1,n-1}) > 0. \tag{6.23}$$

Again, since $0 < U_{n-2,n-2} \leq U_{n-1,n-1}$ a sufficient condition is that $V_{n-2,n-2} > V_{n-1,n-1} > V_{nn}$ which is the case when V is nonsingular. That is, if V is nonsingular M is strictly row diagonally dominant at row n.

(iv) When $n \geq 3$, we define for any $i \in \{2, \cdots, n-1\}$ the set $J = \{i-1, i, i+1\}$. The matrix W_J is nonsingular and its inverse $N(i)$ is a tridiagonal M-matrix, which is row diagonally dominant at rows $1, 3$ (strictly row diagonally dominant at rows 1 and 3 in the case U_J, V_J are nonsingular).

(v) Any of the following two conditions is necessary and sufficient for M to be row diagonally dominant at row $i \in \{2, \cdots, n-1\}$

$$(U_{i+1,i+1} - U_{i-1,i-1})(V_{i-1,i-1} - V_{ii}) \geq (U_{ii} - U_{i-1,i-1})(V_{i-1,i-1} - V_{i+1,i+1}), \tag{6.24}$$

$$N(i) \text{ is row diagonally dominant at row 2.} \tag{6.25}$$

Furthermore, M is strictly row diagonally dominant at row i if there is a strict inequality in (6.24) or equivalently a strict inequality in (6.25), which is equivalent to say that $N(i)$ is strictly row diagonally dominant at row 2.

When U and V are nonsingular (6.24) and (6.25) are equivalent to

$$U_{i,i-1}^{-1} V_{i,i+1}^{-1} \geq U_{i,i+1}^{-1} V_{i,i-1}^{-1}. \tag{6.26}$$

(vi) If U and V are Markov potentials, that is $U^{-1} = \mathbb{I} - P, V^{-1} = \mathbb{I} - Q$ with P, Q substochastic matrices, then the diagonal of M is bounded by one

$$\forall i \quad M_{ii} \leq 1.$$

The proof of this theorem requires the next lemma, which is essentially the result we want to show for dimension $n = 3$.

Lemma 6.39 *Consider the tridiagonal symmetric substochastic matrices*

$$P = \begin{pmatrix} 1-x-z & x & 0 \\ x & 1-x-y & y \\ 0 & y & 1-y \end{pmatrix} \text{ and } Q = \begin{pmatrix} 1-p & p & 0 \\ p & 1-p-q & q \\ 0 & q & 1-q-s \end{pmatrix},$$

where x, y, z, p, q, s are positive parameters that satisfy

$$x + z < 1,\ x + y \leq 1,$$
$$q + s < 1,\ p + q \leq 1.$$

Then $A = (\mathbb{I} - P)^{-1} \in LUM(1)$ and $B = (\mathbb{I} - Q)^{-1} \in LUM(3)$. The matrix $W = A \odot B$ is nonsingular and $N = W^{-1}$ is the tridiagonal M-matrix given by

$$N = \begin{pmatrix} \frac{\beta}{\alpha} & -\frac{\gamma}{\alpha} & 0 \\ -\frac{\gamma}{\alpha} & \frac{\epsilon}{\alpha\delta} & -\frac{\eta}{\delta} \\ 0 & -\frac{\eta}{\delta} & \frac{\kappa}{\delta} \end{pmatrix}, \begin{cases} \alpha = pqz + psz + qsx + qsz; \\ \beta = (x+z)pqsz; \\ \gamma = xpqsz; \\ \epsilon = (pqzx + pqzy + psyx + pszx + pszy + qsyx + qszx + qszy)zqsx; \\ \delta = qxz + syx + sxz + syz; \\ \kappa = (q+s)sxyz; \\ \eta = qsxyz. \end{cases}$$

The following inequalities hold

$$0 < \gamma < \beta < \alpha,\ 0 < \epsilon < \alpha\delta \text{ and } 0 < \eta < \kappa < \delta.$$

Thus, N is strictly row diagonally dominant at rows $1, 3$ and the diagonal elements of N are smaller than 1. The sum of the second row is

$$\frac{s^2 z^2 qx(qx - py)}{\alpha\delta}.$$

Hence, N is a tridiagonal row diagonally dominant M-matrix if and only if

$$qx - py \geq 0, \tag{6.27}$$

and in this case we have $N = \mathbb{I} - R$, *for a substochastic matrix* R, *which is stochastic at the second row if and only if equality holds in (6.27).*

Proof Direct using MAPLE.

Proof (Theorem 6.38) We first assume that $U \in \mathrm{LUM}(1)$ and $V \in \mathrm{LUM}(n)$ are nonsingular. Recall that U, V are inverse M-matrices and therefore they are positive definite matrices. This shows that W is positive definite and a fortiori nonsingular, proving the last claim in (i). We postpone to the end of this first part to show formula (6.21), which will finish the proof of (i).

Let us now show (ii)–(vi). Both U, V are determined by their diagonal values, which satisfy $0 < U_{11} < U_{22} < \cdots < U_{n-1,n-1} < U_{nn}$ and $V_{11} > V_{22} > \cdots > V_{n-1,n-1} > V_{nn} > 0$.

(ii)–(iv) We show by induction on n, the size of W, that $M = W^{-1}$ is a tridiagonal M-matrix, which is strictly row diagonally dominant at rows $1, n$. When the size is 1 or 2 the result is trivial. The case of size 3 is just Lemma 6.39.

So, assume the result is true up to size $n - 1$ and we shall prove it for size $n \geq 4$. Without loss of generality we assume that $U^{-1} = \mathbb{I} - P$ and $V^{-1} = \mathbb{I} - Q$, where P, Q are symmetric substochastic matrices, which are also irreducible and tridiagonal. Moreover, we suppose P loses mass only at 1 and Q only at n. We decompose the matrices W and M in the following blocks

$$W = \begin{pmatrix} X & V_{nn}u \\ V_{nn}u' & V_{nn}U_{nn} \end{pmatrix} \text{ and } M = \begin{pmatrix} \Omega & -\zeta \\ -\zeta' & \alpha \end{pmatrix}$$

where $X = W_{\{1,\cdots,n-1\}}$, $u' = (U_{n1}, \cdots, U_{n,n-1})$.

The basic computation we need is, for $i = 1, \cdots, n - 1$

$$(Xe)_i = U_{n-1,i} V_{n-1,i} = V_{n-1,n-1} U_{i,n-1},$$

where $e = e(n - 1)$ is the last vector of the canonical basis of $\mathbb{R}^{n-1}$. On the other hand $U_{i,n} = U_{n,i} = f^U_{n,n-1} U_{n-1,i} = U_{n-1,i} = U_{i,n-1}$. This means that

$$Xe = V_{n-1,n-1}u. \tag{6.28}$$

Then, using the formula for the inverse by blocks we get that

$$\Omega = \left(X - \frac{V_{nn}}{U_{nn}} uu' \right)^{-1} = X^{-1} + \gamma ee',$$

with

$$\gamma = \frac{V_{nn}}{V_{n-1,n-1}(V_{n-1,n-1}U_{nn} - V_{nn}U_{n-1,n-1})} > 0.$$

Notice that γ is well defined and positive because $U_{n-1,n} = f^U_{n-1,n} U_{nn} < U_{nn}$ and $V_{n-1,n} = f^V_{n-1,n} V_{nn} = V_{nn}$, but $V_{n,n-1} = f^V_{n,n-1} V_{n-1,n-1} < V_{n-1,n-1}$.

The induction hypothesis implies that X^{-1} is a tridiagonal M-matrix strictly row diagonally dominant at rows 1 and $n-1$. We conclude that Ω is a tridiagonal M-matrix that is strictly row diagonally dominant at rows 1 and $n-1$, because the nonnegative term $\gamma e e'$ modifies just the diagonal element $\Omega_{n-1,n-1}$.

The equation for ζ is $V_{nn} \Omega u - V_{nn} U_{nn} \zeta = 0$ and therefore

$$U_{nn} \zeta = X^{-1} u + \gamma U_{n-1,n} e = \left(\frac{1}{V_{n-1,n-1}} + \gamma U_{n-1,n} \right) e,$$

which gives

$$\zeta = \frac{1}{V_{n-1,n-1} U_{nn} - V_{nn} U_{n-1,n-1}} e. \tag{6.29}$$

Finally we compute $\alpha = M_{nn}$ which is given by

$$\alpha = \left(V_{nn} U_{nn} - V^2_{nn} u' X^{-1} u \right)^{-1} = \frac{V_{n-1,n-1}}{V_{nn} (V_{n-1,n-1} U_{nn} - V_{nn} U_{n-1,n-1})}.$$

Since $V_{n-1,n-1} > V_{nn}$ we obtain: The sum of the row n is positive and n is only connected to $n-1$ and n, which is $M_{n,n-1} < 0$, $M_{nn} > 0$. The connections of M on $\{1, \cdots, n-1\}$ are the same as in X^{-1} (including the connection $(n-1, n-1)$, because $X^{-1}_{n-1,n-1} > 0$) and then the induction hypothesis shows that M is tridiagonal. With respect to the row sums, the only one that can change sign in $\{1, \cdots, n-1\}$, with respect to the ones in X^{-1}, is that associated with row $n-1$ because the vector ζ is null out of the node $n-1$. Here we point out that $\zeta_{n-1} > \gamma$.

So we have proved that M is a tridiagonal M-matrix, and the row sum of row n is positive (by symmetry the row sum of the row 1 is also positive). This proves in particular (iii). Since W is a positive matrix and its inverse is an M-matrix, we deduce that W^{-1} is irreducible. We also have proved (iv) because $W_J = U_J \odot V_J$ and according to the extra hypothesis the three matrices are nonsingular (they are principal matrices of inverse M-matrices).

(v) To investigate the other row sums we have to give a look more closely to the previous induction and use the fact that the row sums of M, in $\{1, \cdots, n-2\}$ and those of X^{-1} are the same.

Take now $i \in \{2, \cdots, n-1\}$ we shall prove that condition (6.26) is necessary and sufficient to have a nonnegative row sum at row i in M. Since $n \geq 4$ we choose $J = \{i-1, i, i+1\}$. By Lemma 6.33 we have

$$(U_J^{-1})_{2,k-i+2} = U_{ik}^{-1}, \quad (V_J^{-1})_{2,k-i+2} = V_{ik}^{-1},$$

holds for $k \in J$. Lemma 6.39 gives a necessary and sufficient condition for $N = (U_J \odot V_J)^{-1}$ to be a row diagonally dominant M-matrix, which written in terms of U, V is

$$U_{i,i-1}^{-1} V_{i,i+1}^{-1} \geq U_{i,i+1}^{-1} V_{i,i-1}^{-1}.$$

We conclude that (6.26) and (6.25) are equivalent (also the equivalence between their corresponding strict counterparts).

As we add states to this initial set J, the only row sums that can be modified are the ones associated with nodes $i-1$ and $i+1$ (depending on which side we add nodes). Hence, the row sum associated with node i at the final stage on the matrix M is the same as the row sum of the second row in N showing that (6.26) is necessary and sufficient for M to be row diagonally dominant at row i (again there is a correspondence between their strict counterparts).

Now, we show that conditions (6.26) and (6.24) are equivalent. For that purpose consider $\lambda \in \mathbb{R}^3$ the unique solution of $(U_J \odot V_J)\lambda = \mathbb{1}$. This solution is

$$\lambda = \begin{pmatrix} \lambda_1 \\ \lambda_2 \\ \lambda_3 \end{pmatrix} = N\mathbb{1},$$

which implies that λ_2 is the sum of the second row of N. Thus, condition (6.26) is equivalent to $\lambda_2 \geq 0$. According to Cramer's rule we get

$$\lambda_2 \det(U_J \odot V_J) = \begin{vmatrix} U_{i-1,i-1}V_{i-1,i-1} & 1 & U_{i-1,i-1}V_{i+1,i+1} \\ U_{i-1,i-1}V_{i,i} & 1 & U_{i,i}V_{i+1,i+1} \\ U_{i-1,i-1}V_{i+1,i+1} & 1 & U_{i+1,i+1}V_{i+1,i+1} \end{vmatrix}. \tag{6.30}$$

Since $U_J \odot V_J$ is a positive definite matrix, the sign of λ_2 is the same as the sign of

$$(U_{i+1,i+1} - U_{i-1,i-1})(V_{i-1,i-1} - V_{i,i}) - (U_{i,i} - U_{i-1,i-1})(V_{i-1,i-1} - V_{i+1,i+1}),$$

and the equivalence is shown.

(vi) The result is straightforward when $n = 1, 2$. The case $n = 3$ is done in Lemma 6.39. By Lemma 6.33, we obtain

$$\begin{cases} M_{11} = (W_{\{1,2,3\}})_{11}^{-1}; \\ M_{ii} = (W_{\{i-1,i,i+1\}})_{22}^{-1} \text{ when } 2 \leq i \leq n; \\ M_{nn} = (W_{\{n-2,n-1,n\}})_{33}^{-1}. \end{cases}$$

Now that we know $M = W^{-1}$ is a tridiagonal matrix, formula (6.20) shows that (6.21) holds. The theorem is proven under the extra hypothesis that U and V are nonsingular.

We now show (i)–(v) without the assumption that U and V are nonsingular.

Denote by $0 < x_1 \leq \cdots \leq x_n$ and $y_1 \geq \cdots \geq y_n > 0$ the numbers associated with U and V respectively. For $0 < \epsilon < y_n/n$, we consider $x_i^\epsilon = x_i + i\epsilon$ and $y_i^\epsilon = y_i - i\epsilon$. Then, $0 < x_1^\epsilon < \cdots < x_n^\epsilon$ and $y_1^\epsilon > \cdots > y_n^\epsilon > 0$.

So, the associated ultrametric matrices $U(\epsilon), V(\epsilon)$ are nonsingular. The properties we have proved, applied to the matrix $W(\epsilon) = U(\epsilon) \odot V(\epsilon)$ gives that $W^{-1}(\epsilon)$ is a tridiagonal M-matrix, which is strictly diagonally dominant at rows $1, n$.

(i) Assume first W is nonsingular. Since $W(\epsilon) \to W$ as $\epsilon \downarrow 0$, we conclude that $M = W^{-1}$ is a tridiagonal M-matrix. Then, again from formula (6.20) we conclude that (6.21) holds.

Conversely, if (6.21) holds then formula (6.20) applied to $W(\epsilon)$ gives that $W^{-1}(\epsilon)$ converges to a certain matrix that we call Z. From the fact $W(\epsilon)W^{-1}(\epsilon) = \mathbb{I}$, it is straightforward to check that Z is the inverse of W. Hence, W is nonsingular and (i) is proven.

For the rest of the proof we assume that W is nonsingular and $M = W^{-1}$. We already know that M is a tridiagonal M-matrix, which is row diagonally dominant at rows $1, n$, proving (ii) and (iv).

(v) In order to prove that (6.24) is equivalent to W^{-1} is row diagonally dominant at row i, we notice that the proof we have done in the restricted case, relays in two facts. First, (6.24) is necessary and sufficient to have a row diagonally dominant at the second row of W_J^{-1}, where $J = \{i-1, i, i+1\}$. This was done in (6.30).

The second ingredient is that the row sum of node i does not change as we add more nodes until we arrive to I. Assume that $i < n - 1$. As we have done before we decompose W and W^{-1} in blocks as

$$W = \begin{pmatrix} X & V_{nn}u \\ V_{nn}u' & V_{nn}U_{nn} \end{pmatrix} \text{ and } W^{-1} = \begin{pmatrix} \Omega & -\zeta \\ -\zeta' & \alpha \end{pmatrix}.$$

The fact that W^{-1} is tridiagonal implies that $\zeta = ae(n-1)$ for some $a > 0$. a has to be positive, otherwise node n cannot connect to any other node and $W_{jn} = 0$ for $j < n$ which is not possible. Moreover

$$a = \frac{1}{V_{n-1,n-1}U_{nn} - V_{nn}U_{n-1,n-1}},$$

where we notice that $(V_{n-1,n-1}U_{nn} - V_{nn}U_{n-1,n-1})U_{n-1,n-1}V_{nn} = det(W_{\{n-1,n\}}) > 0$. Also we have

$$\Omega = \left(X - \frac{V_{nn}}{U_{nn}}uu' \right)^{-1} = X^{-1} + \frac{V_{nn}}{V_{n-1,n-1}}a\, e(n-1)e(n-1)'. \qquad (6.31)$$

The row sums of W^{-1} and X^{-1} are the same on $\{1, \cdots, n-2\}$. The rest is done by induction. If $i = n-1$, we decompose the matrices in blocks indexed by $\{1\}$ and $\{2, \cdots, n\}$ and proceed as before.

(*iii*) To show that (6.22) is equivalent to the fact that M is strictly row diagonally dominant at row 1, we proceed as in the proof of (*iv*). This condition is exactly that the inverse of $W_{\{1,2,3\}}$ is strictly row diagonally dominant at the first row. This row sum does not change as we add more states proving the desired equivalence. Similarly, (6.23) is equivalent to: The inverse of $W_{\{n-2,n-1,n\}}$ is strictly row diagonally dominant at the third row. The rest of the argument is analogous.

As consequence of these theorems, we obtain the following result.

Corollary 6.40 *Assume W is a symmetric nonsingular matrix with inverse $M = W^{-1}$. The following two conditions are equivalent:*

(*i*) *M is a tridiagonal irreducible row diagonally dominant M-matrix.*
(*ii*) *$W = U \odot V$, where $U \in LUM(1)$, $V \in LUM(n)$ and (6.24) is satisfied for all $i \in \{2, \cdots, n-1\}$.*

Under any of these two equivalent conditions we have

$$M_{ij} = \begin{cases} (W_{\{1,2,3\}})^{-1}_{ij} & \text{if } i = 1, j = 1, 2 \\ (W_{\{i-1,i,i+1\}})^{-1}_{2,j-i+2} & \text{if } 2 \le i \le n-1, |i-j| \le 1 \\ (W_{\{n-2,n-1,n\}})^{-1}_{3,j-n+3} & \text{if } i = n, j = n-1, n \end{cases} \qquad (6.32)$$

If in addition, U and V are Markov potentials then W is also a Markov potential.

Proof (*i*) $\Rightarrow$ (*ii*) follows from Theorem 6.34. (*ii*) $\Rightarrow$ (*i*) follows from Theorem 6.38. The fact that (6.32) holds under each one of these conditions follows from the proof we have done of Theorem 6.38. In particular, we have for all $3 \le p \le n$ (see (6.31))

$$(W_{\{1,2,3\}})^{-1}_{11} = (W_{\{1,\cdots,p\}})^{-1}_{11} ;$$

$$(W_{\{1,2,3\}})^{-1}_{12} = (W_{\{1,\cdots,p\}})^{-1}_{12} .$$

The other cases in (6.32) are shown in the same way. Finally, if U and V are Markov potentials then by Theorem 6.38 (*vi*) $M_{ii} \le 1$ for all i, which together with the fact that M is a row diagonally dominant M-matrix show that $M = \mathbb{I} - S$ for some substochastic matrix S. Thus W is a Markov potential.

Remark 6.41 It may happen that $W = U \odot V$ is nonsingular while $U \in$ LUM(1), $V \in$ LUM(n) are singular. Indeed, consider the example

$$W = \begin{pmatrix} 2 & 1 & 1 \\ 1 & 1 & 1 \\ 1 & 1 & 2 \end{pmatrix} = \begin{pmatrix} 1 & 1 & 1 \\ 1 & 1 & 1 \\ 1 & 1 & 2 \end{pmatrix} \odot \begin{pmatrix} 2 & 1 & 1 \\ 1 & 1 & 1 \\ 1 & 1 & 1 \end{pmatrix}.$$

Moreover, if $W = A \odot B$ with $A \in$ LUM(1), $B \in$ LUM(n) then both A, B are singular. Notice that $W \in$ LUM(2).

The following result gives conditions to have U and V nonsingular.

Proposition 6.42 *Assume that $U \in LUM(1)$, $V \in LUM(n)$ and consider $W = U \odot V$, which is of course a symmetric matrix.*

(i) *A necessary and sufficient condition to have U and V nonsingular is that for all $i = 1, \cdots, n - 1$*

$$W_{i,i+1} < \min\{W_{ii}, W_{i+1,i+1}\}. \tag{6.33}$$

Hence, using Theorem 6.38 (i), this condition implies that W is nonsingular.

(ii) *Assume that W is nonsingular and $M = W^{-1}$ is a tridiagonal irreducible row diagonally dominant M-matrix. Then, U is nonsingular if and only if M is strictly diagonally dominant at row 1. Similarly, V is nonsingular if and only if M is strictly diagonally dominant at row n.*

Proof (i) Consider $0 < x_1 \le x_2 \le \cdots \le x_n$ and $y_1 \ge y_2 \ge \cdots \ge y_n > 0$ the numbers defining U and V. Given that $W_{i,i+1} = x_i y_{i+1}, W_{i+1,i+1} = x_{i+1} y_{i+1}, W_{ii} = x_i y_i$, condition (6.33) is equivalent to saying that x's are strictly increasing and y's are strictly decreasing, which is equivalent to U and V being nonsingular (see Theorem 6.32).

(ii) If U is nonsingular then from Theorem 6.38 (*iii*), we get that M is strictly diagonally dominant at row 1. To prove the converse we assume that M is strictly diagonally dominant at row 1. Without loss of generality we also assume that $M = \mathbb{I} - P$ with P symmetric, substochastic, irreducible, tridiagonal and strictly substochastic at row 1. The Markov chain associated with P loses mass at least at node 1, which together with the fact that P is tridiagonal imply that $f_{i,i+1}^W < 1$ for all $i = 1, \cdots, n - 1$.

We denote by $0 < x_1 \le x_2 \le \cdots \le x_n$ and $y_1 \ge y_2 \ge \cdots \ge y_n > 0$ the numbers defining U and V. Since $x_i y_{i+1} = W_{i,i+1} = f_{i,i+1}^W W_{i+1,i+1} < W_{i+1,i+1} = x_{i+1} y_{i+1}$ we conclude the x's are strictly increasing and therefore U is nonsingular. The conclusions for V follow similarly.

Remark 6.43 If $W = U \odot V$ then condition (6.33) is sufficient for W being nonsingular. On the other hand, if W is nonsingular, then the 2×2 matrix $W_{\{i,i+1\}}$

is positive definite, because W is an inverse M-matrix. In particular its determinant is positive, or equivalently

$$W_{i,i+1} < \sqrt{W_{ii}W_{i+1,i+1}}.$$

So, condition (6.33) is a strengthening of the necessary condition for W to be positive definite, namely that each principal minor of size 2 must be positive.

The aim of the next results is to show that if W is the potential of a symmetric irreducible random walk, then it is either a linear ultrametric potential or the Hadamard product of two **nonsingular** linear ultrametric potentials.

Theorem 6.44 *Assume that $M = W^{-1}$ is a symmetric, irreducible, tridiagonal and row diagonally dominant M-matrix. Then*

(i) $\mathscr{R}(W) = \{k\}$ if and only if $W \in LUM(k)$;
(ii) $|\mathscr{R}(W)| \geq 2$ if and only if $W = U \odot V$ where $U \in LUM(k)$, $V \in LUM(m)$ are nonsingular and $k = \min \mathscr{R}(W), m = \max \mathscr{R}(W)$.

Proof In what follows we assume that $M = \mathbb{I} - P$ for a substochastic matrix P.

(*i*) Follows immediately from Theorem 6.32.
(*ii*) Here we take $J = \{k \leq j \leq m\}$ the smallest interval containing $\mathscr{R}(W)$, which by hypothesis has size at least 2. According to Lemma 6.33 $X = W_J$ is a potential matrix. Its inverse is an irreducible tridiagonal row diagonally dominant M-matrix and $\mathscr{R}(X) + (k - 1) \supseteq \mathscr{R}(W) \cap J \supseteq \{k, m\}$.

The case $k = 1$ and $m = n$, that is $X = W$, follows from Theorem 6.34 and Proposition 6.42. So, without loss of generality, for the rest of the proof, we can assume that $k > 1$. Again Theorem 6.34 and Proposition 6.42 implies that $X = R \odot S$ for nonsingular $R \in LUM(1)$, $S \in LUM(m-k+1)$ of size $m-k+1$. Now we look for extending these matrices to a decomposition of W. We shall give the idea how to extend this decomposition to $K = \{k-1 \leq j \leq m\}$. Let us consider W_K, the restriction of W to K, that is

$$W_K = \begin{pmatrix} W_{k-1,k-1} & w' \\ w & X \end{pmatrix},$$

with $w = (W_{k-1,k}, \cdots, W_{k-1,m})' = (W_{kk}, \cdots, W_{km})'$, because $f^W_{k-1,k} = 1$. Using that $w' = X_{1\bullet}$ is the first row of X and $X_{\bullet 1}$ the first column of X, we rewrite W_K as

$$W_K = \begin{pmatrix} W_{k-1,k-1} & X_{1\bullet} \\ X_{\bullet 1} & X \end{pmatrix}.$$

Since $f^W_{k,k-1} < 1$ we have that $W_{k-1,k-1}$ strictly dominates the values in $X_{1\bullet}$. Indeed, for $j = k, \cdots, m$ we have

$$W_{kj} = W_{k-1,j} = W_{j,k-1} = f^W_{j,k-1} W_{k-1,k-1} < W_{k-1,k-1}.$$

Let us now introduce the numbers associated with R and S

$$0 < x_k < x_{k+1} < \cdots < x_m \text{ and } y_k > y_{k+1} > \cdots > y_m > 0,$$

respectively. In particular $X_{1\bullet} = x_k(y_k, \cdots, y_m)$. Hence, if we take $x_{k-1} > x_k$ and $y_{k-1} > y_k$ such that $x_{k-1} y_{k-1} = W_{k-1,k-1} > W_{kk} = x_k y_k$ we get

$$W_K = \begin{pmatrix} x_{k-1} & x_k \mathbb{1}'_{m-k+1} \\ x_k \mathbb{1}_{m-k+1} & R \end{pmatrix} \odot \begin{pmatrix} y_{k-1} & (y_k, \cdots, y_m) \\ (y_k, \cdots, y_m)' & S \end{pmatrix}.$$

The rest of the proof is done by an argument based on induction. The matrix U constructed (from the x's) belongs to $\mathrm{LUM}(k)$ and V belongs to $\mathrm{LUM}(m)$.

Remark 6.45 When $|\mathscr{R}(W)| \geq 2$ the decomposition given by this theorem is not unique. We shall see that there are

$$1 + (k - 1) + (n - m)$$

degrees of freedom in such decomposition.

As a converse of the previous theorem, we have the following result.

Theorem 6.46 *Assume that $U \in \mathrm{LUM}(k)$, $V \in \mathrm{LUM}(m)$ are nonsingular of size n (without loss of generality we assume $k \leq m$). Then $W = U \odot V$ is nonsingular and $M = W^{-1}$ is an irreducible tridiagonal M-matrix. The sum of row i is zero for $i \notin \{k \leq j \leq m\}$ and it is strictly positive for $i \in \{k, m\}$. When $k < i < m$, M is row diagonally dominant at row i if and only if (6.24) holds, that is*

$$(U_{i+1,i+1} - U_{i-1,i-1})(V_{i-1,i-1} - V_{i,i}) \geq (U_{i,i} - U_{i-1,i-1})(V_{i-1,i-1} - V_{i+1,i+1}).$$

There is a strict inequality in this formula if and only if M is strictly diagonally dominant at row i.

Proof W is a positive definite matrix and therefore it is nonsingular. On the other hand, there exist $U1, U2 \in \mathrm{LUM}(1)$ and $V1, V2 \in \mathrm{LUM}(n)$ such that $U = U1 \odot V1$ and $V = U2 \odot V2$. Then $W = (U1 \odot U2) \odot (V1 \odot V2)$ is the Hadamard product of $U1 \odot U2 \in \mathrm{LUM}(1)$ and $V1 \odot V2 \in \mathrm{LUM}(n)$. Hence, M is an irreducible tridiagonal M-matrix (see Theorem 6.38).

Consider $J = \{i : k \leq i \leq m\}$. Then, $W_J = U_J \odot V_J$ is again an inverse M-matrix and $U_J \in \mathrm{LUM}(1), V_J \in \mathrm{LUM}(m - k + 1)$ are nonsingular of size $m - k + 1$ (see Remark 6.31). In particular $R = W_J^{-1}$ is an M-matrix, which is strictly row diagonally dominant at the first and last row. Take now the vector

$$v = R \mathbb{1}_{m-k+1} \in \mathbb{R}^{m-k+1},$$

the signed equilibrium potential of W_J. We know that $v_1 > 0$ and $v_{m-k+1} > 0$, because of Theorem 6.38 (*ii*). Consider $\lambda \in \mathbb{R}^n$ given by

$$\lambda_i = \begin{cases} 0 & \text{if } i \notin J \\ v_{i-k+1} & \text{if } i \in J \end{cases}.$$

Let us prove that $W\lambda = \mathbb{1}$. For that purpose we compute

$$(W\lambda)_i = \sum_j U_{ij}V_{ij}\lambda_j = \sum_{j \in J} U_{ij}V_{ij}v_{j-k+1}.$$

There are three cases to analyze: $i < k$, $i \in J$ and $i > m$. In the first case we use the fact that for $i < k$ and $j \in J$

$$U_{ij} = U_{kj}, \quad V_{ij} = V_{kj},$$

and therefore this case is reduced to the second one. Similarly, the third case is reduced to the second one. Hence, we need to analyze the case $i \in J$

$$(W\lambda)_i = \sum_{i \in J} U_{ij}V_{ij}v_{j-k+1} = \sum_{j \in J}(W_J)_{i-k+1,i-k+1}v_{j-k+1} = (W_J \, v)_{i-k+1} = 1.$$

Therefore, $\lambda = W^{-1}\mathbb{1} = M\mathbb{1}$ is the right signed equilibrium potential of W. In particular, the row sums of M are 0 at rows $i \notin J$. Also the row sums at rows k, m are strictly positive. Finally, M is row diagonally dominant at row $i : k < i < m$ if and only if $v_{i-k+1} \geq 0$ or equivalently R is diagonally dominant at row $i - k + 1$. By Theorem 6.38 (*v*) this is equivalent to

$$(U_{i+1,i+1}-U_{i-1,i-1})(V_{i-1,i-1}-V_{i,i}) \geq (U_{i,i}-U_{i-1,i-1})(V_{i-1,i-1}-V_{i+1,i+1}).$$

Similarly, M is strictly diagonally dominant at row i if there is a strict inequality in the last formula.

We summarize some of these results in the next corollary.

Corollary 6.47 *Assume that W is a positive nonsingular matrix and denote by M its inverse. Then, M is an irreducible tridiagonal M-matrix if and only if there exist two positive diagonal matrices D, E such that $X = DWE$ is a symmetric potential and X is a linear ultrametric matrix or the Hadamard product of two nonsingular linear ultrametric matrices U, V. The first case occurs when X has one root. In the second case the roots of X are contained in the convex set determined by the roots of U and V. If M is row diagonally dominant we can take $D = \mathbb{I}$ and if M is symmetric we can take $D = E$.*

Theorem 6.48 *Assume that W^{-1} is an irreducible tridiagonal M-matrix of size n. Then, for all positive real numbers $r > 0$, the r-th Hadamard power of W, is nonsingular and its inverse is an irreducible tridiagonal M-matrix. If W is a potential (respectively Markov potential) and $r \geq 1$ then $W^{(r)}$ is also a potential (respectively Markov potential).*

For $r < 0$ the matrix $W^{(r)}$ is nonsingular and its inverse $C(r)$ is an irreducible tridiagonal matrix and the following properties hold

(i) $sign(\det(W^{(r)})) = (-1)^{n+1}$;
(ii) if $n \geq 2$ then for all i, j we have

$$C(r)_{ij} \text{ is } \begin{cases} < 0 & \text{if } i = j \\ > 0 & \text{if } |i - j| = 1 \; ; \\ = 0 & \text{otherwise} \end{cases}$$

(iii) If W is symmetric then the eigenvalues of $W^{(r)}$ are negative, except for the principal one λ_1 which is positive and with maximal absolute value.

In this theorem the *sign* function is given by

$$sign(x) = \begin{cases} 1 & \text{if } x > 0 \\ 0 & \text{if } x = 0 \; . \\ -1 & \text{if } x < 0 \end{cases}$$

Proof If W^{-1} is an irreducible tridiagonal M-matrix, then W is an entrywise positive matrix and there exist two positive diagonal matrices D, E and a nonsingular $U \in \text{LUM}(1)$ such that

$$DWE = U.$$

Hence, for all $r \in \mathbb{R}$ we get

$$D^{(r)} W^{(r)} E^{(r)} = U^{(r)}.$$

For $r > 0$ we have $U^{(r)} \in \text{LUM}(1)$ is nonsingular and therefore $W^{(r)}$ is also nonsingular. Moreover, $(W^{(r)})^{-1} = E^{(r)}(U^{(r)})^{-1} D^{(r)}$. Since $U^{(r)} \in \text{LUM}(1)$, its inverse is a symmetric tridiagonal row diagonally dominant M-matrix. Hence, $W^{(r)}$ is the inverse of an irreducible tridiagonal M-matrix.

When W is a potential and $r \geq 1$, the matrix $W^{(r)}$ is also a potential (respectively Markov potential) see Theorem 2.2 (respectively Theorem 2.3) in [23].

Now, let us assume that W^{-1} is a tridiagonal irreducible M-matrix and $r < 0$. In order to prove $(i), (ii)$ we can assume without loss of generality that $W \in \text{LUM}(1)$. We shall prove the desired properties by induction on n the size of W. The cases $n = 1, 2$ are obtained immediately. So, we assume the properties hold up to dimension

$n - 1$ and we shall prove them for dimension $n \geq 3$. Also we shall assume that $r = -1$. The general case follows from the fact $W^{(r)} = \left(W^{(-r)}\right)^{(-1)}$.

Take $0 < x_1 \leq x_2 \leq \cdots \leq x_n$ the numbers defining W. Let us denote $T = W^{(-1)}$

$$
T = \begin{pmatrix}
\frac{1}{x_1} & \frac{1}{x_1} & \cdots & \frac{1}{x_1} \\
\frac{1}{x_1} & \frac{1}{x_2} & \cdots & \frac{1}{x_2} \\
\vdots & \vdots & \ddots & \vdots \\
\frac{1}{x_1} & \frac{1}{x_2} & \cdots & \frac{1}{x_n}
\end{pmatrix}.
$$

Now, we partitioned T and T^{-1} (here for the moment we assume it exists) in blocks of sizes $n - 1$ and 1 as

$$
T = \begin{pmatrix} A & a \\ a' & \frac{1}{x_n} \end{pmatrix}, T^{-1} = \begin{pmatrix} \Omega & \zeta \\ \zeta' & z \end{pmatrix},
$$

where $a' = (\frac{1}{x_1}, \frac{1}{x_2}, \cdots, \frac{1}{x_{n-1}})$. The equations for Ω, z and ζ are

$$
\begin{aligned}
\Omega &= (A - x_n a a')^{-1}, \\
z &= (\tfrac{1}{x_n} - a' A^{-1} a)^{-1}, \\
\zeta &= -z A^{-1} a.
\end{aligned}
$$

The induction hypothesis implies that A is nonsingular. The important relation in order to solve these equations is $Ae = a$, where $e = e(n - 1)$ is the last vector of the canonical basis in $\mathbb{R}^{n-1}$. This fact implies that

$$
\frac{1}{x_n} - a' A^{-1} a = \frac{1}{x_n} - a'e = \frac{1}{x_n} - \frac{1}{x_{n-1}} < 0,
$$

and $z = \frac{x_n x_{n-1}}{x_{n-1} - x_n} < 0$. On the other hand, we have

$$
\Omega = \left(A - x_n a a'\right)^{-1} = A^{-1} + \alpha e e',
$$

with $\alpha = \frac{x_n x_{n-1}}{x_{n-1} - x_n} = z < 0$. Finally, we get

$$
\zeta = -ze.
$$

The induction hypothesis implies that A^{-1} is an irreducible tridiagonal matrix, with negative diagonal elements and positive elements in the upper and lower next diagonals. The same is true for Ω and also for T^{-1} since $(T^{-1})_{nn} < 0$. Finally, from the well known formula for the determinant of a matrix by blocks

$$
\det(W^{(-1)}) = \det(A) \left(\frac{1}{x_n} - a' A^{-1} a \right) = \frac{\det(A)}{z},
$$

we get that $sign(\det(W^{(-1)})) = -sign(\det(A))$ and the proof of (i), (ii) is complete.
(iii) The proof is done by induction on n the size of W. When $n = 1, 2$ the proof
is straightforward. So we assume the property holds up to dimension $n - 1$. Let
$\lambda_1 \geq \lambda_2 \geq \cdots \geq \lambda_n$ be the eigenvalues of $W^{(r)}$. We take $X = W_{\{1,\cdots,n-1\}}$ a
principal submatrix of W and consider the ordered set of eigenvalues for $X^{(r)}$ given
by $\mu_2 \geq \mu_3 \geq \cdots \geq \mu_n$. Since $X^{(r)}$ is a principal submatrix of $W^{(r)}$ we have by the
Cauchy's Interlace Theorem for eigenvalues

$$\lambda_1 \geq \mu_2 \geq \lambda_2 \geq \cdots \geq \lambda_{k-1} \geq \mu_k \geq \lambda_k \geq \cdots \geq \mu_n \geq \lambda_n,$$

where $k = 2, \cdots, n$. The matrix X, of size $n - 1$, satisfies the induction hypothesis
and therefore $\mu_2 > 0 > \mu_3$ which implies that $\lambda_n \leq \lambda_{n-1} \leq \cdots \leq \lambda_3 < 0 < \lambda_1$.
We need to prove that $\lambda_2 < 0$. For that purpose we use that

$$\det(W^{(r)}) = \prod_{k=1}^{n} \lambda_k, \quad \det(X^{(r)}) = \prod_{k=2}^{n} \mu_k$$

and $sign(\det(W^{(r)})) = -sign(\det(X^{(r)}))$. This implies that necessarily $\lambda_2 < 0$. That
λ_1 is maximal in absolute value follows from the Perron-Frobenious Theorem.

Remark 6.49 Notice that $W^{(0)}$ is the constant matrix of ones and so it is singular
(unless the dimension of W is one). On the other hand, if W is a potential and
$0 < r < 1$ the matrix $W^{(r)}$ is not in general a potential as the following example
shows. Take the matrices

$$U = \begin{pmatrix} 1 & 1 & 1 & 1 \\ 1 & 4 & 4 & 4 \\ 1 & 4 & 9 & 9 \\ 1 & 4 & 9 & 16 \end{pmatrix} \in \text{LUM}(1), \quad V = \begin{pmatrix} 25 & 16 & 9 & 1 \\ 16 & 16 & 9 & 1 \\ 9 & 9 & 9 & 1 \\ 1 & 1 & 1 & 1 \end{pmatrix} \in \text{LUM}(4).$$

The matrix $W = U \odot V$ is a potential (checked numerically, or use Theorem 6.38)
but $T = W^{(1/2)}$ is an inverse M-matrix that is not row diagonally dominant (the
sum of the third row of T^{-1} is negative).

Remark 6.50 For $r < 0$ the inverse of $W^{(r)}$, that we denoted by $C(r)$, has the sign
pattern opposite to an M-matrix, but clearly $-C(r)$ is not an M-matrix because its
inverse is an entrywise negative matrix.

Theorem 6.51 *Assume that W and X are nonsingular and W^{-1}, X^{-1} are two
irreducible tridiagonal M-matrices. Then the Hadamard product $W \odot X$ is again
the inverse of an irreducible tridiagonal M-matrix.*

Proof There exist diagonal matrices $D, \tilde{D}, E, \tilde{E}$ and nonsingular matrices $U, \tilde{U} \in$
LUM(1), such that

$$DWE = U, \quad \tilde{D} X \tilde{E} = \tilde{U}.$$

Then

$$U \odot \tilde{U} = (DWE) \odot (\tilde{D}X\tilde{E}) = D\tilde{D}(W \odot X)E\tilde{E},$$

is again a nonsingular matrix in LUM(1) and the result follows.

6.4.3   Potentials Associated with Random Walks That Lose Mass Only at the Two Ends

The aim of this section is to characterize the potentials associated with random walks on $I = \{1, \cdots, n\}$ that lose mass exactly at $1, n$. As we know from Theorem 6.44 these potentials (at least the symmetric ones) are the Hadamard product of two nonsingular ultrametric matrices, one of them in LUM(1) and the other one in LUM(n). We shall see that this decomposition has a very special form.

Consider the simplest random walk on $\{1, \cdots, n\}$ losing mass at 1 and reflected at n, that is $P^{01} = P^{01}(n)$ is the $n \times n$ matrix given by

$$\forall i, j \quad P_{ij}^{01} = \begin{cases} 1/2 & \text{if } |i - j| = 1 \text{ or } i = j = n \\ 0 & \text{otherwise} \end{cases}. \tag{6.34}$$

This matrix is stochastic except at $i = 1$. According to Theorem 6.32 the matrix $U^{01} = (\mathbb{I} - P^{01})^{-1}$ is ultrametric. Moreover, this matrix is given by

$$\forall i, j \quad U_{ij}^{01} = 2(i \wedge j). \tag{6.35}$$

Indeed, if W is the matrix given by the right hand side of (6.35), then it is straightforward to check that $W = \mathbb{I} + WP^{01}$, proving that $U^{01} = W$. We also notice that the matrix U^{10} obtained from U^{01} by

$$\forall i, j \quad U_{ij}^{10} = U_{n+1-i,n+1-j}^{01} = 2([n+1-i] \wedge [n+1-j]) = 2(n+1-(i \vee j)),$$

is also ultrametric and it is the potential of the random walk P^{10} which loses mass at n and it is reflected at 1. Also we notice that P^{01}, P^{10} are tridiagonal.

The random walk on $\{1, \cdots, n\}$ that loses mass at 1 and n has a transition matrix $P^{00} = P^{00}(n)$ given by

$$\forall i, j \quad P_{ij}^{00} = \begin{cases} 1/2 & \text{if } |i - j| = 1 \\ 0 & \text{otherwise} \end{cases}. \tag{6.36}$$

This matrix is stochastic except at 1 and n and it is tridiagonal. The matrix $W^{00} = (\mathbb{I} - P^{00})^{-1}$ is a potential and it is given by

$$W_{ij}^{00} = \frac{2}{n+1}(i \wedge j)(n+1-(i \vee j)) = \frac{2}{n+1}(i \wedge j)([n+1-i] \wedge [n+1-j]).$$
(6.37)

We notice that W^{00} is not only symmetric but also it is symmetric with respect to the change $(i, j) \to (n+1-i, n+1-j)$.

The main observation is that W^{00} is proportional to the Hadamard product $U^{01} \odot U^{10}$, where U^{01} is the ultrametric potential of the standard random walk that loses mass at node 1 (see (6.35)) and U^{10} is the one that loses mass at node n. Indeed we have

$$W^{00} = \frac{1}{2(n+1)} U^{01} \odot U^{10}.$$

This is a part of a general result that now we state.

Theorem 6.52 *Consider the matrix W indexed by $I = \{1, \cdots, n\}$ and given by*

$$W_{ij} = z_{i \wedge j}(\mathfrak{m} - z_{i \vee j})d_j,$$
(6.38)

where $0 < z_1 < z_2 \cdots < z_n < \mathfrak{m}$ and $d_j > 0$ for all j. We notice that W is symmetric if and only if (d_j) is constant. Then, W is a potential matrix with inverse $W^{-1} = \kappa(\mathbb{I} - P)$, where κ is a constant and P is an irreducible tridiagonal substochastic matrix, which is stochastic except at 1 and n and moreover

$$\begin{cases} W_{ij}^{-1} = 0 \text{ if } |i-j| > 1; \\ W_{11}^{-1} = \frac{1}{d_1 \mathfrak{m}}\left(\frac{1}{z_1} + \frac{1}{z_2-z_1}\right), \quad W_{nn}^{-1} = \frac{1}{d_n \mathfrak{m}}\left(\frac{1}{\mathfrak{m}-z_n} + \frac{1}{z_n-z_{n-1}}\right); \\ \forall i \leq n-1 \quad W_{i,i+1}^{-1} = -\frac{1}{d_i \mathfrak{m}(z_{i+1}-z_i)}, \quad W_{i+1,i}^{-1} = -\frac{1}{d_{i+1}\mathfrak{m}(z_{i+1}-z_i)}; \\ \forall 2 \leq i \leq n-1 \quad W_{ii}^{-1} = -W_{i,i-1}^{-1} - W_{i,i+1}^{-1}. \end{cases}$$
(6.39)

Conversely, assume P is an irreducible tridiagonal substochastic matrix, which is stochastic except at 1 and n, then there exist $z, \mathfrak{m}, d$ as before such that $W = (\mathbb{I} - P)^{-1}$ has a representation like (6.38). In order to compute d, put $a_1 = 1$ and define $a_{i+1} = a_i \frac{P_{i,i+1}}{P_{i+1,i}}$ for $i = 1, \cdots, n-1$. Then, d is obtained from

$$d_i = \frac{a_i}{\max\{a_j : j = 1, \cdots, n\}} \leq 1.$$
(6.40)

Clearly, if P is symmetric we have $d_j = 1$ for all j. Now, $\mathfrak{m}$ and z can be calculated as

$$\mathfrak{m}^2 = \frac{1}{d_1(1-P_{11}-P_{12})} + \sum_{j=1}^{n-1} \frac{1}{d_j P_{j,j+1}} + \frac{1}{d_n(1-P_{nn}-P_{n,n-1})},$$

(6.41)

$$\forall i \geq 1 \quad z_i = \frac{1}{\mathfrak{m} d_1(1-P_{11}-P_{12})} + \sum_{j=1}^{i-1} \frac{1}{\mathfrak{m} d_j P_{j,j+1}}.$$

Proof Assume that W has a representation like (6.38) where $d_j = 1$ for all j, that is

$$W_{ij} = z_{i \wedge j} (\mathfrak{m} - z_{i \vee j}).$$

The first thing to notice is that $W = U \odot V$ where U, V are the ultrametric matrices given by

$$U_{ij} = z_{i \wedge j}, \quad V_{ij} = (\mathfrak{m} - z_i) \wedge (\mathfrak{m} - z_j).$$

It is straightforward to show that both $U \in \text{LUM}(1), V \in \text{LUM}(n)$ are nonsingular. According to Theorem 6.38 the matrix W is nonsingular and its inverse is an irreducible tridiagonal symmetric matrix. It is straightforward to see that U, V satisfy relations (6.24) with equality for all $i = 2, \cdots, n - 1$. Hence, $W^{-1} = \kappa(\mathbb{I} - P)$, where κ is a constant and P is an irreducible tridiagonal substochastic matrix, which is stochastic except at 1 and n (see Theorem 6.38 (*iii*)).

Formula (6.39) for W^{-1} follows from formula (6.20), by using the symmetry of W and that $W_{i1} = z_1(\mathfrak{m} - z_i)$. This shows the result when W is symmetric.

In the general case $W = \tilde{W}D$, where $\tilde{W}$ is symmetric and D is a diagonal matrix, with strictly positive diagonal entries. The result follows from the properties we have proved for $\tilde{W}$, because $W^{-1} = D^{-1}\tilde{W}^{-1}$ is a row diagonally dominant tridiagonal irreducible M-matrix, whose row sums are 0 for rows $2, \cdots, n - 1$, and positive for rows $1, n$.

Now, we assume that P is an irreducible tridiagonal substochastic matrix, that is stochastic except for rows $1, n$. At the beginning we shall assume that P is symmetric. We shall prove that $W = (\mathbb{I} - P)^{-1}$ has a representation like (6.38). For that purpose we use representation (6.39) in order to define $\mathfrak{m}, z$. Consider first $\alpha(1) = \frac{1}{1-P_{11}-P_{12}}$, which is well defined because P is substochastic at 1. Notice that $\alpha(1)$ represents the unknown $\mathfrak{m} z_1$. In general for $i = 2, \cdots, n$ we define

$$\alpha(i) = \alpha(i - 1) + \frac{1}{P_{i-1,i}}.$$

Also notice here that $\alpha(i)$ represents $\mathfrak{m}z_i$. Then, the formula for W_{nn}^{-1} in (6.39) and the symmetry of P suggest that $\mathfrak{m}$ should satisfy

$$\mathfrak{m}^2 = \alpha(n) + \frac{1}{1 - P_{nn} - P_{n,n-1}}.$$

The right hand side is well defined because P is substochastic at n. Then, we obtain

$$\forall\, i \geq 1 \quad \alpha(i) = \frac{1}{1 - P_{11} - P_{12}} + \sum_{j=1}^{i-1} \frac{1}{P_{j,j+1}},$$

$$\mathfrak{m}^2 = \frac{1}{1 - P_{11} - P_{12}} + \sum_{j=1}^{n-1} \frac{1}{P_{j,j+1}} + \frac{1}{1 - P_{nn} - P_{n,n-1}}.$$

Given $\alpha(i), i = 1, \cdots, n$ and $\mathfrak{m}$ we can define $z_i = \alpha(i)/\mathfrak{m}$, which gives the formula (6.41).

It is clear that $0 < z_1 < z_2 \cdots z_n < \mathfrak{m}$. The matrix A defined by

$$A_{ij} = z_{i \wedge j}(\mathfrak{m} - z_{i \vee j})$$

is, according to what we have proved, a potential matrix and its inverse is given by formula (6.39). This shows that $A = W$ and the result is proven in this case.

Assume now P is not symmetric. Take $a_1 = 1$ and define inductively for $i = 1, \cdots, n-1$

$$a_{i+1} = a_i \frac{P_{i,i+1}}{P_{i+1,i}}.$$

We define $d_i = \frac{a_i}{\max\{a_j : j = 1, \cdots, n\}} \leq 1$ and the associated diagonal matrix D. It is straightforward to show that DP is symmetric. The symmetric matrix Q defined by

$$\mathbb{I} - Q = D(\mathbb{I} - P),$$

that is, $Q_{ij} = (1 - d_i)\delta_{ij} + d_i P_{ij}$, is again irreducible tridiagonal and substochastic. Moreover, Q is symmetric and stochastic except at rows $1, n$. Hence, there exists $z, \mathfrak{m}$ such that $T = (\mathbb{I} - Q)^{-1}$ satisfies

$$T_{ij} = z_{i \wedge j}(\mathfrak{m} - z_{i \vee j}).$$

Let $W = (\mathbb{I} - P)^{-1}$, then $W = TD$. Hence, $W_{ij} = T_{ij}d_j$ and then W has a representation like (6.38) showing the result.

Remark 6.53 This result bears some similarities with the case of one dimensional diffusions on the interval $[0, 1]$ killed at both ends. In this context W plays the role of G the Green potential of such diffusions. For example, in the case of Brownian motion killed at $0, 1$, the Green potential is given by

$$G(x, y) = 2(x \wedge y)(1 - (x \vee y)),$$

which is the analog of (6.37).

Remark 6.54 Assume that W is symmetric. The decomposition (6.38) is unique, that is, if for all i, j

$$W_{ij} = z_{i \wedge j}(\mathfrak{m} - z_{i \vee j}) = \tilde{z}_{i \wedge j}(\tilde{\mathfrak{m}} - \tilde{z}_{i \vee j}), \tag{6.42}$$

then $\tilde{\mathfrak{m}} = \mathfrak{m}$ and $z = \tilde{z}$ on I. So, if $J = \{\ell_1 < \cdots < \ell_k\} \subseteq I$ and we define $V = W_J$, then clearly for all $1 \leq s, t \leq k$

$$V_{st} = W_{\ell_s \ell_t} = z_{\ell_s \wedge \ell_t}(\mathfrak{m} - z_{\ell_s \vee \ell_t}) = u_{s \wedge t}(\mathfrak{m} - u_{s \vee t}),$$

with $u_s = z_{\ell_s}$. In particular $\mathfrak{m}$ is the same quantity for all the principal submatrices of W. Thus, $\mathfrak{m}$ may be computed, for example, from $W_{\{1,2\}}$ and it is given by

$$\mathfrak{m}^2 = \frac{(W_{11} W_{22} - W_{12}^2)^2}{(W_{11} - W_{12})(W_{22} - W_{12}) W_{12}}.$$

Appendix A
Beyond Matrices

A good part of what we have done in the preceding chapters in the case of matrices can be done in a more general setting. We will consider three types of generalizations, which can be combined:

(1) Extension in space;
(2) Extension in time;
(3) Nonlinear analogues.

In (1) and (2) "space" and "time" refer to the probabilistic interpretation of potential theory. In Chap. 2 we considered Markov chains (discrete and continuous time) with finite state space. If extensions (1) and (2) are considered together, we meet the general Markov processes and the general potential theory: many old and new books are consecrated to these theories and we will not try to give an outline of these; let us mention nevertheless [9], a nice (old, but still incomparable) place where to gain a precise idea of all aspects of the subject. Here we will have a look only at generalizations which do not go too far away in spirit from what was done for matrices.

There are three natural ways to address the extension in space. The first is to consider a denumerable (infinite) set I of indexes and so look at infinite matrices; in that case the difficulty lies in the fact that counting measure is no more finite. The second is to replace I by a probability space $(\Omega, \mathscr{F}, \mathbb{P})$, where the matrices become continuous operators on the corresponding L^2 space. We will look at these two first extensions when generalizing below the concepts of ultrametric and filtered matrices in Chaps. 3 and 5, where we also make a modest extension in time, in considering filtrations indexed by $\mathbb{R}_+$. The third is to replace I by a compact metrisable space F, the Z- and M-matrices becoming linear or nonlinear (third extension), continuous or not (see below), operators defined on a subspace $\mathscr{D}$ of the space $\mathscr{C}$ of continuous functions. We begin with this generalization (which may implicitly contain a time extension) without straying too far from the Chap. 2: it's amazing to see how the

© Springer International Publishing Switzerland 2014
C. Dellacherie et al., *Inverse M-Matrices and Ultrametric Matrices*, Lecture Notes in Mathematics 2118, DOI 10.1007/978-3-319-10298-6

principal features of Z- and M-matrices can be easily extended, usefully to a wider context.

Extension of Z-Matrices, M-Matrices and Inverses

Instead of the set I of indices for matrices we take a compact metrizable space F; beware that "x" will denote here a generic point of F whereas "i" denotes a generic point of I. Let $\mathscr{C}$ be the space of (real) continuous functions on F equipped with the uniform norm and consider a map A from a subset $\mathscr{D}$ of $\mathscr{C}$ into $\mathbb{R}^F$ verifying

$$\forall u, v \in \mathscr{D} \ \forall x \in F \ [u \leq v \text{ and } u(x) = v(x)] \implies [(Au)(x) \geq (Av)(x)] \quad \text{(A.1)}$$

which, if $\mathscr{D}$ is a vector subspace and A is linear, is equivalent to

$$\forall v \in \mathscr{D} \ \forall x \in F \ [0 \leq v \text{ and } v(x) = 0] \implies [(Av)(x) \leq 0]. \quad \text{(A.2)}$$

Actually (A.2) is a very known "principle" of potential theory where it bears different names, but what matters here is if F is finite, identified to $\{1, \ldots, n\}$, $\mathscr{D}$ is $\mathbb{R}^n$ and A is linear, identified to a matrix, then it can be recognized in (A.2) and so in (A.1) a "functional" way to say A is a Z-matrix, no more, no less. As we will see below, (A.1) is the right statement to acquiring comparison theorems, even in the linear case.

They are numerous examples of operators verifying (A.1). The most sophisticated are probably the so called "fully nonlinear degenerate second order elliptic operators" [16] (in despite of the name, these include all of the usual linear elliptic and parabolic operators) and will be called **differential elliptic operators** subsequently. In that case, F is the closure of a relatively compact open subset E of some $\mathbb{R}^n$, $\mathscr{D}$ is the restriction to F of the $\mathscr{C}^2$-functions on $\mathbb{R}^n$ and A is defined on E by

$$(Au)(x) = \Phi[x, u(x), \nabla u(x), Hu(x)] \quad \text{(A.3)}$$

with $u \in \mathscr{D}, x \in E, \nabla u$ the gradient of u, Hu the hessian matrix of u, Φ any function with appropriate arguments (the last one is a symmetric matrix), and nonincreasing in its last argument w.r.t. the usual order on semidefinite matrices—here is the ellipticity property. One completes the definition of A on F by taking $Au(x) = u(x)$ for $x \in \partial E$, that is, in other words, by the Dirichlet's condition at the boundary. To check that A verifies (A.1) is just a short exercise using Taylor's formula.

We propose to say that operators verifying (A.1) are elliptic operators (they are called "derivers" in [19]).

Moreover, the most simple examples are what we call the **elementary elliptic operators**. In that case, F being an arbitrary metrizable compact space, $\mathscr{D}$ is $\mathscr{C}$ and by definition A can be written $A = \varphi(I - P)$ where φ is (the multiplication by) a nonnegative function, I is the identity and P is a nondecreasing map from $\mathscr{C}$ into $\mathbb{R}^F$; sure, we met often very particular cases in the Chap. 2. It is quite simple to confirm that such an A verifies (A.1) and so is elliptic in our terminology.

A final example before we look at a generalization of an M-matrix. This time, as for matrices, F is a finite set identified with $I = \{1, \ldots, n\}$. A continuous map $\Psi = (\Psi_i)_{i \in I}$ from $\mathbb{R}^n$ into $\mathbb{R}^n$ is called **quasimonotone** (see [57]) if each Ψ_i is nondecreasing in each of its arguments except the ith. So if Ψ is linear, $-\Psi$ can be identified with a Z-matrix, and in the general case $-\Psi$ is a nonlinear elliptic operator (often elementary, but not always). The quasimonotone maps are important ingredients in the study of some dynamical systems called competitive or cooperative systems [55]. An easy and useful result, even in the case of a Z-matrix, is the following: let $-A$ be quasimonotone and v, f two points of $\mathbb{R}^n$; if the system of inequalities $u \geq v$, $Au \geq f$ has at least a solution then it has a smallest one $\tilde{u}$, which verifies $[\tilde{u}_i = v_i$ or $(A\tilde{u})_i = f_i]$ for each $i \in I$. A good part of linear or non linear potential theory is devoted to hard extensions of this kind of result.

Coming back to the general situation (F metrizable compact, A defined on a subset $\mathscr{D}$ of $\mathscr{C}$) we suppose here $\mathscr{D}$ is a vector subspace and now introduce a condition implying that our elliptic operator A is injective and its inverse on $A(\mathscr{D})$ is nondecreasing, namely we suppose the existence of $\chi \in \mathscr{D}$

$$\chi \text{ is positive and } \forall u \in \mathscr{D} \ \forall x \subset F \ t \mapsto (Au + t\chi)(x) \text{ is increasing on } \mathbb{R} \qquad \text{(A.4)}$$

which is, if A is linear, equivalent to

$$\chi > 0 \text{ and } A\chi > 0. \qquad \text{(A.5)}$$

With a nod to mathematical economics we will say that the elliptic operator A is **productive** if it verifies (A.4) and the function χ can be called a **witness of productivity**. One recognizes in (A.5) one of the usual necessary and sufficient condition for a Z-matrix A to be an M-matrix. At the other end of complexity, where A is a differential elliptic operator as in (A.3), one often sees the hypothesis that the function Φ is increasing in its second argument: that implies at once that the positive constant functions are witnesses to productivity. This brings us to a fundamental theorem; here we consider an arbitrary partition $(E, \partial E)$ of F in a subset E and its complementary ∂E, but the notation suggests the possibility of applying it to Dirichlet's problem. The following result is known as the **comparison Theorem**.

Theorem A.1 (Theorem of Comparison) *Let A be a productive elliptic operator, $(E, \partial E)$ a partition of F and u, v two elements of $\mathscr{D}$. Then*

$$[u \leq v \text{ on } \partial E \text{ and } Au \leq Av \text{ on } E] \Longrightarrow u \leq v \text{ everywhere.}$$

Proof Suppose there is $x \in E$ such that $u(x) > v(x)$, let χ be a witness to productivity and $\tau = \inf\{t \geq 0 : u \leq v + t\chi\}$, which is positive and finite. Setting

$w = v + \tau\chi$, we have $w \geq u$, and since we are looking at continuous functions on a compact space, there is $\xi \in F$ such that $w(\xi) = u(\xi)$. From our hypotheses we have $\xi \in E$, from the ellipticity of A we have $Aw(\xi) \leq Au(\xi)$ and from the productivity of A we have $Av(\xi) < Aw(\xi)$, and so finally a contradiction with our hypothesis $Av(\xi) \geq Au(\xi)$.

Corollary A.2 *Any productive elliptic operator A is injective and its inverse A^{-1} is increasing on its domain of definition $A(\mathscr{D})$.*

Proof Take ∂E empty in the theorem: you get $Au \leq Av \Rightarrow u \leq v$, and so the injectivity of A and the nondecreasingness of A^{-1}, even its increasingness by again using (A.1).

So if F is finite, identified to $\{1, \ldots, n\}$ and A is linear identified to a matrix, then A is an M-matrix since here the injectivity implies the surjectivity. If F is just a singleton and $\mathscr{C}$ identified to $\mathbb{R}$, then a (nonlinear) elliptic operator A defined on $\mathscr{C}$ is just a (real) function, and a productive one is just an increasing function: the corollary can be viewed as a generalization of a part of the usual theorem on the inversion of an increasing function; one can delve further in this direction: under mild conditions, it can be quite easily proved that A^{-1} is continuous for the uniform norm on the interior of $A(\mathscr{D})$. The major difficulty is to identify $A(\mathscr{D})$ (remember, the Dirichlet's problem is an instance of our problem of inversion); below we will briefly look at the case when A is continuous from $\mathscr{C}$ into $\mathscr{C}$ (which is certainly not the case for a differential elliptic operator which is not completely degenerated).

Let us now state what is called the **Domination Principle**.

Corollary A.3 (Domination Principle) *Let A be a productive elliptic operator and u, v two elements of $\mathscr{D}$. Then*

$$u \leq v \text{ on } \{Au > Av\} \Longrightarrow u \leq v \text{ everywhere.}$$

Proof Take $\partial E = \{Au > Av\}$ in the theorem, and so $E = \{Au \leq Av\}$ (do not be afraid by the tautology "$Au \leq Av$ on $\{Au \leq Av\}$"!).

We retrieve, in a general setting, the domination principle of Chap. 2, except here it is explicitly written with A, not with A^{-1}. If one supposes that A verifies a "submarkovianity" property, namely $t \mapsto (Au + t)(x)$ is nondecreasing in $t \in \mathbb{R}$ for any $u \in \mathscr{D}$, $x \in F$, one can easily deduce from the corollary a general statement of the complete maximum principle seen in Chap. 2.

Suppose now that A is a continuous productive elliptic operator from $\mathscr{C}$ into $\mathscr{C}$. Adapting modern tools used in the study of Dirichlet's problem [16] (essentially here a modern variant of the classical Perron's method in potential theory), it is possible but not trivial to prove that $A(\mathscr{C})$ verifies an intermediary property, namely if f, g belong to $A(\mathscr{C})$, then any $h \in \mathscr{C}$ verifying $f \leq h \leq g$ belongs also to $A(\mathscr{C})$. So if A is linear, A is surjective since $tA\chi$, $t \in \mathbb{R}$, χ witness, can be as great or as little as you wish. But, if A is linear, there is a more elementary way to look at A^{-1}: A is linear and continuous, so lipschitzian and so A is a elementary elliptic operator and can be written $A = k(I - P)$ where k is a constant, I the identity and P is a nonnegative continuous linear operator. Knowing that $A\chi$, χ a witness, belongs of course to $A(\mathscr{C})$, it is a good exercise on series to prove that the geometric series $\sum_{n \geq 0} P^n$ is normally convergent and that its sum is equal to kA^{-1}, a series we saw for matrices in Chap. 2 and others.

Extension of Ultrametricity to Infinite Matrices

Here I is an infinite denumerable set. As in Definition 3.2, we will say an infinite matrix U indexed by I is **ultrametric** if U is symmetric, the diagonal of U is pointwise dominant and the ultrametric inequality

$$U_{ik} \geq \inf(U_{ij}, U_{jk})$$

is verified for any i, j, k in I. Is such an ultrametric matrix a potential matrix in some way as in Theorem 3.5.

Before providing some elements in response, we need to introduce substochastic matrices and potential matrices into our infinite context.

An infinite matrix P will be said **substochastic** if we have $0 \leq P_{ij} \leq 1$ and $\sum_{j \in I} P_{ij} \leq 1$ for any i; the **potential**-matrix V generated by such a P, when it exists—that is when the series is convergent, is

$$V = \sum_{n \geq 0} P^n$$

(cf Sect. 2.3). There are some important differences with the finite case: suppose V exists; it is a right inverse of $\mathbb{I} - P$, but generally $\mathbb{I} - P$ has many different right inverses; nevertheless it can be proved V is the smallest nonnegative right inverse of $\mathbb{I} - P$. Let us have a closer look. An (infinite) vector v is said **invariant** (or **harmonic**) w.r.t. the substochastic matrix P if $Pv = v$; in the finite case, it is impossible to have a non-null invariant vector and a potential matrix, but it is not the case for infinite matrices—it is even possible to have $\mathbb{1}$ as an **invariant** (P is stochastic) and $\sum_{n \geq 0} P^n < \infty$—and if H is an infinite matrix whose columns are invariant vectors and V the potential of P, then $V + H$ is another right inverse of $\mathbb{I} - P$.

Let us return to the study of an ultrametric matrix U, looking only at the particularly important case where U is the canonical ultrametric matrix associated with a (infinite denumerable) rooted tree T with I as set of nodes and r as root. We suppose T locally finite, that is any node i has a finite number s_i of successors, and, in order to avoid referring to the potential of a continuous time Markov chain, we suppose $\sup_i s_i < +\infty$. We will extend in our context all the vocabulary and notations introduced for a finite rooted tree in the Chap. 3. So our ultrametric matrix

$U = (U_{ij})$ is given by

$$U_{ij} = |i| + 1,$$

where $|i|$ is the length of the geodesic from the root r to the node i. Further the canonical M-matrix M associated to T is given by:

$$\begin{cases} M_{ij} = 0 \text{ if } i \neq j \text{ and } (i, j) \notin T; \\ M_{ij} = M_{ji} = -1 \text{ if } j \text{ is the predecessor of } i; \\ M_{ii} = 1 + s_i \text{ where } s_i \text{ is the number of successors of } i. \end{cases}$$

It is evident that M is diagonally dominant (the sum of each line is 0 except for the line r where the sum is 1) and easy to check as a left inverse of U; moreover M can be written $M = k(\mathbb{I} - P)$ where $k = \sup_i(1 + s_i)$ and P is a substochastic matrix, stochastic at any node except for r which therefore is attached to an added trap ∂. Consequently we have $k(\mathbb{I} - P)U = \mathbb{I}$ and so kU is candidate to be the potential of P, but it often misses the point. For example, if T is the infinite linear tree ($s_i = 1$ for any i), then $2U$ is the potential V of P, but if T is the infinite dyadic tree ($s_i = 2$ for any i), then $3U$ is not the potential V of T: $3U - V$ is a positive harmonic matrix for P. This situation is enlightened by the probabilistic interpretation. Let $\mathbb{P}$ be the probability attached to the Markov chain (X_n) on T starting from r with P as transition matrix, and $\tau = \inf\{n : X_n = \partial\}$ the lifetime of the chain. Then either we have $\mathbb{P}\{\tau < \infty\} = 1$, and kU is the potential of P, or $\mathbb{P}\{\tau = \infty\} > 0$, and kU is not the potential. In this last case a detailed study of the situation requires the addition of T as an "exit" boundary (a kind of Cantor set); as in [24].

Extension of Ultrametricity to L^2-Operators

One can see the finite index set I of our finite matrices as a probability space equipped tacitly with the field $\mathscr{P}$ of all the parts of I and more or less tacitly with the normalized counting measure. Then any (symmetric) matrix is identified with a (symmetric) L^2-operator. More generally, we replace I by any probability space $(\Omega, \mathscr{F}, \mathbb{P})$ and define and study ultrametric operators. In this general context, ultrametric distances and partitions are not good tools, but filtrations (nondecreasing families of conditional expectations) are excellent ones; as said in Definition 5.1 a conditional expectation is a symmetric projection in L^2 which is nonnegative and leaves $\mathbb{1}$ invariant. We know from Chaps. 3 and 5 that in the finite case looking to ultrametric distances or increasing sets of partitions or filtrations are equivalent; in particular following Proposition 5.17 a symmetric matrix is ultrametric if and only if it is filtered. So, in the general case we can say that an operator U is **ultrametric** if it is filtered, that is there exists a finite (for the moment) nondecreasing sequences

$(\mathbb{E}_k : 0 \leq k \leq n)$ of conditional expectations (shortly, a filtration) and a sequence $(a_k : 0 \leq k \leq n)$ of nonnegative bounded random variables such that $\mathbb{E}_k a_k = a_k$ for any k and

$$U = \sum_{k=0}^{n} a_k \mathbb{E}_k$$

where in the formula a_k should be interpreted as the multiplication operator by a_k, which commutes with $\mathbb{E}_k$. Then, if U is such a filtered or ultrametric operator, the operator $U + t\mathbb{I}$ is, for any $t > 0$ (and sometimes for $t = 0$), invertible and its inverse can be written $k(\mathbb{I} - P)$, k, P functions of t, where k is a positive constant and P is a substochastic operator, that is positive and verifies $P\mathbb{1} \leq \mathbb{1}$. As such, we observe the same result as in Chap. 3. In fact, the notion of a filtered operator and its study were first introduced in this probabilistic context [21], and the comments in Chap. 5 on filtered matrices are a restriction of more general statements.

More generally one can study weakly filtered operators. Such an operator U can be expressed

$$U = \sum_{k=0}^{n} a_k \mathbb{E}_k z_k$$

where a_k, z_k are nonnegative bounded random variables such that $\mathbb{E}_{k+1} a_k = a_k$ and $\mathbb{E}_{k+1} z_k = z_k$ for $k < n$. The set forth in Chap. 5 is still valid, except for the elements revolving around the consideration of "non-normalized" conditional expectations (whose entries are just 0 and 1), which do not exist outside the matrix case.

Coming back to an ultrametric operator $U = \sum_{k=0}^{n} a_k \mathbb{E}_k$, let us set $A_i = \sum_{k=0}^{i-1} a_k$ and $A_{-1} = 0$ so that we can write $U = \sum_{k=0}^{n} \mathbb{E}_k (A_k - A_{k-1})$ where (A_k) is an adapted ($\mathbb{E}_k A_k = A_k$) nondecreasing ($A_{k-1} \leq A_k$) process. This suggests an extension of the notion of an ultrametric (or filtered) operator in the case of a continuous time filtration ($\mathbb{E}_t : 0 \leq t$), right continuous, by considering

$$U := \int_0^\infty \mathbb{E}_t \, dA_t$$

where (A_t) is an adapted nondecreasing process and is bounded (A_∞ is a bounded random variable)—or we encounter difficulties as in the case of infinite matrices. A classical approach to the continuous by the discrete (see [21]) permits us to prove that again $U + t\mathbb{I}$, for any $t > 0$, is invertible, with inverse of the form $k(\mathbb{I} - P)$ like above. It is also possible to consider a weakly filtered operator U in this context, but the integral defining U in this case is more involved, so we won't comment further on this subject here.

Appendix B
Basic Matrix Block Formulae

In this appendix we include some basic facts about matrices than we need in our exposition.

Take $n \geq 2$, $1 \leq p < n$ and $q = n - p$. We define $J = \{1, \cdots, p\}$ and $K = \{p + 1, \cdots, n\}$. Consider the following partition of an $n \times n$ matrix U

$$U = \begin{pmatrix} A & B \\ C & D \end{pmatrix},$$

where $A = U_{JJ}$, $D = U_{KK}$. We assume that U and D are nonsingular. Then, if we apply the Gauss algorithm by blocks we obtain

$$\begin{pmatrix} A & B \\ C & D \end{pmatrix} \begin{pmatrix} \mathbb{I} & 0 \\ -D^{-1}C & D^{-1} \end{pmatrix} = \begin{pmatrix} A - BD^{-1}C & BD^{-1} \\ 0 & \mathbb{I} \end{pmatrix}$$

and therefore

$$U = \begin{pmatrix} A & B \\ C & D \end{pmatrix} = \begin{pmatrix} A - BD^{-1}C & BD^{-1} \\ 0 & \mathbb{I} \end{pmatrix} \begin{pmatrix} \mathbb{I} & 0 \\ C & D \end{pmatrix}.$$

In particular, one has the well known formula for the determinant of U given by $|U| = |D| \, |A - BD^{-1}C|$. This shows that $A - BD^{-1}C$, which is called the **Schur complement** of D, is nonsingular.

© Springer International Publishing Switzerland 2014
C. Dellacherie et al., *Inverse M-Matrices and Ultrametric Matrices*, Lecture Notes in Mathematics 2118, DOI 10.1007/978-3-319-10298-6

On the other hand the inverse of U can be computed as

$$U^{-1} = \begin{pmatrix} \mathbb{I} & 0 \\ C & D \end{pmatrix}^{-1} \begin{pmatrix} A - BD^{-1}C & BD^{-1} \\ 0 & \mathbb{I} \end{pmatrix}^{-1}$$

$$= \begin{pmatrix} \mathbb{I} & 0 \\ -D^{-1}C & D^{-1} \end{pmatrix} \begin{pmatrix} [A - BD^{-1}C]^{-1} & -[A - BD^{-1}C]^{-1}BD^{-1} \\ 0 & \mathbb{I} \end{pmatrix}$$

$$= \begin{pmatrix} [A - BD^{-1}C]^{-1} & -[A - BD^{-1}C]^{-1}BD^{-1} \\ -D^{-1}C[A - BD^{-1}C]^{-1} & D^{-1} + D^{-1}C[A - BD^{-1}C]^{-1}BD^{-1} \end{pmatrix}.$$

Thus, the block of U^{-1} associated with J is

$$(U^{-1})_{JJ} = [A - BD^{-1}C]^{-1}.$$

Given a permutation σ of $\{1, \cdots, n\}$, we construct the matrix Π given by

$$\Pi_{ij} = \begin{cases} 1 & \text{if } j = \sigma(i) \\ 0 & \text{otherwise} \end{cases}$$

Then the matrix ΠU is obtained from U by permuting the rows of U according to σ, that is the i-th row of ΠU is the $\sigma(i)$-th row of U. Similarly $U\Pi'$ is the matrix obtained from U by permuting the columns of U according to σ. Finally, the matrix $\Pi U \Pi'$ is similar to U and is obtained from it by permuting the rows and columns according to σ.

Appendix C
Symbolic Inversion of a Diagonally Dominant M-Matrix

Following [26], which is too rarely cited in the bibliography of books on matrices or graphs, we present without proofs (see [8, 26] and [27]) a wonderful method for inverse symbolic diagonally dominant symmetric M-matrices. The examples provided here are for a 3×3 matrix but the results are true for any order.

We start with the following matrix

$$A = \begin{pmatrix} x+a+b & -a & -b \\ -a & y+a+c & -c \\ -b & -c & z+b+c \end{pmatrix}$$

where x, y, z, a, b, c are symbolic variables intended to be replaced by nonnegative reals. In fact, any 3×3 symbolic matrix can be written like this after obvious change in variables, but it does not seems very interesting. The associated graph to A is the complete K_4 graph; it is allowed to annulate (and drop) a variable—if you set $c = 0$, the associated graph is the graph of a Wheatstone bridge—, but it is not allowed to identify some variables (for example $x = y$ or $a = b$) if the intent is to use the Wang algebra (but of course it is allowed in the ordinary algebra).

Wang algebra will be seen here as a commutative algebra $\mathfrak{A}$ such that $x + x = 0$ and $x.x = 0$ for any $x \in \mathfrak{A}$ (after Duffin it is the Grassmann algebra on $\mathbb{Z}/2\mathbb{Z}$, and he uses that in the proofs). Such a property implies spectacular simplifications in algebraic calculations: for example one has $(a + b).(b + c).(c + a) = 0$ for any $a, b, c \in \mathfrak{A}$.

We are going to calculate the determinant and the principal cofactors of A with the help of the Wang algebra generated by x, y, z, a, b, c, and deduce from them the remaining cofactors with the help of the ordinary algebra.

Determinant The determinant of A is merely the product of the diagonal elements, but in the Wang algebra generated by x, y, z, a, b, c. Indeed if one expands $D = (x + a + b)(y + a + c)(z + b + c)$ in this algebra one finds first (recall that $c^2 = 0$)

© Springer International Publishing Switzerland 2014

C. Dellacherie et al., *Inverse M-Matrices and Ultrametric Matrices*, Lecture Notes in Mathematics 2118, DOI 10.1007/978-3-319-10298-6

$$D = (x + a + b)(yz + by + cy + az + ab + ac + cz + bc)$$

and finally one finds D equal to

$$xyz + bxy + cxy + axz + abx + acx + cxz + bcx+$$
$$+ayz + aby + acy + acz + byz + bcy + abz + bcz$$

since the terms $a^2z, a^2b, a^2c, b^2y, ab^2, abc+abc, b^2c$ vanish. This long polynomial of six variables, called DD in the sequel, is what is found if one computes as usual the determinant of A after simplification.

The monomials are linked only by "+" signs, and so DD was called an **unisignant** by Sylvester about 1850, a few years before Kirchhoff discovered this determinant in his study of electrical networks. Both of them, independently, gave a wonderful graphical interpretation of DD: let us consider the graph associated to A and let us label the six edges by x, y, z, a, b, c respectively (which can be interpreted as probabilities, or resistances, or conductances, etc.). Then each triplet in DD can be seen as the name of a tree generating the graph, and DD supplies the list of the generating trees, without omission or repetition. This kind of result, famous in graph theory and electrical networks, is often called the **matrix-tree theorem** and generally ascribed to Kirchhoff alone, with matrices like A being called **Kirchhoff's matrix** when they are not called **laplacian** of the graph. The concept of Wang algebra is more recent, being initiated by Wang and other Chinese electricians between 1930 and 1940 (see [13]) and, despite Duffin's popularization, seems to be alive only among electricians.

Principal Cofactors The computations, via Wang algebra, are analogous but easier since we have only two terms in the diagonals. For example the cofactor C_{11} à la Wang is equal to $(y + a + c)(z + b + c)$; we expanded it already when computing DD and got

$$C_{11} = (y + a + c)(z + b + c) = yz + by + cy + az + ab + ac + cz + bc.$$

In the same way we have

$$C_{22} = (x + a + b)(z + b + c) = xz + bx + cx + az + ab + ac + bz + bc$$

and

$$C_{33} = (x + a + b)(y + a + c) = xy + ax + cx + ay + ac + by + ab + bc.$$

Remaining Cofactors Here we will not use any more the Wang algebra; nevertheless we have a nice result due to Bott and Duffin. For example, in order to compute C_{13}, one looks, in ordinary algebra, at the expanded forms of C_{11} and C_{33} and simply keeps of them their "intersection", that is the common monomials, $by + ab + ac + bc$.

So we have

$$C_{13} = C_{31} = C_{11} \cap C_{33} = by + ab + ac + bc.$$

In the same way we have

$$C_{12} = C_{21} = C_{11} \cap C_{22} = az + ab + ac + bc$$

and

$$C_{23} = C_{32} = C_{22} \cap C_{33} = cx + ab + ac + bc.$$

We leave to the reader the pleasure of writing the inverse of A from the determinant and the cofactors. Note that all of them are unisignant.

References

1. W. Anderson, *Continuous-Time Markov Chains: An Applications Oriented Approach* (Springer, New York, 1991)
2. T. Ando, Inequalities for M-matrices. Linear Multilinear Algebra **8**, 291–316 (1980)
3. R.B. Bapat, M. Catral, M. Neumann, On functions that preserve M-matrices and inverse M-matrices. Linear Multilinear Algebra **53**, 193–201 (2005)
4. J.P. Benzecri et collaborateurs, *LÁnalyse des données* (Dunod, Paris, 1973)
5. R. Bellman, *Introduction to Matrix Analysis*. Classics in Applied Mathematics, vol. 12 (Society for Industrial and Applied Mathematics, Philadelphia, 1995)
6. E. Bendito, A. Carmona, A. Encinas, J. Gesto, Characterization of symmetric M-matrices as resistive inverses. Linear Algebra Appl. **430**(4), 1336–1349 (2009)
7. N. Bouleau, Autour de la variance comme forme de Dirichlet. Séminaire de Théorie du Potentiel 8 (Lect. Notes Math.) **1235**, 39–53 (1989)
8. R. Bott, R.J. Duffin, On the algebra of networks. Trans. Am. Math. Soc. **74**, 99–109 (1953)
9. M. Brelot, H. Bauer, J.-M. Bony, J. Deny, J.L. Doob, G. Mokobodzki, *Potential Theory* (C.I.M.E., Stresa, 1969) (Cremonese, Rome, 1970)
10. D. Capocacia, M. Cassandro, P. Picco, On the existence of ther- modynamics for the generalized random energy model. J. Stat. Phys. **46**(3/4), 493–505 (1987)
11. S. Chen, A property concerning the Hadamard powers of inverse M-matrices. Linear Algebra Appl. **381**, 53–60 (2004)
12. S. Chen, Proof of a conjecture concerning the Hadamard powers of inverse M-matrices. Linear Algebra Appl. **422**, 477–481 (2007)
13. W.K. Chen, Topological network analysis by algebraic methods. Proc. Instit. Electr. Eng. Lond. **114**, 86–87 (1967)
14. G. Choquet, J. Deny, Modèles finis en théorie du potentiel. J. d'Analyse Mathématique **5**, 77–135 (1956)
15. K.L. Chung, *Markov Chains with Stationary Transition Probabilities* (Springer, New York, 1960)
16. M.G. Crandall, Viscosity solutions: a primer. Lect. Notes Math. **1660**, 1–43 (1997). Springer
17. P. Dartnell, S. Martínez, J. San Martín, Opérateurs filtrés et chaînes de tribus invariantes sur un espace probabilisé dénombrable. *Séminaire de Probabilités XXII Lecture Notes in Mathematics*, vol. 1321 (Springer, New York, 1988)
18. C. Dellacherie, Private Communication (1985)
19. C. Dellacherie, Nonlinear Dirichlet problem and nonlinear integration. *From classical to modern probability*, vol. 83–92 (Birkhäuser, Basel, 2003)

© Springer International Publishing Switzerland 2014
C. Dellacherie et al., *Inverse M-Matrices and Ultrametric Matrices*, Lecture Notes in Mathematics 2118, DOI 10.1007/978-3-319-10298-6

20. C. Dellacherie, S. Martínez, J. San Martín, Ultrametric matrices and induced Markov chains. Adv. Appl. Math. **17**, 169–183 (1996)
21. C. Dellacherie, S. Martínez, J. San Martín, D. Taïbi, Noyaux potentiels associés à une filtration. Ann. Inst. Henri Poincaré Prob. et Stat. **34**, 707–725 (1998)
22. C. Dellacherie, S. Martínez, J. San Martín, Description of the sub-Markov kernel associated to generalized ultrametric matrices. An algorithmic approach. Linear Algebra Appl. **318**, 1–21 (2000)
23. C. Dellacherie, S. Martínez, J. San Martín, Hadamard functions of inverse M-matrices. SIAM J. Matrix Anal. Appl. **31**(2), 289–315 (2009)
24. C. Dellacherie, S. Martínez, J. San Martín, Ultrametric and tree potential. J. Theor. Probab. **22**(2), 311–347 (2009)
25. C. Dellacherie, S. Martínez, J. San Martín, Hadamard functions that preserve inverse M-matrices. SIAM J. Matrix Anal. Appl. **33**(2), 501–522 (2012)
26. R.J. Duffin, An analysis ot the Wang algebra of networks. Trans. Am. Math. Soc. **93**, 114–131 (1959)
27. R.J. Duffin, T.D. Morley, Wang algebra and matroids. IEEE Trans. Circ. Syst. **25**, 755–762 (1978)
28. M. Fiedler, V. Pták, Diagonally dominant matrices. Czech. Math. J. **17**, 420–433 (1967)
29. M. Fiedler, Some characterizations of symmetric inverse M-matrices. Linear Algebra Appl. **275/276**, 179–187 (1998)
30. M. Fiedler, Special ultrametric matrices and graphs. SIAM J. Matrix Anal. Appl. **22**, 106–113 (2000) (electronic)
31. M. Fiedler, H. Schneider, Analytic functions of M-matrices and generalizations. Linear Multilinear Algebra **13**, 185–201 (1983)
32. L.R. Ford, D.R. Fulkerson, *Flows in Networks* (Princeton University Press, Princeton, 1973)
33. D. Freedman, *Markov Chains* (Holden-Day, San Francisco, 1971)
34. F.R. Gantmacher, M.G. Krein. *Oszillationsmatrizen, Oszillationskerne und kleine Schwingungen mechanischer Systeme* (Akademie-Verlag, Berlin, 1960)
35. R.E. Gomory, T. C. Hu, Multi-terminal network flows. SIAM J. Comput. **9**(4), 551–570 (1961)
36. E. Hille, R. Phillips, *Functional Analysis and Semi-Groups*, vol. 31 (AMS Colloqium, Providence, 1957)
37. R. Horn, C. Johnson, *Matrix Analysis* (Cambridge University Press, Cambridge, 1985)
38. R. Horn, C. Johnson, *Topics in Matrix Analysis* (Cambridge University Press, Cambridge, 1991)
39. C. Johnson, R. Smith, Path product matrices and eventually inverse M-matrices. SIAM J. Matrix Anal. Appl. **29**(2), 370–376 (2007)
40. C. Johnson, R. Smith, Inverse M-matrices, II. Linear Algebra Appl. **435**(5), 953–983 (2011)
41. J. Kemeny, J.L. Snell, A. Knapp, *Denumerable Markov Chains*, 2nd edn. (Springer, New York, 1976)
42. R. Lyons, Y. Peres, *Probability on Trees and Networks* (Cambridge University Press, Cambridge, 2014). In preparation, current version available at http://mypage.iu.edu/~rdlyons/
43. T. Markham, Nonnegative matrices whose inverse are M-matrices. Proc. AMS **36**, 326–330 (1972)
44. S. Martínez, G. Michon, J. San Martín, Inverses of ultrametric matrices are of Stieltjes types. SIAM J. Matrix Anal. Appl. **15**, 98–106 (1994)
45. S. Martínez, J. San Martín, X. Zhang, A new class of inverse M-matrices of tree-like type. SIAM J. Matrix Anal. Appl. **24**(4), 1136–1148 (2003)
46. S. Martínez, J. San Martín, X. Zhang, A class of M-matrices whose graphs are trees. Linear Multilinear Algebra **52**(5), 303–319 (2004)
47. J.J. McDonald, M. Neumann, H. Schneider, M.J. Tsatsomeros. Inverse M-matrix inequalities and generalized ultrametric matrices. Linear Algebra Appl. **220**, 321–341 (1995)
48. J.J. McDonald, R. Nabben, M. Neumann, H. Schneider, M.J. Tsatsomeros, Inverse tridiagonal Z-matrices. Lincar Multilinear Algebra **45**, 75–97 (1998)

49. C.A. Micchelli, R.A. Willoughby, On functions which preserve Stieltjes matrices. Linear Algebra Appl. **23**, 141–156 (1979)
50. R. Nabben, On Green's matrices of trees. SIAM J. Matrix Anal. Appl. **22**(4), 1014–1026 (2001)
51. R. Nabben, R.S. Varga, A linear algebra proof that the inverse of a strictly ultrametric matrix is a strictly diagonally dominant Stieltjes matrix. SIAM J. Matrix Anal. Appl. **15**, 107–113 (1994)
52. R. Nabben, R.S. Varga, Generalized ultrametric matrices – a class of inverse M-matrices. Linear Algebra Appl. **220**, 365–390 (1995)
53. R. Nabben, R. Varga, On classes of inverse Z-matrices. Linear Algebra Appl. **223/224**, 521–552 (1998)
54. M. Neumann, A conjecture concerning the Hadamard product of inverses of M-matrices. Linear Algebra Appl. **285**, 277–290 (1998)
55. H. Smith, *Monotone Dynamical Systems. An Introduction to the Theory of Competitive and Cooperative Systems.* Mathematical Surveys and Monographs, vol. 41 (American Mathematical Society, Providence, 1995)
56. R.S. Varga, Nonnegatively posed problems and completely monotonic functions. Linear Algebra Appl. **1**, 329–347 (1968)
57. W. Walter, *Ordinary Differential Equations.* Graduate Texts in Mathematics, vol. 182 (Springer, New York, 1998)
58. B. Wang, X. Zhang, F. Zhang, On the Hadamard product of inverse M-matrices. Linear Algebra Appl. **305**, 2–31 (2000)
59. X. Zhang, A note on ultrametric matrices. Czech. Math. J. **54**(**129**)(4), 929–940 (2004)

Index of Notations

A

$A^{(\alpha)}$: Hadamard power of A 166
$\hat{a}^{\mathbb{E}}$: Envelop of a 136

C

CBF : Constant block form   63
CMP : Complete maximum principle   8

D

D_x : Diagonal matrix, $x \in \mathbb{R}^n$ 121
$d_{\mathscr{G}}(i, j)$: Geodesic distance   67

E

$\mathbb{E}$: Conditional expectation   120
$e(i)$: The i-th element of the canonical basis   5

F

$\mathbb{F}$: Incidence matrix   129
$\mathbf{F}, \mathbf{G}$: Filtration   126
$\mathfrak{F}$: Finest partition   58
$f(A)$: Hadamard function   166

f_{ij}^{U} : Probability that the chain ever visits j starting from i 40

G

γ : Path   66
GUM : Generalized ultra. matrix   64
Geod(i, j) : The geodesic in a tree   67
$\mathscr{G}$: Graph   66
$\mathscr{G}^{W}$: Incidence graph of W^{-1} 86

H

$H(T)$: Height of T 67

I

$\langle\,,\,\rangle_{\mu}$: Inner product in $\mathbb{R}^n$ 121
$\mathbb{I}$: Identity matrix   5
$\mathbb{1}_B$: Indicator of B 5
$\mathbb{1}$: Vector of ones   5
Im(A) : Image of A 123
i^- : Immediate predecessor   67

L

$\bar{\lambda}$: Total mass of λ 7
LUM : Linear ultrametric matrix   188

© Springer International Publishing Switzerland 2014
C. Dellacherie et al., *Inverse M-Matrices and Ultrametric Matrices*, Lecture Notes
in Mathematics 2118, DOI 10.1007/978-3-319-10298-6

λ^U : Right equilibrium potential   7
$\mathscr{L}(T)$: The set of leaves of T 67

M

$\mathbb{M}_n$: Mean value matrix   120
$\min\{U\}$: minimum value of U 60
μ^U : Left equilibrium potential   7

N

NBF : Nested block form   63
$\mathfrak{N}$: Coarsest partition   58

O

$\odot$: Hadamard product   121

P

$bi\mathscr{P}$: Bi-potential   7

R

$\mathfrak{R}$: Partition   58
$\mathscr{R}(V)$: Roots of V 7

S

$\mathfrak{S}(i)$: Immediate successors of i 67

T

T : Tree   67
$\mathscr{T}$: Class   183
$\tau(W)$: Upper limit in $bi\mathscr{P}$ 175

X

$x \cdot A$: Product of D_x and A 130

Index

A

Adapted   130

B

Balayage principle   12
Block form
   constant   63
   nested   63

C

Class $\mathcal{T}$ 183
Completely monotone function
   162
Conditional expectation   120
   $\mathbb{E}$-envelop   136
   incidence matrix   129
Constant block form
   positive   63
Counting vector   129

D

D-matrix   188
Domination principle   12

E

Equilibrium potential
   left   7
   right   7
   signed   7
Equilibrium principle   12

F

Filtered   130
   strongly   129
   weakly   136
      normal form   138
Filtration
   dyadic   126
   expectations   126
   maximal   126
   strict   126

G

Geodesic   66
   distance   67
Graph   66
   incidence   86
   path   66
      cycle   67
      length   66
      loop   67

© Springer International Publishing Switzerland 2014
C. Dellacherie et al., *Inverse M-Matrices and Ultrametric Matrices*, Lecture Notes
in Mathematics 2118, DOI 10.1007/978-3-319-10298-6

H

Hadamard function   166
Harmonic function   53

I

Infinitesimal generator   24
Irreducible   22

M

Markov   28
    killed chain   22
Markov semigroup   28
    substochastic   28
Matrix
    Dirichlet   156
    Dirichlet-Markov   156
Maximum principle   12
    complete   8
Measurable   126
M-matrix   6
    irreducible   27

P

Partition   58
    chain   58
        dyadic   58
    finer   58
    measurable   58
Potential   7
    bi-potential   7
    Green   24
    Markov   8

R

Reversibility condition   53

Roots of a matrix   7
Row diagonally dominant   7
Row pointwise diagonally dominant   7

S

Skeleton   31
Stochastic matrix   10
    substochastic   10
Strong Markov property   25
Supermetric   83

T

Tree   67
    branch   67
    geodesic   67
    hight   67
    immediate predecessor   67
    immediate successor   67
    leaf   67
    level function   67
    linear   67
    matrix   68
        weight function   68
    root/rooted   67
    weighted   68
        extension   69

U

Ultrametric
    generalized   64
    geometry   82
        triangle equilateral   82
        triangle isosceles   82
    inequality   58
    matrix   58
        linear   188
    strictly   58
    tree matrix   68

LECTURE NOTES IN MATHEMATICS  Springer

Edited by J.-M. Morel, B. Teissier; P.K. Maini

Editorial Policy (for the publication of monographs)

1. Lecture Notes aim to report new developments in all areas of mathematics and their applications - quickly, informally and at a high level. Mathematical texts analysing new developments in modelling and numerical simulation are welcome.

   Monograph manuscripts should be reasonably self-contained and rounded off. Thus they may, and often will, present not only results of the author but also related work by other people. They may be based on specialised lecture courses. Furthermore, the manuscripts should provide sufficient motivation, examples and applications. This clearly distinguishes Lecture Notes from journal articles or technical reports which normally are very concise. Articles intended for a journal but too long to be accepted by most journals, usually do not have this "lecture notes" character. For similar reasons it is unusual for doctoral theses to be accepted for the Lecture Notes series, though habilitation theses may be appropriate.

2. Manuscripts should be submitted either online at www.editorialmanager.com/lnm to Springer's mathematics editorial in Heidelberg, or to one of the series editors. In general, manuscripts will be sent out to 2 external referees for evaluation. If a decision cannot yet be reached on the basis of the first 2 reports, further referees may be contacted: The author will be informed of this. A final decision to publish can be made only on the basis of the complete manuscript, however a refereeing process leading to a preliminary decision can be based on a pre-final or incomplete manuscript. The strict minimum amount of material that will be considered should include a detailed outline describing the planned contents of each chapter, a bibliography and several sample chapters.

   Authors should be aware that incomplete or insufficiently close to final manuscripts almost always result in longer refereeing times and nevertheless unclear referees' recommendations, making further refereeing of a final draft necessary.

   Authors should also be aware that parallel submission of their manuscript to another publisher while under consideration for LNM will in general lead to immediate rejection.

3. Manuscripts should in general be submitted in English. Final manuscripts should contain at least 100 pages of mathematical text and should always include

   – a table of contents;
   – an informative introduction, with adequate motivation and perhaps some historical remarks: it should be accessible to a reader not intimately familiar with the topic treated;
   – a subject index: as a rule this is genuinely helpful for the reader.

   For evaluation purposes, manuscripts may be submitted in print or electronic form (print form is still preferred by most referees), in the latter case preferably as pdf- or zipped ps-files. Lecture Notes volumes are, as a rule, printed digitally from the authors' files. To ensure best results, authors are asked to use the LaTeX2e style files available from Springer's web-server at:

   ftp://ftp.springer.de/pub/tex/latex/svmonot1/ (for monographs) and
   ftp://ftp.springer.de/pub/tex/latex/svmultt1/ (for summer schools/tutorials).

Additional technical instructions, if necessary, are available on request from lnm@springer.com.

4. Careful preparation of the manuscripts will help keep production time short besides ensuring satisfactory appearance of the finished book in print and online. After acceptance of the manuscript authors will be asked to prepare the final LaTeX source files and also the corresponding dvi-, pdf- or zipped ps-file. The LaTeX source files are essential for producing the full-text online version of the book (see http://www.springerlink.com/openurl.asp?genre=journal&issn=0075-8434 for the existing online volumes of LNM). The actual production of a Lecture Notes volume takes approximately 12 weeks.

5. Authors receive a total of 50 free copies of their volume, but no royalties. They are entitled to a discount of 33.3 % on the price of Springer books purchased for their personal use, if ordering directly from Springer.

6. Commitment to publish is made by letter of intent rather than by signing a formal contract. Springer-Verlag secures the copyright for each volume. Authors are free to reuse material contained in their LNM volumes in later publications: a brief written (or e-mail) request for formal permission is sufficient.

Addresses:
Professor J.-M. Morel, CMLA,
École Normale Supérieure de Cachan,
61 Avenue du Président Wilson, 94235 Cachan Cedex, France
E-mail: morel@cmla.ens-cachan.fr

Professor B. Teissier, Institut Mathématique de Jussieu,
UMR 7586 du CNRS, Équipe "Géométrie et Dynamique",
175 rue du Chevaleret
75013 Paris, France
E-mail: teissier@math.jussieu.fr

For the "Mathematical Biosciences Subseries" of LNM:

Professor P. K. Maini, Center for Mathematical Biology,
Mathematical Institute, 24-29 St Giles,
Oxford OX1 3LP, UK
E-mail: maini@maths.ox.ac.uk

Springer, Mathematics Editorial, Tiergartenstr. 17,
69121 Heidelberg, Germany,
Tel.: +49 (6221) 4876-8259

Fax: +49 (6221) 4876-8259
E-mail: lnm@springer.com